W9-CAJ-135

# FEWER, RICHER, GREENER

# FEWER, RICHER, GREENER

## PROSPECTS FOR HUMANITY IN AN AGE OF ABUNDANCE

LAURENCE B. SIEGEL

WILEY

Cover image: © grynold/Shutterstock, © grakozy/Unsplash
Cover design: Wiley



Published by John Wiley & Sons, Inc., Hoboken, New Jersey.

Published simultaneously in Canada.



***Library of Congress Cataloging-in-Publication Data:***

Name: Siegel, Laurence B., author.
Title: Fewer, richer, greener : prospects for humanity in an age of
abundance / Laurence B. Siegel.
Description: Hoboken, New Jersey : John Wiley & Sons, Inc., [2020] |
Includes bibliographical references and index.
Identifiers: LCCN 2019035276 (print) | LCCN 2019035277 (ebook) | ISBN
978-1-119-52689-6 (cloth) | ISBN 978-1-119-52693-3 (adobe pdf) | ISBN
978-1-119-52692-6 (epub)
Subjects: LCSH: Economic development—Forecasting. | Economic
history—Forecasting. | Progress—Forecasting.
Classification: LCC HC59.15 .S5564 2020 (print) | LCC HC59.15 (ebook) |
DDC 330.9001/12—dc23
LC record available at https://lccn.loc.gov/2019035276
LC ebook record available at https://lccn.loc.gov/2019035277

Printed in the United States of America

V10015075_110719

# *Contents*

# Foreword

## On Capitalism and Humane True Liberalism

Adam Smith, the eighteenth-century founder of economics, and Milton Friedman, two centuries later, promoted the idea that capitalism improves humanity in a way that cannot be accomplished through any other means. Deirdre McCloskey, the great modern economist, historian, and philosopher, wrote a book that was at first called *Humane True Liberalism.* (It is now in print under the more prosaic, but still admirable, title of *Why Liberalism Works.*) It connects capitalism to the ideals of freedom and human autonomy. Rajendra Sisodia and I have written about conscious capitalism in much the same vein.

In this spirit, Larry Siegel, an author trained in finance and economics but now branching out into demography and environmentalism, has written the optimistic book you now hold in your hands. Observing that the end of the population explosion is relatively close at hand, Siegel argues that the whole world (and especially the developing world) has been getting richer and will continue to do so, and that the future will be greener as more resources can be devoted to preserving our precious natural heritage. I think he is right. The future will be better than the present, just as the present is better than the past.

## Apocalypse Now . . . or Apocalypse Not?

This is deeply contrarian stuff. People just love to be told that the world is coming to an end. And they often believe it. Apocalyptic thinking has been a feature of every religion since recorded history began, and that habit of mind persists even in these irreligious times. As a result, a book arguing, as this one does, that economic and living conditions have been steadily improving for hundreds of years and will continue to do so if we don't foul things up, swims against the current. Optimists sound like they are trying to sell you something, while pessimists sound wise and well intentioned.

But *Fewer, Richer, Greener* is not a statement of blind optimism. It treats challenges like overpopulation, economic stagnation, and climate change as risks to be managed and problems to be solved, rather than existential threats that we can't do anything about. It proposes solutions that may be uncomfortable—urbanism, nuclear power, and environmental engineering among them—but will work. Panicking and saying "we're all going to die" will not.

## Free Markets and Free Minds

Perhaps the most important fact of which Siegel reminds us is that *free markets and free minds* are the key to solving the most fundamental problem of human life, which is allocating scarce resources to unlimited wants. This is usually called the economic problem. As the great economist Gary Becker showed, however, this challenge extends far beyond ordinary economic considerations to infiltrate almost every aspect of life. Applying economic thinking to a wide variety of problems opens the mind to insights that cannot be obtained in any other way.

Here are some questions that are amenable to economic thinking although they do not appear to be at first blush: Why do some animals (say, rabbits) have a lot of offspring and devote few resources to each, while others (say, whales) do the opposite? Which kind of animal are we, and is that equilibrium changing? What are the unintended consequences of zoning laws, of gas mileage regulations, of government transfer payments to individuals? Were the Germans who invaded Rome in the fifth century climate refugees? How about the Europeans, from whom I'm descended, who invaded the Americas a little over a millennium later?

These are the kinds of questions Siegel raises in a book that asks the reader to remove the constraints on his or her reasoning that come from a lifetime of mainstream thinking and uninspired education. He asks you to engage, instead, in the opposite: unconventional thinking—bolstered by quotes from adventurous philosophers, scientists, and humanists through the millennia—and passionate self-education.

## A Big and Rich Africa?

His chapter on education, for example, proposes that "we could be on the cusp of a worldwide golden age of education," spurred by access, through the Internet, to books, teachers, and ideas that will reach those previously barred from such knowledge by custom and geography. Siegel quotes Deirdre McCloskey as saying that "a big and rich Africa will yield a crop of geniuses unprecedented in world history. In a century or so the leading scientists and artists in the world will be black."

Have you thought about that? You might want to contemplate such an outcome, because it's very likely if what we've observed about genetic diversity on that huge continent is correct.

## Conscious Capitalism

I've written extensively about free markets. While I believe that the single-minded pursuit of profit by businesses is misguided and should be tempered with (or replaced by) concern for all the stakeholders, the first lesson that Rajendra Sisodia and I teach in *Conscious Capitalism* is that capitalism is "marvelous, misunderstood, [and] maligned." From a starting point 200 years ago of more than 90 percent of the world living in desperate poverty, how did we get to be as rich as we are?

We wrote,

> In a mere 200 years, business and capitalism have transformed the face of the planet and the complexion of daily life for the vast majority of people. The extraordinary innovations that have sprung from this system have freed so many of us from much of the mindless drudgery that has long accompanied ordinary existence and enabled us to lead more vibrant and fulfilling lives. Wondrous technologies have shrunk time and distance, weaving us together into a seamless fabric of humankind extending to the remotest corners of the planet.

Siegel's narrative fills in the details of this transformation, drawing on modern authors as varied as Matt Ridley, Stewart Brand, Hans Rosling, Johan Norberg, and of course McCloskey—but also a zoo of historical characters, including Charles Dickens (who, it turns out, appears to have been a capitalist free trader, entirely consistent with his Victorian liberalism and compassion for the poor); the Flemish priest Ferdinand

Verbiest, who developed a self-propelled "car" in the 1600s as a gift to the emperor of China; the anonymous builders of the great French cathedrals in the Middle Ages; and the wonderfully named Isambard Kingdom Brunel, England's greatest engineer.

## Perils and Promise of the Future

But the last 200 years were not a straightforward march from the benighted past into the sunlit meadows of liberal capitalist prosperity. Nor will the future be free of problems—far from it! In *Conscious Capitalism*, we cautioned:

> So much has been accomplished, yet much remains to be done. The promise of this marvelous system for human cooperation is far from being completely fulfilled. . . . [A]s a result, we are collectively far less prosperous and less fulfilled than we could be.

And, at this time, we are experiencing a reaction against classical liberal principles in pursuit of the security of borders and trade restrictions, and autocratic rule is in a bit of a revival. These tendencies are almost worldwide and concern me. But, in the long term, I do not think they will be more than a blip on the path to greater productivity, exchange, and freedom.

## Population and the Environment

We also face environmental challenges (not just climate change) and demographic patterns to which we are unaccustomed. Among the demographic patterns are low birth rates, not only in the rich world but in most of the rest of the world; longer and longer lives, due to medical technology and better public health practices; and the aging of the population due to both causes. Only in Africa are birth rates anywhere near their traditionally high levels, and they, too, are falling, although perhaps not fast enough.

At any rate, the world's population will peak around 11 billion later in this century, about a 50 percent increase from the current level. That is, in percentage terms, no more than the increase from 1950 to 1970! And then the population will begin to decline.

Thus we are, finally, getting our population under control. This is wonderful news for the human race and for the environment. We just have to take advantage of this circumstance and use the opportunity to preserve the natural environment and avoid further degradation of it.

## Rational Optimism

Larry Siegel is, of course, not alone in making his contrarian case—although few of his peers have used art, architecture, poetry, philosophy, and humor as extensively as he has, making this book an exceptionally fun read. The genre that I'd call rational optimism (after Matt Ridley's 2012 book, *The Rational Optimist*) has been flourishing. In addition to the books that Larry's volume refers to, I recommend the works of Robert Bryce, Byron Reese, and Thomas Sowell. To gather the courage and resources needed to solve the problems of the future, we need to wash away the hopelessness and alarmism that has permeated public discourse and the world of education. These books, in addition to Larry's, will provide the desperately needed restorative.

To sum up, don't fear the future. Embrace it with all your heart and mind, and work to improve it.

John P. Mackey
Founder and CEO, Whole Foods Market, Inc.

# *Preface*

## How I Came to Write This Book[1]

Thirteen years ago, I read Ben Wattenberg's book, *Fewer*, his elegy for the population explosion. Showing that the explosion was effectively finished in the developed world and quickly coming to an end in the rest of the world, Wattenberg wrote in funereal tones, lamenting the end of an age of youth, exuberance, and innovation.

I thought it was the best news I had ever heard.

(And I'm not one of those antipeople people. Some such folks exist. More about that later.)

For decades we had heard that the population explosion would impoverish and perhaps kill us all. Now that it's almost finished, it's time to reexamine our outlook: we will have *fewer* people than we were expecting, we will become *richer,* and the planet will become *greener.*

This betterment will take place not just in the advanced industrial societies, where self-perpetuating economic growth has been taking place for more than two centuries, but in China, India, and the rest of the developing world. It is the first decent break that three-quarters of the world's population has ever gotten. I don't want it to stop.

[1] A note on language: English does not have a neutral gender embracing both the female and male of the species. In German it's *mensch*, human being, and the word is neuter in meaning (although grammatically masculine, don't ask why). As the best of a bunch of bad choices, when I'm referring to human beings collectively, I often use "mankind," "the ascent of man," "the rights of man," and so forth in this Germanic sense. "Humankind," the repetitive use of "the human race" (which is not a race but a species), and other half-solutions just don't sound right to my ear, although I use them occasionally. I apologize to those who object.

Yet many people are afraid. We have been told, consistently and repeatedly by people who seem to be well informed, that the world is running out of resources; that the economy used to be kinder and more humane; that the costs of advanced technology are greater than the benefits.

These concerns, although mostly misplaced, should not be dismissed as the foolish thinking of Luddites who romanticize the past. Change always produces winners and losers, and we need to understand and respect the concerns of people who believe they are likely to suffer rather than benefit from it. Progress is not a military march from debased to elevated ways of living. It is confusing and messy.

Life has improved tremendously in the last 250 years; this book argues that it will continue to improve in almost every dimension: health, wealth, longevity, nutrition, literacy, peace, freedom, and so forth. Without overlooking the many obstacles on the path of progress, my aim is to reinforce and help restore people's faith in the future—and help them understand why optimism is amply justified.

▪ ▪ ▪

But this is not a book about population, and I am not a demographer. I've spent my adult life studying the economy, businesses, and investments, so that is the lens through which I tend to observe human action. As a result, economic progress is my principal focus, although it's tempting—and I've succumbed to the temptation—to write about closely related noneconomic topics such as population and the environment.

This book, then, is mostly about "richer." More food, better food, less hunger. Longer lives, healthier lives, happier lives. Less work, easier work. Technology that satisfies and makes life more pleasant and interesting. A culture worth enjoying.

The amount of betterment that has taken place over the last 250 years—the Great Enrichment—in what we call the developed world, and which is rapidly spreading to the rest of the world, is almost unbelievable. Humans have lived on Earth for many tens of thousands of years, but only in the last quarter-millennium has any large number of them enjoyed the fruits of economic growth and technology on a sustained basis. There were exceptions—Leonardo's Florence and Mozart's Vienna come to mind—but most human lives until the Great Enrichment were lived under conditions of deprivation that are almost unimaginable today.[2]

How we got out of that morass and how the whole world can and will enjoy the fruits of the Great Enrichment is the real topic of this book.

Of course, like anything else, economic output cannot grow forever—if, by output, we mean piles of stuff. But economic growth can, and will, mean other things too: products that are more useful or economical, services and experiences that did not exist before, methods of delivery and disposal that improve upon existing practices. That kind of progress can proceed as far as human ingenuity takes it.

■ ■ ■

I am also not a naturalist, but I care about nature. I'd better, since we're an essential part of it.

It is possible to imagine a future in which humans appear to prosper at the expense of other living species and the rest of the physical world; this is unacceptable and, in the long run, self-defeating. I am old enough to remember filthy air and water in the United States. Through new and necessary laws, but more importantly through economic growth and improvements in technology, we've largely resolved that problem. The air and water are still filthy in some other places.

There are many environmental dragons to slay, and to do that we need an abundance of fiscal and intellectual resources. Continued economic growth is the key to the solution. It will not be cheap or easy, but we will largely prevail.

Fewer, richer, greener.

■ ■ ■

This book is a guide to the past and the likely (but not guaranteed) future of human betterment, but it is also a reader's guide. In the reader's

---

[2] The phrase "Great Enrichment" is from the works of Deirdre McCloskey, from whom we will hear much in this book. She said, "It was a stunning Great Enrichment, material and cultural, well beyond the classic Industrial Revolution, 1760–1860, which merely doubled income per head. Such doubling revolutions as the Industrial had been rare but not unheard of, as in the surge of northern Italian industrialization in the Quattrocento" (McCloskey 2019). Unlike previous doublings, the doubling of income from 1760 to 1860 led to yet another doubling, then another, continuing today and spreading across the world. That is what makes the modern era unique in human history.

guide, I briefly describe the 20 or so books that have most directly influenced my thinking on these topics, plus data sources.

Luck is the most important factor in life, and I've been blessed with it. I dedicate this book to Connie O'Hara, who, on account of my inexplicable good fortune, became my wife almost four decades ago and is sitting right next to me at this moment.

# Acknowledgments

This book literally could not have been written without the efforts of David L. Stanwick and Joanne Needham. Dave Stanwick, research assistant extraordinaire, is the kind of colleague who knows the answers to a question before I've finished asking it. He is also a skilled web designer, digital marketer, literary editor, and food vendor (Blazing Bella oil and vinegar). Joanne Needham, my permissions editor, solved some of the most difficult permissions riddles I could dream up, investigating sources behind sources and so on *ad infinitum*. She even commissioned artists and photographers to create a number of illustrations. Thanks!

I want to express deep appreciation to John Mackey, who graciously volunteered to write the foreword to this book, and who founded what I think is the world's finest chain of grocery stores, Whole Foods Market, Inc.

My delightful literary agent, Lucinda Karter, is someone else without whom this book would not have been written. She persuaded me to turn my old (2012) magazine article on this topic into a full-length book. I also want to thank my publisher, Bill Falloon at Wiley, and his excellent staff, for their help and support.

Lavish thanks are due to those who made helpful comments, including Stephen C. Sexauer and M. Barton Waring—both frequent co-authors of mine whose ideas made it into the book in various ways—as well as David E. Adler, Ted Aronson, the late Peter Bernstein, Erik Brynjolfsson, Thomas Coleman, Elizabeth Hilpman, Michael Falk, Michael Gibbs, William Goetzmann, Walter (Bud) Haslett, Gary Hoover, Bob Huebscher, Lee Kaplan, Robert Kiernan, Marty Leibowitz, Deirdre McCloskey, John O'Brien, Ben Rudd, Thomas Totten, and Wayne Wagner. Roger Ibbotson introduced me to the nascent New Optimism genre a very long time ago, pointing me to the works of Julian Simon and Petr Beckmann. He and I have had a long and fruitful collaboration, and I look forward to further efforts with him.

I am of course deeply indebted to all the people whose research and writing I referred to, and to those who gave permission for their work to be cited or illustrations reproduced. They are noted in the footnotes, illustration credits, and the reader's guide.

Finally, I wish to thank my family for putting up with me during this time-intensive wild adventure. My wife, Connie, turns up in the preface with a fuller expression of my appreciation of her.

# PART I
# THE GREAT BETTERMENT

# 1

# *Right Here, Right Now*

Not too long ago, President Barack Obama said, "If you had to choose any time in the course of human history to be alive, you'd choose this one. Right here . . . right now."[1]

In a world suffering from war and rumors of war, environmental destruction, poverty, disease, inequity, and every other form of misery known to man, what was he talking about? What is so special, so desirable, about this time in history?

First of all, we are richer, far richer, than any human population that has ever preceded us. Richer not just in money and goods, but in food, health, longevity, education, culture, safety, and just about everything else that people need and crave.

Second, we are going to continue to get richer. The present state of human achievement is not a brief, shining moment that will all too quickly pass. We are on the verge of the greatest democratization of wealth and well-being that the world has ever known.

How do I know this? We mostly understand the future by studying the past. Since this is not just an economics book but also a picture book, one that tries to convey its message visually, let's start right away with a picture of income per person in the United States from 1820 to the present and for the whole world from 1870 to the present, as shown in Figure 1.1.

---

[1] Obama (2016).

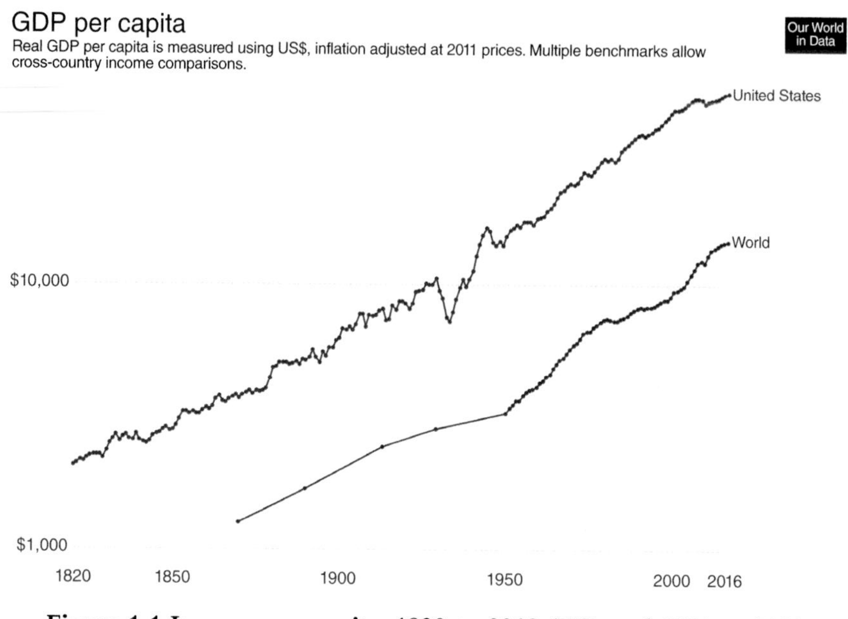


**Figure 1.1 Income per capita, 1820 to 2018 (US) and 1870 to 2018 (world), in today's dollars.**

*Note:* "Today's" dollars are 2011 dollars. To convert to 2019 dollars, multiply by 1.12.

*Source:* https://ourworldindata.org/economic-growth.[2] Maddison Project Database, 2018. Our World in Data. Licensed under CC BY 4.0.

Third, and most importantly, the growth shown for the United States in the top line of Figure 1.1 is finally being experienced by the world at large, including what are historically some of the world's poorest countries. We in the United States are not alone in our prosperity and in the promise of our future. We do not want to be alone. Economic opportunity, cultural achievement, and human betterment are a precious resource to be shared.

This book is concerned with the past, present, and future of the whole world. If I appear to be US-centric, it's because much of the available data come from the United States, and I am an American. The United Kingdom, where careful records have been kept for longer than in the United States, is also an important source of information. But these are just examples. The forces that have enriched these countries can enrich anyone, and will.

[2] Bolt et al. (2018).

Looking at Figure 1.1 a little more closely reveals a remarkable fact. In 2016, the average income of all the people in the world was $14,574.[3] How far back do you have to go in history to find a year when the average income in the United States was that low? Take a guess. 1800? 1875? 1920?

It was 1949. The United States and its allies had just won World War II, we were enjoying the fruits of a postwar economic boom, and that time in our history is generally remembered fondly. We were the richest country in the world. *Yet the world today is as rich as we in America were then.*

That is what Barack Obama was talking about. In the Obama quote, I left out "in America"—he actually said, "right here in America, right now"—but I wish he had said, "right here in the world, right now." Though the developed world has been rich for a while, the Great Fact of our time is that the developing world is catching up with us.[4] It's about time.

## The Charming Little House

How well did the average American live in 1949, with an income *per capita* of about $14,500 in today's money?[5] One very visible measure of well-being is housing. Figure 1.2 shows a five-room house that a typical American family might have lived in at that time. I did not select a new house, since most people live in houses that have been around for a while; the house in the drawing was offered for sale two decades earlier. Also, everyone shared one bathroom and all the children would have had to share the one bedroom not occupied by the parents. But it was certainly adequate and, to modern tastes, charming.

[3] Again, in "2011 US dollars"—that is, the income that each person earned in 2016 was equal to what $14,574 would have bought in the United States in 2011. This number has been adjusted for "purchasing power parity," that is, the differences in the cost of living between various countries. Also, a minor technical point, these numbers are gross domestic product (GDP) per capita, not personal income, though the two concepts are very closely related.

[4] I am being a little cute with the use of Deirdre McCloskey's term, "the Great Fact." She uses it to refer to the fact of massive overall economic development since about 1800; I use it (at least in this instance) to refer to the catch-up of less developed countries to the more developed ones.

[5] Again, strictly speaking, a GDP *per capita* of that amount. Households do not get all of it, the government and capital providers receiving the rest.

THE WAYNE

The Wayne is one of our most popular designs. One looking at the exterior of this house is impressed with its stability. A study of its floor plan reveals the unusual care taken by the architect to give the largest rooms possible. Special attention is drawn to the large living room and large bedroom on the second floor and the well-arranged stairway.

**Details and features:** Five rooms and one bath. Full-width front porch supported by square columns; gabled dormer. Swinging door between kitchen and dining room; open stairs.

**Years and catalog numbers:** 1925 (13210); *1926* (P13210); 1928 (C13210); 1929 (P13210)

**Price:** $1,994 to $2,121



**Figure 1.2 A house that an American family of average means in 1949 might have lived in.**

*Source:* Ward and Stevenson (1996). © John Wiley & Sons, Inc. Reproduced with permission.

Now, here is what I've called the single most amazing fact in economic history: *the average person in the world lives about this well now.*[6]

That does not mean the average person lives in this particular way. Patterns of human settlement have changed. An average (that is to say, moderately comfortable by 1949 American standards) family in a developing country is more likely to live in a high-rise, or in a multifamily unit with less green space around it. Today's average householder might also not live in a home that is quite as charming (according to my old-fashioned, American biases)—or they might love high-rise living; everybody's different. But they'd have an income consistent with this general level of living, which is the point I'm really trying to make.

## A Tale of Three Authors

This book will fill in the details of how this remarkable story of human betterment came to be true, what the problems and shortcomings are, and what solutions one might consider. I am emphatic about thinking as an economist does: about the existence of limits, the importance of trade-offs, the power of incentives, and the tyranny of unforeseen consequences.

I rely heavily on researchers and authors who have tread this path already:

- Matt Ridley, who described "how prosperity evolves."
- The late Hans Rosling, who documented the transition, achieved in different countries at various rates and at various times, from "poor and sick" to "rich and healthy."
- Deirdre McCloskey, who teaches that culture and values are forces even more powerful than pure economic calculation in determining human futures.

There are, of course, many other thinkers whom we'll meet along the way. The people who produce ideas worth thinking about are as important as the ideas themselves.

The remainder of this chapter introduces the work of Ridley and Rosling—little pieces of each, since their work is so voluminous. Their ideas are basic to understanding the book's main themes. Their

---

[6] To be more precise, the house is one that an American of statistically average income could have bought in 1949 if he or she had access, then, to what is the statistically average income in the world now.

contributions make it clear how *betterment*—McCloskey's term—has been the grand trend of the last 250 years, emerging at first in north-western Europe and its overseas offshoots and then only quite recently spreading to much of the rest of the world.

I'll get to McCloskey later in the book, but her insights were as much an inspiration to me in writing this book as Ridley's and Rosling's, and I wanted to give early mention to all three because of their influence in setting me along this path. In the subsequent sections, we'll begin to explore the idea of *Fewer*—the fact that the population explosion is ending and why it's happening. Then *Richer*, then *Greener*.

## Matt Ridley: How Prosperity Evolves

It's a little surprising for an optimistic book on the future of the human race to start by citing the work of Matt Ridley (shown in Figure 1.3), more properly fifth Viscount Ridley, member of the House of Lords, banker, and owner of land on which coal is mined.

But Ridley is also one of the world's finest popular science writers, an explainer of complicated things in simple terms (but not too simple). The variety of his interests is revealed by a few of his book titles: *The Origins of Virtue*, *The Evolution of Everything*, and *Genome: The Autobiography of a Species in 23 Chapters*.

For us, the book of Ridley's that is most relevant is *The Rational Optimist: How Prosperity Evolves.*[7] It's a tour de force that explores how ideas "have sex" (examples: carriage + engine = automobile; telephone + computer = Internet), creating prosperity out of, well, almost nothing.

Although ideas having sex is a headline grabber, Ridley's main point is that prosperity evolves through specialization and trade. This idea is hardly new: it's also the principal theme of Adam Smith's 1776 book, *The Wealth of Nations*. But Ridley has new ways of explaining old concepts:

> In the two hours since I got out of bed I have showered in water heated by North Sea gas, shaved using an American razor running on electricity made from British coal, eaten a slice of bread made from French wheat, spread with New Zealand butter and Spanish marmalade, then brewed a cup of tea using leaves grown in Sri Lanka,

[7] All the quotes in this section are from Ridley (2010).

**Figure 1.3 Matt Ridley**

*Source:* Chris McAndrew, 2018, https://upload.wikimedia.org/wikipedia/commons/2/2c/Official_portrait_of_Viscount_Ridley_crop_2.jpg. Licensed under CC 3.0.

> dressed myself in clothes of Indian cotton and Australian wool, with shoes of Chinese leather and Malaysian rubber, and read a newspaper made from Finnish wood pulp and Chinese ink . . .

How does this exotic bounty link back to Adam Smith and specialization? Ridley explains,

> More to the point, I have also consumed minuscule fractions of the productive labour of many dozens of people. . . . They were all, though they did not know it, working for me. In exchange for some fraction of my spending, each supplied me with some fraction of their work. They gave me what I wanted just when I wanted it—as if I were the Roi Soleil, Louis XIV, at Versailles in 1700 . . .

But I'm not the Roi Soleil, the Sun King, who had 498 servants at his disposal *just to prepare meals*. How does my life differ from his?

> [While Louis XIV] . . . was the richest of the rich . . . , the cornucopia that greets you [a person of average means] as you enter the supermarket dwarfs anything that [Louis] ever experienced . . . beef, chicken, pork, lamb, fish, prawns, scallops, eggs, potatoes, beans, carrots, cabbage, aubergine, kumquats, celeriac, okra, seven kinds of lettuce, cooked in olive, walnut, sunflower or peanut oil and flavoured with cilantro, turmeric, basil or rosemary . . .

It's not just delicious food, of course. Ridley reminds you that although "you have no woodcutters to bring you logs for the fire, . . . the operators of gas rigs in Russia are clamouring to bring you clean central heating." You can select from affordable clothes made "all over Asia." A light switch will bring you the power of a thousand candles. In other words,

> [Y]ou have far, far more than 498 servants at your immediate beck and call. Of course, unlike the Sun King's servants, these people work for many other people too, but from your perspective what is the difference? That is the magic that exchange and specialisation have wrought for the human species.

Okay, you've gotten the flavor of Matt Ridley's way of thinking; enough for now. You wish your high school history teacher could have explained things like that. But, through technology that began with Gutenberg in the 1450s, he *can* be your history teacher. Now that the Internet exists, he can do so without your lifting a finger. Well, maybe one finger.

## Hans Rosling: Poor and Sick to Rich and Healthy

Whereas Matt Ridley is a consummate armchair intellectual, Hans Rosling (Figure 1.4) was a man of action—too much action for his own good, although he did great good for others. A medical doctor, Rosling died in 2017 of a disease he might have contracted while helping the suffering poor in East Africa.[8] He was known for his startling lectures (distributed as TED talks) on improvements in health and wealth over time, and his posthumous book, *Factfulness,* co-authored with Ola Rosling and Anna Rosling Rönnlund, has become a bestseller.[9]

---

[8] It is not definitively known whether the hepatitis that apparently led to his death from liver failure and pancreatic cancer at age 68 was contracted through his medical work.

[9] Rosling et al. (2018).

**Figure 1.4 Hans Rosling**

*Source:* Photo by Elisabeth Toll. Courtesy of Gapminder.

## The World in 1800: Mostly Poor and Sick

As visual as Ridley is literary, Rosling tells his story in pictures (which were, when he presented them, movies). Figure 1.5, a composite of two of Rosling's famous bubble diagrams, shows the world in 1800 and 2018, with the area of each country's bubble drawn in proportion to its population and its location on the grid showing the income ($x$-axis) and life expectancy ($y$-axis) in that country. The colors denote geographic areas: Asia (dark gray circles), Europe (light gray circles), and so forth. Magnified versions, so you can see the individual countries and the data, will come a little later.

Note that, in 1800, all the countries in the world are squeezed into the lower left corner, with low incomes and short lives. The impact of using the same scale for 1800 and 2018 is very powerful—even more powerful when you realize the income (horizontal) axis is logarithmic,

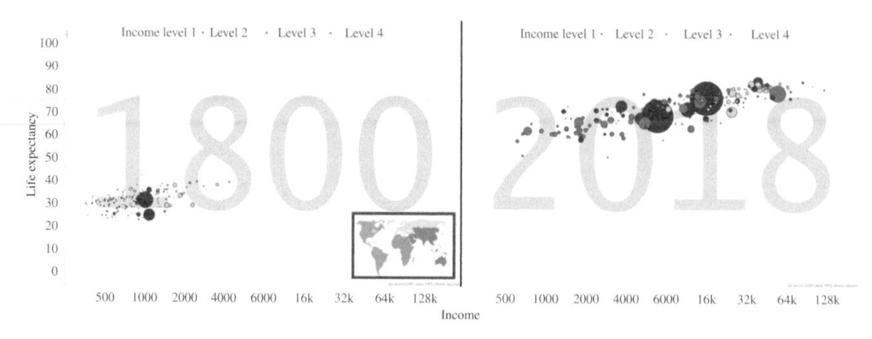


**Figure 1.5 The world in 1800 and 2018: Poor and sick to rich and healthy.**
*Source:* Created with Gapminder.org's Tools Offline.

doubling with each marker. Few people, if any, dreamed in 1800 that the upper and rightmost reaches of this diagram would be filled with people (actually, bubbles representing people) two centuries later.

As Rosling exclaimed with great emotion in his TED talk, "All the countries were sick and poor!" In 1800, even the richest countries, the UK and the Netherlands, had life expectancies of about 40. Of course, infant mortality was responsible for much of the low life expectancy, and a great many people lived beyond 40. But the average person's life was pretty miserable, even in the more advanced countries.

To illustrate, Nathan Mayer Rothschild (1777–1836), the richest man in the world in the early 1800s, died at 58 of an infection that could have been cured with about a dollar's worth of penicillin if that substance had been discovered. There were many things that money couldn't buy.[10]

Figure 1.6 shows the world in 1800 (the left panel of the previous figure) on a magnified scale, so you can see each country, where it was "located" in terms of income and life expectancy, and how big its relative population was.

## The World in 1948: Maximum Inequality, the End of the Great Divergence

Skip ahead a century and a half, to 1948, which Rosling describes as "a great year." World War II had finally come to an end, the economy was booming, a number of countries such as India and Israel were newly independent, and, Rosling adds, "I was born."

[10] Rothschild's fortune, expressed in today's money, has been estimated at a quarter trillion dollars—about as much as the net worth of Jeff Bezos and Warren Buffett combined as of October 2018.

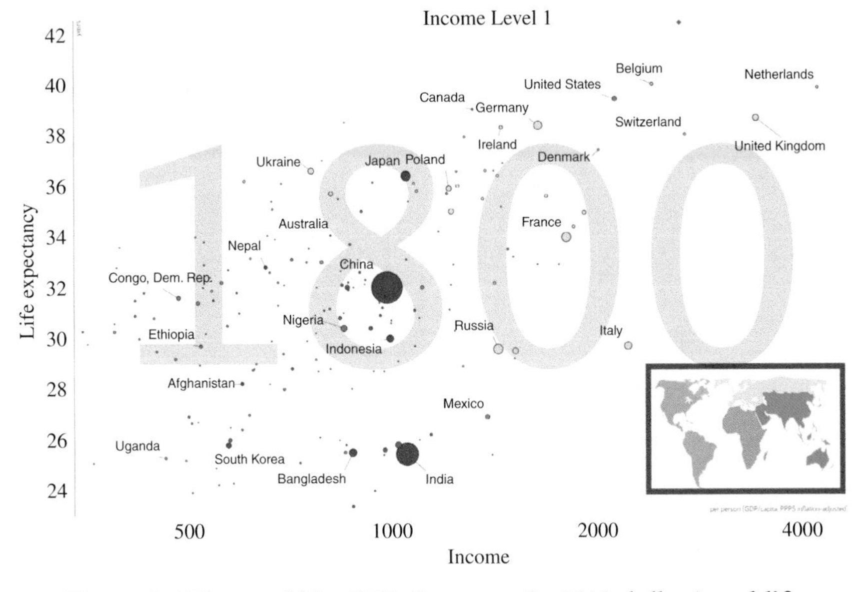


**Figure 1.6 The world in 1800: Incomes (in 2011 dollars) and life expectancy at birth.**

*Source:* Created with Gapminder.org's Tools Offline.

But, as Figure 1.7 shows, 1948 was also roughly the year of peak inequality in the world. The rich world was already pretty rich: the United States is the large gray circle way off to the right. In terms of income, the poor world, much larger, had either made modest gains or had actually lost ground. (Life expectancies had risen everywhere.) Astonishingly, China, the largest bubble, was considerably *poorer* than it had been in 1800, and India, the second largest bubble, was a little poorer than in 1800. Asia was poorer than Africa.

This splitting of the world into rich and poor spheres is sometimes called the *Great Divergence*.

## The World in 2018: The Great Convergence Finally Takes Hold

After World War II, the Great Divergence began to reverse. At first, the poor countries that started to get ahead quickly were in southern Europe, then the Asian Tigers and parts of Latin America. But the really big change in the world picture came with the opening of China to market reforms in 1978 and of India in 1991 to 1993. Figure 1.8 is the bubble chart as of today, a magnified version of the right panel in Figure 1.6.

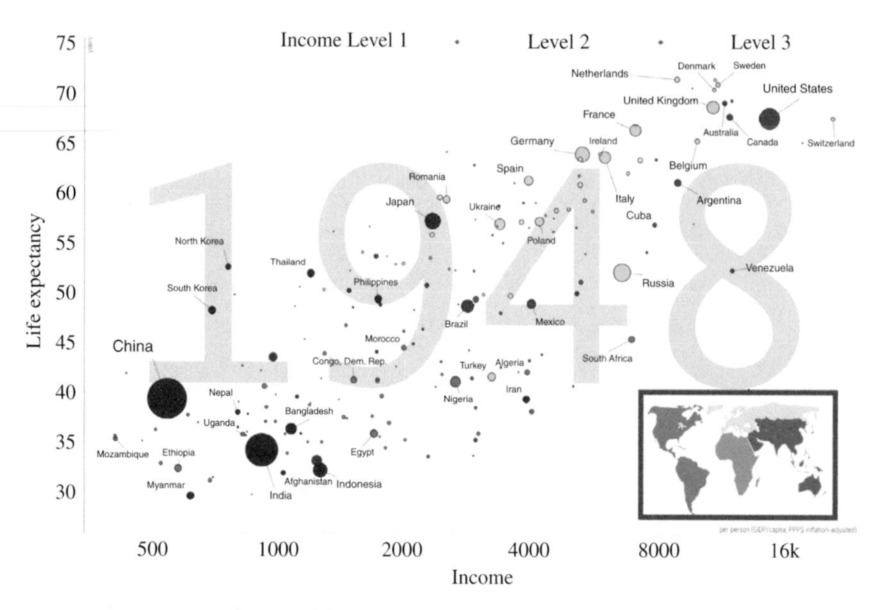


**Figure 1.7 The world in 1948: Some are rich, most are poor.**

*Source:* Created with Gapminder.org's Tools Offline.

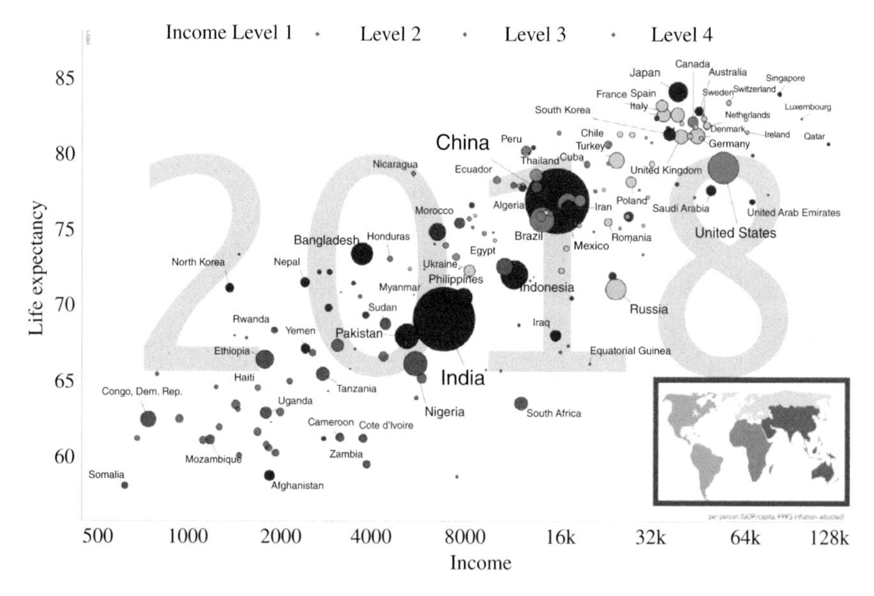


**Figure 1.8 The world in 2018: Approaching a middle-income world.**

*Source:* Created with Gapminder.org's Tools Offline.

In Rosling's TED talk, he practically jumps off the stage as he sees China's bubble rise rapidly in the last quarter century—he exclaims that the poor countries are "getting healthier and healthier!" His enthusiasm for the modern period contrasts with his view of World War I and the 1918 influenza epidemic—"what a catastrophe!"—which is revealed as a dramatic, short-lived plunge (not shown here) in length of life.

What does Rosling think of the current state of the world? For the first time in history,

> Most people today live in the middle, [although] some Africans were stuck in civil war and some were hit by HIV. But there are huge differences . . . between the best of countries and the worst of countries. And there are also huge inequalities within countries . . . Take China—I can split it into provinces. Shanghai . . . has the same wealth and health as Italy . . . [but] the rural parts are like Ghana.
>
> And yet, despite the enormous disparities today, we have seen 200 years of remarkable progress. That huge historical gap between the West and the rest is now closing. We have become an entirely new *converging* world. And I see a clear trend into the future: with aid, trade, technology, and peace, it is fully possible that everyone can make it into the healthy, wealthy corner.[11]

## Conclusion

We'll revisit these authors as we document, in Part III of this book, the improvement over the last 250 years in food, health, wealth, energy, education, technology, safety, and just about every other aspect of life. But first, we explore the key reason why we're so optimistic about the future: *the population explosion is coming to an end,* massively reducing pressure on the environment, the food supply, and our living space. Part II, "Fewer," explores this unique hinge event in the world's history.

[11] Although Rosling produced many TED talks, his best-known one, which really should not be missed, is at www.youtube.com/embed/jbkSRLYSojo. The quotes are from this talk.

# PART II
# FEWER

# 2

# *The Population Explosion, Malthus, and the Ghost of Christmas Present*

The population explosion is, in the broad sweep of history, just about over. It will almost certainly end in this century, with a world population about 11 times larger than when the Great Enrichment started about 250 years ago.

Let's introduce the idea with a picture, which in this case is worth way more than a thousand words. Figure 2.1 shows the past and projected world population from 1750 to 2100. The annual rate of growth is also shown, represented by the black line.

This graph depicts one of the two central ideas of this book. The other is Figure 1.1 in Chapter 1, showing the Great Betterment over the last 200 years. Between them, they convey the essence of "fewer, richer, greener." The rest is elaboration.[1]

The graph in Figure 2.1 is on an arithmetic scale; if it were on a logarithmic scale, showing equal percentage changes as equal vertical distances, the slowing of population growth would be much more

[1] Apologies to Hillel.

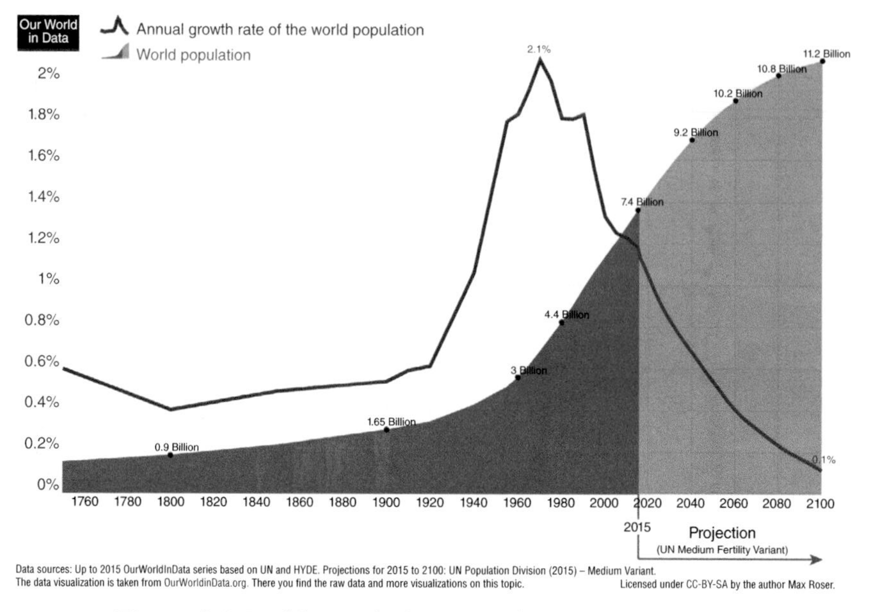


**Figure 2.1 World population growth, 1750 to 2015 with projections to 2100.**

*Data sources:* Up to 2015 OurWorldInData series based on UN and HYDE. Projections for 2015 to 2100: UN Population Division (2015)—Medium Variant. The data visualization is taken from OurWorldinData.org. There you find the raw data and more visualizations on this topic

*Source:* Max Roser, Our World in Data, https://ourworldindata.org/wp-content/uploads/2013/05/updated-World-Population-Growth-1750-2100.png. Licensed under CC BY-SA; a version is in Pinker (2018), p. 125.

dramatic. But the arithmetic scale has the advantage of giving one person exactly as much visual weight as another, regardless of when he or she lived, so we'll stick with that scale for now.

Peak population growth (as a rate, the black line) was around 1970. No wonder the radical pessimists Paul Ehrlich, John Holdren, and the Club of Rome made such terrible forecasts around that time. If current trends had continued, in other words if people did not respond to changing incentives but just kept having as many children as they did when life expectancies were short and death rates high, their forecasts might have been right.[2]

[2]See www.theatlantic.com/personal/archive/2008/12/the-insights-of-paul-ehrlich/55882/. It is interesting to note that John Holdren, who said he wanted to "de-develop the United States," was appointed senior science adviser to President Barack Obama—although, to be fair, he said it 40 years before he took office.

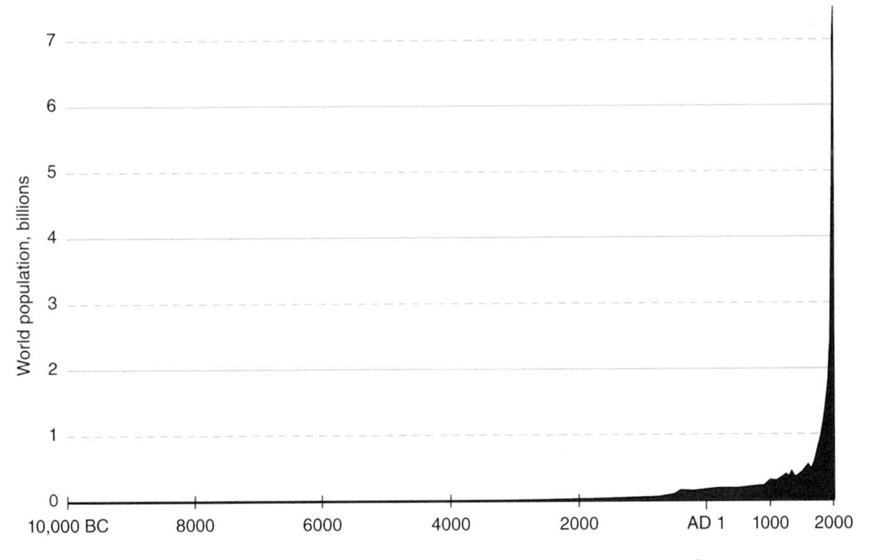


**Figure 2.2 World population from prehistoric times to the present.**
*Source:* https://upload.wikimedia.org/wikipedia/commons/b/b7/Population_curve.svg.

## What Happened to the Population Hockey Stick?

The story told by Figure 2.1 contrasts with the more commonly seen, panic-inducing population hockey stick in Figure 2.2. Both are numerically correct, but focusing on the distant past, when a small group of human beings could barely feed itself, is not particularly helpful in thinking about and planning for the future.

Looking again at Figure 2.1, the big slowdown in the population growth rate came around 1990, as "emerging markets" became more than a marketing slogan and market economies in less developed countries actually began to emerge. China's period of hypergrowth began with the ascent of Deng Xiaoping to power in 1978. India's began with the 1991 rise of the finance minister Manmohan Singh. Thus, the two most populous and influential nonwestern countries started their march upward from poverty within a generation of each other. They both did so by pursuing variations on market themes. Is it any wonder that their populations began to stabilize around that time too?[3]

[3] The one-child policy in China that prevailed from 1979 to 2015 or 2016 is an exception to the rule that people voluntarily reduce their fertility in response to economic incentives; in that case the reduction was involuntary. Researchers differ on whether the policy was effective (that is, how many births were prevented) or if it was justified.

## Falling Birth Rates: Boon or Disaster?

Ben Wattenberg, in *Fewer* (2005), was less than enthusiastic about the dramatic fall in birth rates. The United Nations Population Division had just come out with a set of forecasts that were stunning in the speed at which they expected the population explosion to end—and reverse. (They later revised the forecasts upward somewhat.)

Wattenberg meticulously documents the nearly global baby bust, as we'll do in some detail in the next chapter, but comes to the wrong conclusions. I'll focus on a humorous example so that I don't have to summarize his whole dreary book:

> [A] February 2004 article in the newspaper *The Scotsman* [says], "The population of Scotland will fall below 5 million by 2009 . . . More worrying . . . is . . . the fact that . . . it's also getting older . . . All this suggests that by the year 3573 there'll be two people left in Scotland, probably a married couple in their 90s living in Beardsden." I don't know where Beardsden is, but I know that something strange and unnatural is happening.[4]

This isn't going to happen, and it isn't a new concern. In 1934, the British demographer Enid Charles wrote, in almost exactly the same panicked tone, "in place of the Malthusian menace of over-population there is now a real danger of under-population."[5] Harvard professor Kenneth Hill says, "Charles blamed the low fertility of the time on what she called the 'Acquisitive Society,' in which children were 'regarded as a form of capital expenditure which brings the parent no return commensurate with its investment value to society as a whole. . . . [T]he choice between a Ford and a baby is usually made in favour of the Ford.'"[6]

It is just too tempting to point out that babies are occasionally conceived in Fords. At any rate, the baby bust will self-correct because incentives will change yet again, this time in favor of more children. We will not run out of people. We explore the data and mechanisms of the decline in the birth rate in some detail, along with speculations about the future, in Chapters 3, "The Demographic Transition," and 4, "Having Fewer Children."

---

[4] Wattenberg (2005), p. 224.

[5] Charles (1934).

[6] As quoted in McDonald (2004).

## "The Paragon of Animals"

As I said in the preface, I thought—upon reading Wattenberg's *Fewer*, a dozen years ago—that it conveyed wonderful news, and I'm not one of those antipeople people. Some writers have referred to the human race as a cancer.[7] Actually, we *cure* cancer. Nature gives it to us and, increasingly, we make it stop. No metaphor could be less fitting. My view of the human race is better captured by Hamlet: "What a piece of work is man! how noble in reason! how infinite in faculties! in form and moving how express and admirable . . . the paragon of animals."

And when Hamlet says that we are nevertheless only dust, he is talking about *his* failings, not ours.[8]

I thought the eventual (not that far off) end of the population explosion was welcome news because the alternative is hard to think about. No physical thing can grow *forever* at a rate other than zero, so it's just a question of when, not if, the growth will stop. Sooner is almost certainly better than later. There are some Shakespeares and Einsteins who will never be born, but a great many people will live better lives.

I do not agree with those who say we are facing an ecological catastrophe, although there's some chance of it—but I concede that the planet is getting crowded. (It was getting crowded when the world's population crossed three billion, around 1960.) A lot of the problems we've been facing come from trying to bring to rapidly growing numbers of people

---

[7] "Growth for the sake of growth is the ideology of the cancer cell." —Edward Abbey (1977). Abbey was a prolific and gifted writer on many topics, including nature and Western living; one of his novels was made into the classic film *Lonely Are the Brave*. He may have been referring to economic growth, not population growth, but they're related. Sir David Attenborough more explicitly said we're a disease: "We are a plague on the Earth" (www.telegraph.co.uk/news/earth/earthnews/9815862/Humans-are-plague-on-Earth-Attenborough.html). In 2013, when Attenborough said it, he should have known better; in 1960 or 1970 the comment might have been forgivable, since the population explosion was in full flower.

[8] Although this question is sometimes debated by Shakespeare scholars, it is clear (to me) that this is the correct interpretation given the immediately preceding line: "I have of late,—but wherefore I know not, —lost all my mirth, forgone all custom of exercises; and indeed it goes so heavily with my disposition that this goodly frame, the earth, seems to me a sterile promontory; this most excellent canopy, the air, look you, this brave o'erhanging firmament, this majestical roof fretted with golden fire, why, it appears no other thing *to me* but a foul and pestilent congregation of vapours" [my italics]. Source: www.bartleby.com, which incidentally uses the alternate wording, "What a piece of work is a man," found in some of Shakespeare's folios but not others.

the benefits of increasing wealth, more advanced technology, and other aspects of modern life. When some of the pressure from population growth is removed, those problems become much less severe—and it becomes easier to envision, in the only moderately distant future, a whole world as prosperous, healthy, and pleasant as the developed world is now.

## Malthus's Dismal Theory

Population grows when birth rates exceed death rates. (In-migration and out-migration changes this equation, but let's focus on "natural increase," births minus deaths.) Under what demographers call Malthusian conditions, named after the English demographer and Anglican preacher Thomas Malthus (1766–1834), population expands to consume all available food and other resources. Thus any economic progress made by a group is temporary, and is wiped out by a subsequent increase in population made possible by the transient surplus. He wrote,

> The power of population is so superior to the power of the earth to produce subsistence for man, that premature death must in some shape or other visit the human race. . . . [S]ickly seasons, epidemics, pestilence, and plague advance in terrific array, and sweep off their thousands and tens of thousands. Should success be still incomplete, gigantic inevitable famine stalks in the rear, and with one mighty blow levels the population.[9]

Over most of the world's history, it's widely believed that population growth proceeded in a Malthusian fashion, with world population increasing from somewhere between 200 and 400 million in the year 1 to a little under a billion in 1750. This was all made possible, the Malthusian theory says, by increases in the amount of food that could be grown (or gathered, or hunted, or fished) and with almost no economic growth per capita.

A gifted economic blogger calling himself Pseudoerasmus notes that we moderns often misunderstand Malthus's theory:

> [It] does *not* necessarily mean that the average person was living on the edge of starvation. . . . To the contrary, the neo-Malthusian model implies that *anything* which lowers the birth rate and increases the death rate will raise the living standards of the average person. This is

[9] Malthus ([1798] 1993).

why different societies with different fertility practises and mortality conditions had very different income levels.[10]

At various times, then, certain societies flourished for a while and the average person could live at better than subsistence levels, but, until the Great Enrichment began, those periods always ended sooner or later. Then, something changed.

## Malthus's Mistake

It is hard to fault Malthus for setting forth a theory that explained so much of what was, for him, the recent past and even the present. But he had the misfortune to be born at a time when the past was an almost useless forecast of the future. The Industrial Revolution started just as he came into the world, and would be powerfully underway in his native England by the time he left it. "Malthusian" is now an epithet used to describe a world that is, in the political philosopher Thomas Hobbes's words, "solitary, poor, nasty, brutish, and short."[11] Malthus did not foresee that world receding behind him.

His mistake was to think that there were only two forces at work in restraining population growth: the "negative checks" such as famine, disease, and war that Malthus feared, and the "positive checks" of birth control and celibacy by which he hoped mankind would avoid the negative ones. He did not envision the third force, economic growth changing the incentives to procreate so that having fewer, rather than more, children would convey an advantage to the parents. The third force is related to the second—if you decide to have fewer children, you need a mechanism for doing so—but without taking into account changing incentives, it's hard to see why people would voluntarily restrain their reproduction.

## From Malthus to Modernism

Having taken a stand that is essentially economic, Malthus and his work attracted the attention of the other economists of Malthus's day.

[10] "Pseudoerasmus" (2014).

[11] Hobbes (1651). The waggish, conservative *American Spectator* magazine claims that an organization called Solitary, Poor, Nasty, Brutish & Short is its legal counsel.

**Figure 2.3 Charles Dickens.**

*Source:* Nicku/Shutterstock.

Jean-Baptiste Say (pronounced *sigh*), a French disciple of Adam Smith, wrote a book in opposition to Malthus. David Ricardo, the British economic giant of his time, became Malthus's friend and failed to see the flaw in Malthus's forecast. But all these men admired Adam Smith, the founder of modern economics. A roughly contemporary historian observed:

> If Malthus and Ricardo differed, it was a difference of men who accepted the same first principles. They both professed to interpret Adam Smith as the true prophet, and represented different shades of opinion rather than diverging sects.[12]

This intellectual kerfuffle among friends is a good segue to an examination of Adam Smith and *The Wealth of Nations*, the dawning of the age of improvement, and the reasons Malthus was so wrong. It turns out Malthus had crossed (if intellectual fights can reach beyond death) no less a master than Charles Dickens (see Figure 2.3). Herein lies a tale well worth telling, but it is a bit of a side trip relative to our main path, which is to the demographic transition and the eventual end of the

[12] Stephen (1900).

population explosion and the reasons for it. I hope you will be patient, and entertained.

## The Modern Voice of the Ghost of Christmas Present

The idea that Malthus might be wrong about the future, and that per capita economic growth could be sustained and living conditions improved, had been in the wind for some time when Jean-Baptiste Say published his letters contra Malthus.[13] Rather than getting into the weeds of Say's work, let's look at the Forbes.com columnist Jerry Bowyer's extraordinary and influential 2012 article linking the topic to Charles Dickens' *A Christmas Carol.*

Dickens had Scrooge say:

> "I help to support the [poorhouses] I have mentioned: they cost enough: and those who are badly off must go there."
>
> "Many can't go there; and many would rather die" [Scrooge's visitor replied].
>
> "If they would rather die," said Scrooge, "they had better do it, and decrease the surplus population."[14]

Surplus population, hmm. Bowyer observes,

> That phrase—surplus population—is what first tipped me off to Dickens' philosophical agenda. . . . He was weighing in on one of the central economic debates of his time, the one that raged between Thomas Malthus and one of the disciples of Adam Smith.
>
> Malthus famously argued that in a world in which economies grew arithmetically and population grew geometrically, mass want would be inevitable. . . . Jean-Baptiste Say, Smith's most influential disciple, argued on the other hand, as had his mentor, that the gains from global population growth, spread over vast expanses of trading, trigger gains from a division of labor which exceed those ever thought possible before the rise of the market order.
>
> Guess whose ideas Charles Dickens put into the mouth of his antagonist Ebenezer Scrooge.[15]

---

[13] Say (1821).

[14] Dickens (1843).

[15] This quote and the next one are from Bowyer (2012).

And guess who speaks in the voice of Adam Smith, Jean-Baptiste Say, and modernity? Bowyer writes:

> The Ghost of Christmas Present is the key to understanding Dickens' political and economic philosophy. He is the symbol of abundance. He literally and figuratively holds a cornucopia, a horn of plenty. . . .
>
> [The Ghost] takes Scrooge to the market, and shows him the abundance there, especially the fruits (sometimes literal) of foreign trade.

Dickens describes the fruits:

> There were great, round, pot-bellied baskets of chestnuts, shaped like the waistcoats of jolly old gentlemen, lolling at the doors, and tumbling out into the street in their apoplectic opulence. There were ruddy, brown-faced, broad-girthed Spanish Friars. . . . There were pears and apples . . . , bunches of grapes, made, in the shopkeepers' benevolence to dangle from conspicuous hooks, . . . piles of filberts, mossy and brown, . . . there were Norfolk Biffins, squab and swarthy, setting off the yellow of the oranges and lemons, and, in the great compactness of their juicy persons, urgently entreating and beseeching to be carried home in paper bags and eaten after dinner.[16]

(A Spanish Friar is a kind of onion; a Norfolk Biffin is an apple.) How else, wonders Bowyer, could the poor Cratchits enjoy all this bounty without international trade, and enterprise, and capital? "Surely, Christmas Present, and his creator Mr. Dickens, and his teacher Mr. Say, are true disciples of [Adam] Smith," Bowyer concludes.

## Conclusion

And so, the terms of the battle between Malthusians and modernists (sometimes wrongly called cornucopians) were set. The battle still rages today. Neo-Malthusians anticipate lower living standards and possibly ecological catastrophe. Modernists make the case for continued (possibly even accelerating) growth in well-being, and regard the environmental challenges as problems to be solved through technology and the wise deployment of economic resources. This tension—and my strong modernist bias—will be a recurring theme in this book.

[16] Dickens (1843).

In Chapter 4, "Having Fewer Children," we look at the demographic transition, in which birth rates fall in response to (or in anticipation of) improved economic conditions, and document the unprecedented fall in birth rates around the world, including in surprising places that we think of as undeveloped. We also (briefly) look at the question of whether this is a good idea: some people, including Ben Wattenberg but also, notably, Julian Simon, whom we'll meet in Chapter 3, "The Demographic Transition," thought that policies should deliberately encourage continued population growth. I disagree, and think the best strategy is to adjust to the downsides of a stabilized population.

# 3

# *The Demographic Transition: Running Out of and Into People*[1]

How do population explosions begin? How do they end? Why do people start having fewer children when they start to get ahead economically? Population dynamics involves a number of mysteries, some of which we'll explore in this chapter.

In Chapter 2, we heard from the British demographer Thomas Malthus, who lived about two centuries ago. The environmentalist Ted Nordhaus describes Malthus's view as follows: "In response to abundance, humans would respond with more—more children and more consumption. Like protozoa or fruit flies, we keep breeding and keep consuming until the resources that allow continuing growth are exhausted."[2] Looking backward at history, Malthus was not wrong to arrive at this conclusion.

[1] My subtitle is inspired by David Greene of Oak Ridge National Laboratory, whose April 4, 2004, presentation at the Institute for Quantitative Research in Finance in Key Largo, Florida, was entitled "Running Out of and Into Oil."

[2] Nordhaus (2018).

But we are not protozoa or fruit flies. We are intelligent beings that respond to incentives; we can shape our environment. Nordhaus explains,

> In reality, human fertility and consumption work nothing like this. Affluence and modernisation bring falling, not rising fertility rates. As our material circumstances improve, we have *fewer* children, not more. The explosion of human population over the past 200 years has not been a result of rising fertility rates but rather falling mortality rates.[3]

As families either consciously or instinctively respond to the lower death rates by lowering the birth rate, the population stabilizes at a new, higher level—and will do so, according to the best forecasts, at a level less than 50% greater than the population we have now. The mechanism by which this operates is called the *demographic transition*.

We'll meet Ted Nordhaus again in Chapter 25, "Ecomodernism." To answer the question posed by the chapter title, no, we are not running out of people. If you're in one of the many places where the population is old and shrinking, it sometimes feels that way. But we have plenty of people. Expanding the economy and the carrying capacity of the Earth to accommodate a still growing and ambitious population should be our main concern, not starting a new population explosion.

## The Demographic Transition and Population Momentum

Once again, a picture—Figure 3.1—is the best way to introduce the idea of demographic transition. This transition has taken place (or started to) at different rates, with various starting dates, in different geographic areas. It was first observed in the United Kingdom and the United States.

In Stage 1, birth rates are high and death rates are just as high, so the population does not grow. I believe death rates are *exogenous*—nature

[3] Nordhaus (2018).

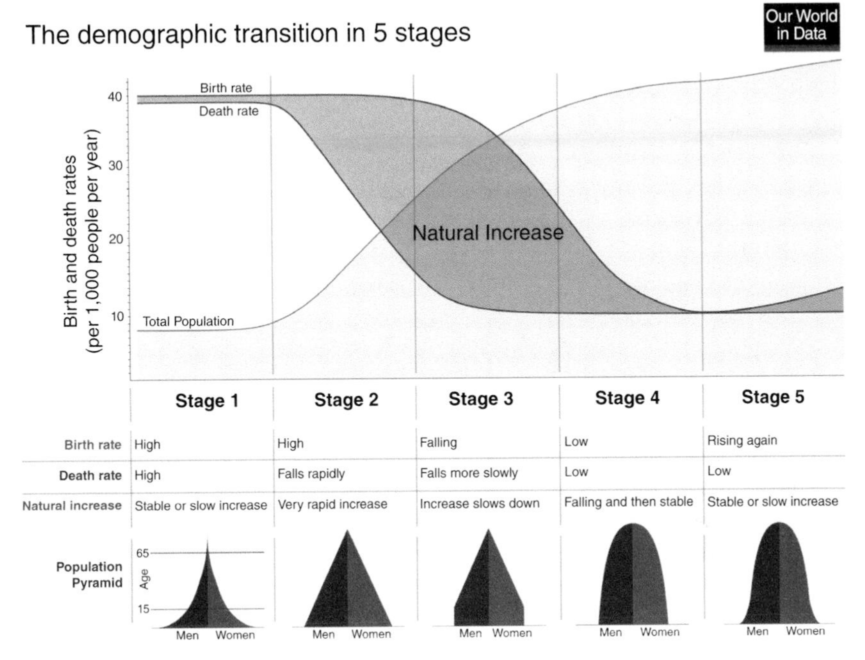


**Figure 3.1 Diagram of demographic transition.**

*Source:* Roser and Ortiz-Ospina (2017). Licensed under CC BY-SA 4.0.

decides. At least until recently, death rates arose in the natural conditions in which we find ourselves.[4]

On the other hand, it's best to treat birth rates as *endogenous*—we decide. Birth rates are the result of conscious human decisions that are made in response to, among other things, death rates.

## Stage 1: Shoemaker and Morning Star

When Stephen Jay Gould, the evolutionary biologist and dean of popular science writers, toured the Middle Amana cemetery in Iowa, he was shocked and saddened by a sight we are no longer used to:

> Then I came upon the Great Reminder . . . so freely available in any town as the ultimate antidote to . . . romantic nostalgia for a

[4] I'm counting "manmade" war and violence, which account for a substantial fraction of mortality in primitive societies, as natural causes because these are the natural outcomes of the struggle for extremely scarce resources.

> simpler past: gravestones of dead children. . . . The longer names do not fit across the small stone . . . Morgenstern (morning star), and Schuhmacher (shoemaker).[5]

In an aside, Gould reflected on the simple beauty of German names, ranging from humble occupations to celestial bodies—Shoemaker and Morning Star. He continues:

> I was looking for more . . . names when I came upon a particularly stark example of the Great Reminder. The stone read, simply: "Emil Neckwinder, died 23 Nov. 1897, 1 day olt" (a conflation of the German "alt" and the English "old"—an example of languages and cultures in transition). Emil's twin sister Emma lies just beside him—"died 11 Dez. (again the German spelling), 3 weeks olt."

Almost unbelievably, this is not the end of the story:

> The loss of twins, though tragic, would not mark an unusual event. But . . . directly in front of Emil and Emma's last resting place . . . [I saw that] in 1904, Frau Neckwinder bore another set of twins. . . . Again, they both died—Evaline on May 23 "0 week alt" (fully in German this time), Eva on September 27. . . . [B]oth pairs of infant twins . . . lie together in death.

This private catastrophe did not unfold in an ancient age. It did so in the prosperous United States, in some of the richest farmland in the world, in modern times. In 1904, Teddy Roosevelt was president; the Wright brothers had taken their first flight; Daimler and Benz had been crafting horseless carriages for almost two decades.

The Neckwinders, then, had to produce a great many children if they were going to leave behind any offspring at all. It appears that they did: there were Neckwinders in Amana, Iowa, until at least 1964; other descendants may have moved away or changed their names, so it's possible some grandchildren or great-grandchildren are still living.

This was the pattern in Stage 1 of the demographic transition: a great many born, few survive to adulthood, but more often than not the genetic line survives.[6] High death rates are the cause of high birth rates. The population, by and large, remains stable at a low level.

---

[5] All the quotes in this section are from Gould (1993).

[6] Sometimes the genetic line doesn't survive. It's easiest to keep track of descendants of famous people since hobbyists and historians do that for us. Shakespeare had four grandchildren who all died without issue, so there are no direct descendants of the Bard today. I could cite many other examples without resorting to people who were themselves childless, such as Mozart and Beethoven.

## Stage 2: The Population Explosion Begins

Returning to the demographic transition diagram in Figure 3.1 (I hope you didn't mind the long, sad detour), look at Stage 2. Death rates come down quickly as rudimentary public health measures are introduced, beginning in the late 1700s or early 1800s in the most advanced countries. (This is largely what made them advanced. Not dying conveys quite a large economic benefit.) Some of the same public health measures—basic sanitation, clean water, access to vaccines—are still being introduced in the least developed countries, which are still experiencing their Stage 2 population growth.

Meanwhile, birth rates come down more slowly. One reason is that, for a while, people don't know about the lower death rates, or don't believe that the new environment will apply to them. So they keep having as many babies as they can. Other reasons include tradition and religion. During this part of the demographic transition, the population expands very rapidly.

## Stage 3: Toward a New Equilibrium

Later, in Stage 3, death rates have fallen about as far as they are going to. This fact becomes widely known. In response, sometimes consciously but perhaps more often unwittingly, people have begun to reduce their fertility rate dramatically to a much smaller number. They are responding to a new environment (described in Chapter 4, "Having Fewer Children") where most children survive, their labor is not needed for survival on the farm, and raising even a small number of children has become expensive. Yet, despite the much lower birth rate, the population continues to expand, due to a phenomenon called *population momentum,* the seemingly paradoxical tendency of populations to keep growing after fertility rates have come way down.

Population momentum is best illustrated by example. My wife's paternal grandparents, the O'Haras, were born around 1880 and had 15 children (they're Irish). Like the Neckwinders, they were among the hardworking poor, but conditions in their town were not quite as harsh as in Amana, and 13 of the 15 survived to adulthood and reproductive age. This was something truly new and different, and could not really be foreseen.

The family did adjust fairly quickly to the new, healthier environment by reducing its fertility. My wife's father, born in 1927 near the end of a long line of siblings, only had two children, as did my wife and I; we all did our part to bring the population explosion to a rapid end. But

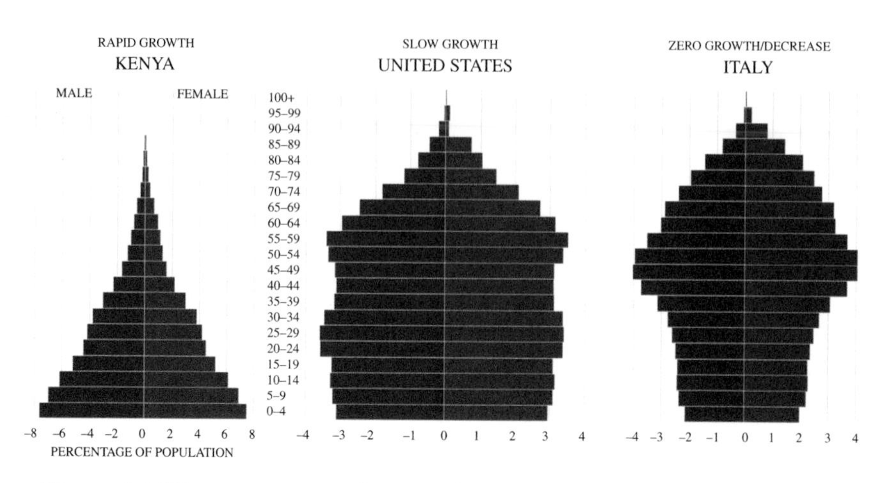


**Figure 3.2 Three patterns of age and sex distribution in a population.**
*Source:* Chart constructed with 2017 data from www.populationpyramid.net/united-states-of-america/2017/.

it did not end quickly. The number of O'Haras continued to grow for the better part of the century. The reasons are twofold: the unexpected survival rate of that first generation of children and their unexpected subsequent longevity. As a result—at least until recently—her father and his many siblings were still alive, as were all the descendants. That's a lot of O'Haras.

This population momentum—the persistence of a growing population in the face of low birth rates—begins in Stage 3 of the demographic transition and continues beyond that. The effect is most pronounced when the age "pyramid," shown in Figure 3.2, is broad at the bottom and narrow at the top: there are a lot of young people who will both live longer lives and someday have children and a few old ones who are going to die.

## Stage 4: Population Stability at a High Level

The left panel of Figure 3.2 shows current demographic conditions in Kenya, which is still in Stage 2 of the demographic transition, its population growing rapidly. The United States (middle panel) is in Stage 4, where birth rates have come down to meet the death rate and, immigration aside, the population has stabilized. Italy (right panel) is in a predicament that isn't even enumerated as one of the stages in the demographic transition—it's a Stage 4 situation gone too far, where there aren't enough young people to maintain the population. As a result, Italy's population, not counting immigration, is declining and will continue to for a while. And Italy is not alone in this quandary.

If "fewer" people are a good thing, why is Italy's situation, and that of other negative-population-growth countries, problematic? First, there will be too few young people to produce the services needed by old folks; we'll explore that problem further in "The Low-fertility World" section of Chapter 4, "Having Fewer Children."

But, more importantly, "fewer" in my title is shorthand for "fewer people than we thought we were going to have." I don't really want the population to decline, at least not at a rapid rate. To speed the process of betterment and to take pressure off the natural environment, I just want the population to stabilize. And it will.[7]

## Stage 5: The Fertility of the Very Rich

As ordinary high-income countries transition to *very*-high-income countries, something we've just begun to see in a few places in the last generation, there's a demographic surprise: fertility turns up again. I guess it's because children provide some pleasure and a sense of continuity with the future, and the very rich can afford to have more than one or two of them. The demographers Max Roser and Esteban Ortiz-Ospina write,

> Mikko Myrskylä, Hans-Peter Kohler, and Francesco Billari [found], at very high levels of development, "a fundamental change in the well-established negative relationship between fertility and development as the global population entered the twenty-first century. . . ." [W]e can see the strong negative association between a country's level of development and the fertility level.
>
> But at very high levels of development—[a human development index, or HDI] over 0.85 or even 0.9—*this association is reversed*. While causality cannot be established in this relationship, it is evident that, after a given point, higher development is associated with increasing fertility.[8]

[7] A 300-year forecast of world population prepared by the United Nations Population Division in 2004 showed world population shrinking to 3.4 billion people by the year 2300. This is their low estimate; the high estimate is 36.4 billion. These numbers mostly demonstrate that 300-year forecasts are ridiculous. Although the UNPD usually does good work, this is not an example of it. We have no idea what will happen in 300 years. See www.un.org/en/development/desa/population/publications/pdf/trends/WorldPop2300final.pdf.

[8] Roser and Ortiz-Ospina (2017), https://ourworldindata.org/fertility-rate#fertility-is-first-falling-with-development-and-then-rising-with-development. The Human Development Index (HDI) is compiled by the United Nations Development Programme at http://hdr.undp.org/en/content/human-development-index-hdi.

Myrskylä et al. note that this fertility upturn "has the potential to slow the rates of population ageing, thereby ameliorating the social and economic problems that have been associated with the emergence and persistence of very low fertility." It's nice to know that there will be a next generation of successful, comfortable people.

Who are these very rich? The upturn is most noticeable in countries with a United Nations Human Development Index (HDI) above 0.9—New Zealand, the United Kingdom, the United States, Norway, Denmark, the Netherlands, and Switzerland. Admittedly some of these fertility upticks are from low levels, but they're still increases. If one could look *within* the income distribution of a country, the higher-fertility-with-very-high-income relationship might be even more apparent.

What incentives are people responding to that cause them to become more fertile at the highest levels of human development? We'll get to that in the next chapter.

## An Aging and Stabilizing World

Stages 4 and 5 in Figure 3.1 bring us back to Ben Wattenberg's concern—or, in my rendition, mostly good news—about rapidly declining rates of population growth, Stage 4 of the demographic transition, and then either slowly falling or slowly rising populations thereafter, depending on how rich we get. How far have birth rates fallen? How far will they fall, and where? What is behind the United Nations projection of a peak world population of 11 billion (give or take) late in this century, then possibly even a decline?

Figure 3.3 gives some answers; it shows *total fertility rates* in various countries. The total fertility rate (TFR) is the number of children to whom a woman gives birth over her lifetime. The rate that yields a stable population, in the long run, is about 2.1 children per woman, with each couple thus exactly reproducing itself. It's 2.1 instead of 2.0 because some children are sterile or don't survive to reproduction age.[9]

This map makes mincemeat out of some of our most cherished prejudices. Mexicans have a lot of children . . . well, no. Indians? Same

[9] The TFR is an estimated rate that cannot be measured exactly. That is because it is supposed to capture the total number of children born to each woman on average, so we'd have to wait for all women in the country to reach the end of their reproductive years and then look back and count the number of children born to them. Since that cannot be done in real time, a statistical method is used to estimate TFR.

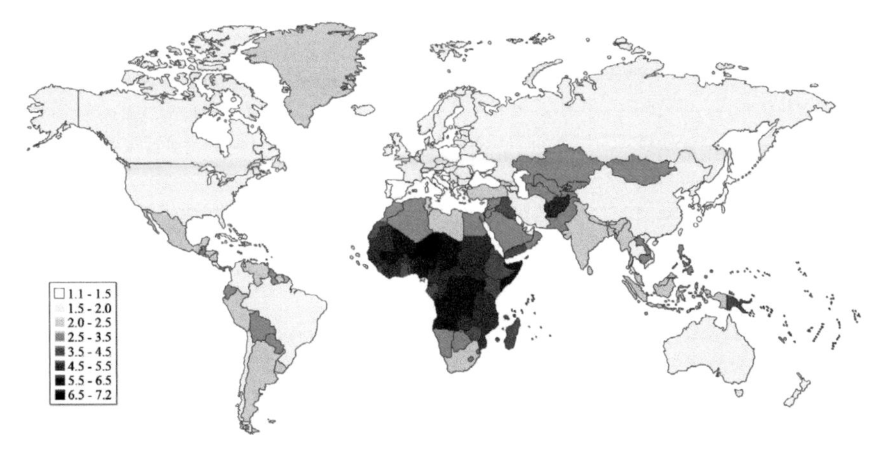


**Figure 3.3 Total fertility rates by country as of 2013.**

*Note:* I chose this graphic because the TFR gradations are finer than in most other available graphics. The differences between <1.5, 1.5–2.0, and 2.0–2.5 are very important.

*Source:* Created using data from the World Bank, https://data.worldbank.org/indicator/sp.dyn.tfrt.in.

thing. We've all heard that birth rates are down—way down—in Europe, Japan, and China, but Iran? Brazil? Thailand? Really?

Yes, really. The outliers are almost all in Africa, mostly sub-Saharan Africa, where death rates also began to decline later than in other areas. As a result, that part of the world is in Stage 2 of the demographic transition and its population is still growing rapidly. (A small cluster of high-fertility countries in Central Asia doesn't have much population, so it doesn't have a big impact on the world picture.)

The TFR for the whole world is 2.42. Just enough above the replacement rate to be significant.[10] It has halved in just over 50 years. Figure 3.4 shows the world fertility rate since 1950, by major regions, demonstrating the transformation of most of the world from a condition of very high fertility—not that long ago—to one that is approaching replacement rate.[11]

There are also major regional differences *within* countries. Of the 32 Mexican administrative units (31 states plus Mexico City), 16 have a TFR below the 2.1 replacement-rate benchmark.[12] Something similar

[10] As of 2016, from the *CIA Factbook*.

[11] This revolution in family planning could also be shown by comparing two maps (Figure 3.3 and a similar one drawn with data from, say, 1950) but we thought that change over time was more vividly portrayed in the style of Figure 3.4, that is, by a "time series."

[12] https://en.wikipedia.org/wiki/List_of_Mexican_states_by_fertility_rate.

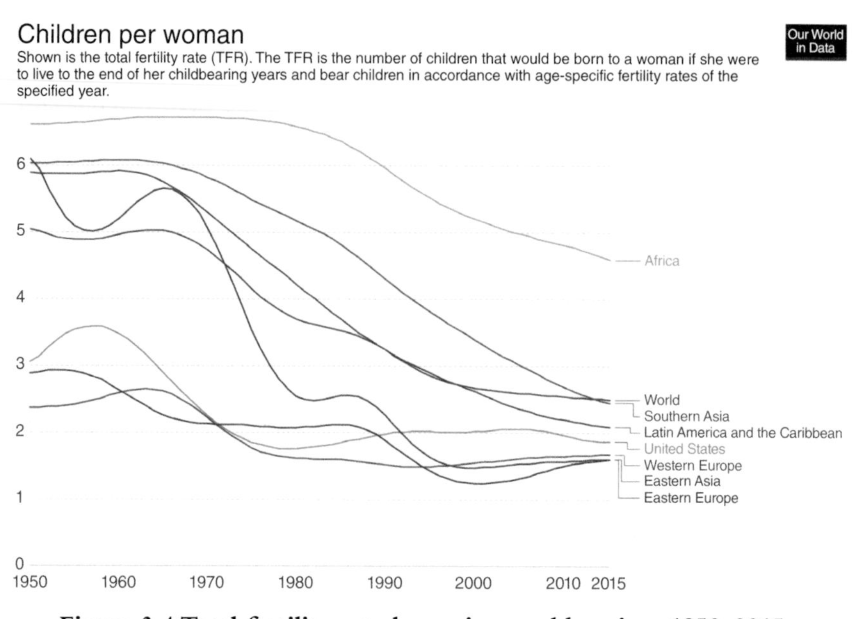


**Figure 3.4 Total fertility rate by major world region, 1950–2015.**

*Source:* UN Population Division (2017 Revision), Our World in Data, https://ourworldindata.org/fertility-rate. Licensed under CC BY.

is true in India, soon to be the world's most populous country (due to population momentum that China is not experiencing); 21 out of India's 36 states and union territories have a TFR below 2.1. The TFRs in the remaining 15 range from 2.11 to 2.91, except for Bihar (3.34), and all are declining. The country as a whole has a TFR of 2.18, a hair's breadth above the benchmark.[13]

## The High-fertility World: A Shrinking Geography, a Growing Population

In case you need a little more convincing that "fewer" is the wave of the future, here's a list of *all the countries in the world* that are (1) not in Africa, (2) have a TFR higher than the pretty-good rate of 2.5, and (3) are have more than two million people so they're big enough to make a difference: Afghanistan, Bolivia, Cambodia, Guatemala, Haiti, Honduras,

[13] https://en.wikipedia.org/wiki/List_of_states_and_union_territories_of_India_by_fertility_rate.

Iraq, Israel, Jordan, Kyrgyzstan, Laos, Malaysia, Oman, Pakistan, Palestine, Philippines, Tajikistan, and Yemen.[14]

If your first impression is that most of them are tiny, you're right. Pakistan and the Philippines are the only large (population above 100 million) countries among them. Notice that Bangladesh, the poster child for misery in the 1970s, is not on the list (and has one of the world's fastest growing, albeit still poor, economies).

## Sub-Saharan Africa: The Last Fertility Frontier

Now, let's look at Africa. Current fertility levels are high, but the decline has begun. In reducing the TFR, sub-Saharan Africa is about 30 or 40 years behind formerly desperately poor Asia, and the rate of decline is slower than in Asia but still impressive.

There are large differences between African countries. Despite being politically and racially troubled, South Africa, a middle-income country,

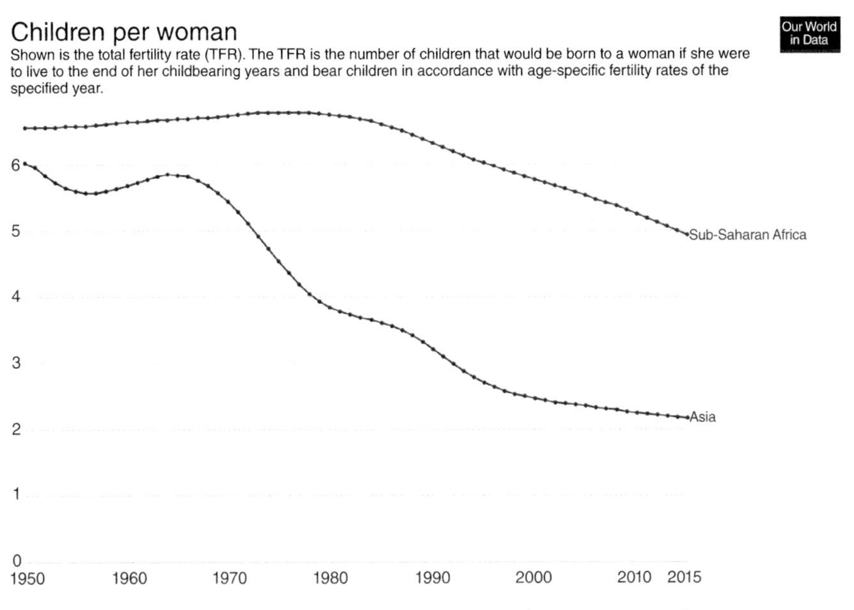


**Figure 3.5 Fertility decline in sub-Saharan Africa versus Asia, 1950–2015.**

*Source:* UN Population Division (2017 Revision), Our World in Data, https://ourworldindata.org/fertility-rate. Licensed under CC BY.

[14] Syria is right on the line (2.50) but has experienced massive emigration. Israel and Malaysia are high-income and don't "count" if you're worried about poverty, but they do count if you're worried about the environmental effects of consumption.

has reduced its TFR to 2.29. Niger's is highest at 6.49. The average for the region (the entire African continent excluding the countries with a Mediterranean coastline) is 4.85, more than twice the world average.

Despite some significant progress in fertility reduction, then, Africa's population is expected to increase massively—before that continent, too, stabilizes by the end of this century. This is, of course, due to population momentum. The UN's median projection for Nigeria's population in 2100 is an incredible 752 million, up from 186 million now, and its various high-variant projections are even more incredible. (Nigeria is a little smaller than Texas and New Mexico combined.)

That's unlikely to happen, because Nigerian birth rates will probably fall much faster than expected, and emigration will also be a factor. (The same organization has low-variant projections of 230 million to 275 million; it's hard to interpret forecasts with such a large known margin of error.)

These trends run the risk of creating a two-speed world: central Africa and everyone else.[15] We've already had a two-speed world, where Western Europe, Japan, and the Anglosphere were rich and everyone else was poor; good riddance.[16] The new two-speed world, if it materializes (and it looks like it's already happening), is a challenge for our children and grandchildren.

However, the always-optimistic Deirdre McCloskey points out a delightful upside to population growth in Africa, the most genetically diverse continent:

> Genetic diversity in a big and rich Africa will yield a crop of geniuses unprecedented in world history. In a century or so the leading scientists and artists in the world will be black. It will be a splendid irony on the history of racism in Europe and Asia.
>
> Today a Mozart in Nigeria follows the plow; a Bashō in Mozambique was recruited as a boy soldier; a Tagore in East Africa tends his father's cattle; a Jane Austen in Congo spends her illiterate days carrying water and washing clothes. . . . But not in 2100.[17]

---

[15] By "central Africa," in this case, I mean all of sub-Saharan Africa (including what is sometimes called West Africa and East Africa), except for the southernmost part of the continent, that is, South Africa, Botswana, and Namibia, which are economically better off and have much lower TFRs.

[16] By the Anglosphere I mean the United States, Canada, Australia, and New Zealand, as well as the United Kingdom, which is in both the Anglosphere and Western Europe. Ireland can also be included.

[17] McCloskey (2016).

# 4

# *Having Fewer Children: "People Respond to Incentives"*

"Most of economics can be summarized in four words: 'People respond to incentives.' The rest is commentary."

Channeling Rabbi Hillel of Babylon, so says Steven Landsburg, professor at the University of Rochester and author of *The Armchair Economist: Economics and Everyday Life*.[1] If there are a few simple precepts to be learned from a study of economics, they are:

- Needs and wants are unlimited.
- Resources are limited.
- All decisions involve trade-offs.
- People respond to incentives.

The first three precepts are a way of saying "you can't have everything." (Steven Wright, the heady comic, added, "If you could, where would you put it?") The fourth is the topic of this chapter, which

[1] Landsburg ([1993] 2012).

argues that people responding to incentives are the cause of the population explosion coming to an end in our time, just as they caused the population explosion to start in the first place. People decide how many children to have based on a conscious or unconscious calculation about what is best for themselves and the children they do or do not have.

The idea that choosing how many children to have is (or can be) a utilitarian calculation discomforts some people. The resistance comes from believing that having children is a deeply personal, even spiritual, decision that economics should have little part of. Yet there is a rich body of respected research, led by that icon of twentieth-century economics, Nobel Prize winner Gary Becker of the University of Chicago, placing economic considerations (cost-benefit analysis, to be specific) at the center of this decision.

Before delving into insights into the ways families choose how many children to have—insights that rely heavily on the work of Becker and his colleagues—let's pick up where we left off in the last chapter. We ended with hope for poor people in countries with rapidly declining birth rates and growing economies. What can we say about today's low fertility rates generally?

## The Low-Fertility World: No Longer the Exception, but the Rule

Low-fertility countries now make up a majority of the world's population. Figure 4.1 illustrates fertility declines over time for some traditionally rich societies where death rates, and then birth rates, fell early. All of these economically advanced, Western countries reached replacement rate in the mid-1970s and fertility kept falling from there.

Today, southern and eastern Europe have joined—and surpassed—Western Europe and North America in getting their populations to stabilize or even decline. Some would say they've gone too far, guaranteeing themselves a stagnant economy in which a small number of young people are stuck supporting a large number of old ones. China looks like it's going to face similar problems, plus a shortage of women due to the persistence, until recently, of a one-child policy wherein boys had been favored. Extreme levels of failure to reproduce are a legitimate concern and are not to be minimized.

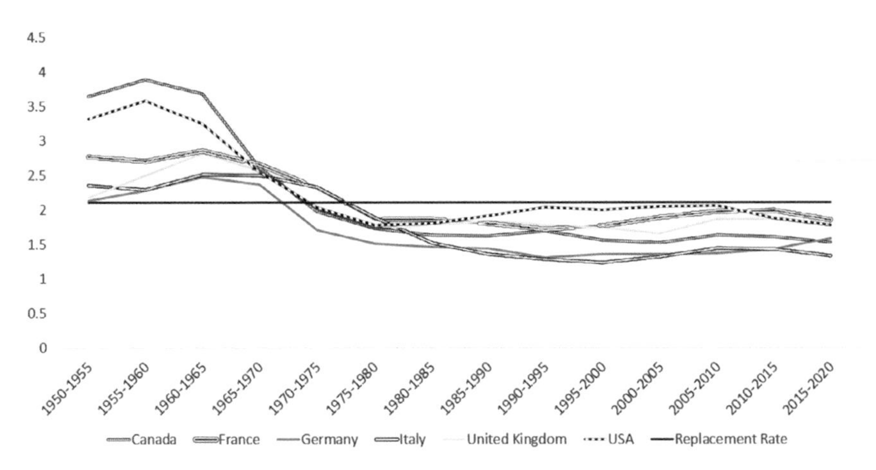


**Figure 4.1 Total fertility rates in western Europe and the United States and Canada, by country, 1960–2017.**

*Source:* Constructed by the author using data from the United Nations Population Division, http://data.un.org/Data.aspx?d=PopDiv&f=variableID%3A54.

## Not Too Hot, Not Too Cold—Just Right

But, in the Goldilocks countries—the United States, Canada, Australia, New Zealand, and I'd add France and the UK—fertility levels are hovering just below replacement rate, a situation that the respected demographers Ronald Lee and Andrew Mason, writing in the *Los Angeles Times*, consider "almost ideal."[2] All of these countries attract intensive immigration, so their populations are still growing, and so are their economies.

It doesn't feel like a Goldilocks situation sometimes. The late Ben Wattenberg and former US Secretary of Commerce Peter G. Peterson (who died at 91 just as I was writing this) are among those who have been ringing the bell for demography-driven fiscal disaster in the United States and Europe for decades. But, in my view, that problem has more to do with overpromised social benefits and political unwillingness to adjust them to changing circumstances than to any fundamental weakness caused by a stable or slowly growing population.[3]

[2] Lee and Mason (2014).

[3] We've had about 20 years of advance notice, at the very least, about this problem. Everyone who will be working and paying taxes 20 years from now has already been born; mid-career workers who will be paying the big bucks to the government 20 years from now were born 30 or more years ago. We know how many of them exist and roughly what countries they'll be in. If you can't plan for a problem you can see coming that far in advance, it's your fault.

## Newcomers to the Low-fertility World

In the mid-1970s, when first-world fertility rates were well on their way to their modern, low levels, the picture did not look so good in the rest of the world. Understandably, that was about the time when panic about the population explosion was peaking.

Figure 4.2 shows fertility rates in six countries that were quite poor when the chart begins and, to a large extent, still are. It's interesting that fertility rates declined very quickly in these poorer locales; this contrasts with the slower rate of decline in the rich countries a generation earlier (Figure 4.1). Note that most of the decline in the poorer countries was between 1980 and 2005.

Iran, which had the most dramatic fertility reduction of the countries shown in Figure 4.2, trimmed the number of children born in a typical woman's lifetime by 80% in one generation! And, to the best of my knowledge, Iran did not have a one-child policy (or any policy).[4] People don't need policies to reduce their birth rate; they need the natural incentives provided by an economy that rewards the "quality" (educational attainment and such) of children instead of the quantity.

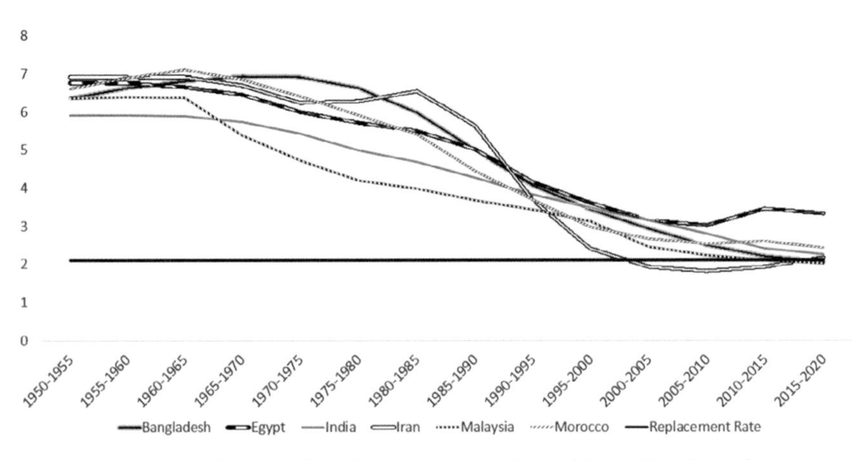


**Figure 4.2 Total fertility rates in selected less-developed countries, 1960–2017.**

*Source:* Constructed by the author using data from the United Nations Population Division, http://data.un.org/Data.aspx?d=PopDiv&f=variableID%3A54.

[4] Do I believe the data coming out of the autocratic Iranian regime? Sure. I have no reason to believe they're wrong, and neutral observers have checked them. I'm also told it is a more open society than some other countries that express antidemocratic values. Still, it's a fair question, given that the Soviet Union, and possibly Nazi Germany, produced fake economic data by the truckload.

## East Asia, Epicenter of Childlessness

Did you know that the most dramatic fertility declines in large populations (setting aside Iran, which may be a unique case) have been in East Asia, where the population explosion was roaring just two generations ago?

The lowest total fertility rate (TFR) in the world is in Singapore, where the number of children in an average woman's lifetime is 0.83.[5] At such a rate, population declines rapidly; a TFR below 1 means that, absent any other effect, each generation is *less than half as large* as its predecessor. South Korea and Taiwan are runners-up in the no-kid sweepstakes, with TFRs of 1.13 and 1.26, respectively.

This is not your grandfather's East Asia. Averaged over 1950 to 1955, Taiwan had a TFR of 6.72, higher than any *African* country today; and South Korea's TFR was 5.65. Japan, currently at 1.41, is the only East Asian country that had anything like a First World fertility rate 50 years ago. The massive drop in East Asian (ex-Japan) fertility rates in the last half-century is one of the most rapid and profound changes in any society in history.

Asians are feeling the pinch. Ronald Lee and Andrew Mason, the demographers whom we met earlier and who generally celebrate lower fertility rates, warn that Japan, South Korea, [and] Taiwan "would benefit from higher fertility [because] the balance between workers and those they support is just too disadvantageous."[6]

Japan, which has had a baby bust for longer than other East Asian countries, is a bellwether of the social changes caused by low fertility. A *Business Insider* report by journalist Chris Weller blames the Japanese woes of overwork, loneliness, and economic stagnation on the aging of the population.[7] And, Weller writes, overworked couples struggling to make ends meet are unlikely to be enthusiastic baby makers. It's a vicious circle, but it may be more cultural than demographic.

## The United States: Baby Boom, Baby Bust

But being overworked and broke, as some Japanese claim to be, does not necessarily mean a baby bust. World War II–era Americans were overworked and broke in their young years. Remember *The Honeymooners*?

---

[5] Central Intelligence Agency, *CIA World Factbook*, 2017 edition showing 2016 data. Other sources show a higher number.

[6] Lee and Mason (2014).

[7] Weller (2017).

Their apartment, furnishings, and appliances were modest in the extreme. Yet Ralph and Alice Kramden were prototypes of the young, postwar middle class. And although the Kramdens did not produce children in their one year of TV episodes (they were too busy fighting), their age cohort sure did.

The US baby boom that lasted roughly from 1946 to 1964 involved fertility rates around 3.5, higher than Pakistan's today. We are still feeling the reverberations from this event, as are other countries that experienced postwar baby booms, such as Britain and France. One possible cause, other than pent-up demand from the war years when most young men were absent, is the expectation of future prosperity, something that is a tough sell for today's young First World populations (but shouldn't be).

But then, in the late 1960s, the US fertility rate fell sharply, reaching below-replacement-rate levels in the 1970s and never really recovering after that. This phenomenon brought population pressure under control but created new problems: the stresses of an aging society, a plethora of only-child and childless households, and a concern that young would-be parents weren't seeing a bright future for the next generation.

All this fertility restraint took place against a background of sometimes slow, sometimes robust, but almost always positive economic growth. Our standard of living is much higher by any measure than that of the Kramdens.

Thus, it is very difficult to disentangle economic influences from other factors—social and cultural—in explaining population trends. The 1970s and subsequent decades saw a boom in career opportunities open to women, a massive increase in the earnings premium to higher education, and a labor force that was growing unusually rapidly. (The rapid labor force growth was due to the earlier baby boom, amplified by the movement of women into market work.) As a result, despite overall increases in prosperity, entry-level wages became depressed. All these factors put a dent in the birth rate.

But these influences just don't quite feel like they're sufficient to explain a fall by half in total fertility in just a few years. Culture matters (Figure 4.3). In this chapter, we'll focus on changing incentives, the costs and benefits of having children as perceived by parents. (As we'll see, the attractive young people in Figure 4.3 are responding to incentives, not just fooling around and having fun.) In Chapter 5, "Age Before Beauty," we'll delve further into the cultural and social aspects of low fertility and ask what life will be like in an aging, population-stable world.

**Figure 4.3 The children who did not have children.**

*Source:* Nejron © 123rf.com.

## Why People Have Children: The Beginnings of Demography

"Demography is a science short on theory, but rich in quantification," writes the Stanford University demographer Dudley Kirk in a 1996 article recounting the history of demographic science.[8] Despite an abundance of data, he was saying, little thought had been given to *why* demographic patterns emerged the way they did. This was the case even though Malthus, Ricardo, Say, and other classical (eighteenth- and nineteenth-century) economists had commented on population—in Malthus's case to great, and largely misplaced, acclaim.

Falling death rates were obvious to everyone as the nineteenth century proceeded. Kirk writes,

> [Although] [m]odern mortality decline was unprecedented in human history . . . [i]t is much easier to explain than fertility decline: the

[8] Kirk (1996), pp. 362–363.

> reduction of epidemics by vaccination and better hygiene, improved diagnosis and treatment of disease, reduction of famines, fewer deaths from violence and civil wars, reductions in infant mortality, and improved standards of living all played a part.

But falling birth rates, documented by the same censuses as those that revealed the plunging death rates, were "harder to explain, except in the proximate sense of the increased use of contraception," writes Kirk. And finding an explanation took a while. In a landmark 1934 work, *a full century* after the demographic transition had actually started, a Frenchman named Adolphe Landry not only called it *la transition démographique*—the first recorded use of the term—but began to craft an explanation.[9] Landry was the first to contemplate the transition in *economic* terms.

**Figure 4.4 Adolphe Landry (1874–1956), French economist, politician, and pioneer of the demographic transition.**

*Source:* https://en.wikipedia.org/wiki/Adolphe_Landry#/media/File:Adolphe_Landry.jpg. Public domain.

[9] Landry (1934). Landry had some odd views. He believed that a possible decline in the population of France would expose it to the threat of foreign domination, and that the then-booming population of Japan could best be accommodated by territorial expansion.

Kirk asks,

> What is the motive for birth regulation? Landry believes [it] to be largely "egotistical": the cost of children, their ability to cause pain and distress to their parents, the limitation of parents' activities and relaxation, and, of course, the problems women experience in pregnancy and child care. His analysis preceded much of the later discussions of individualism and "self-fulfilment" as a cause of declining fertility.[10]

Frank Notestein had similar thoughts only a little later (1945 and 1953), and wrote in English so he found a wider audience. He presciently wrote:

> The new ideal of the small family arose typically in the urban industrial society. . . . Urban life stripped the family of many functions in production, consumption, recreation, and education. . . . The new mobility of young people and the anonymity of city life reduced the pressure toward traditional behavior. . . . Education and a rational point of view became increasingly important. As a consequence, the cost of child-rearing grew and the possibilities for economic contributions by children declined. Falling death rates . . . lowered the inducements to have many births. Women, moreover, found . . . new economic roles less compatible with childbearing.[11]

Incentives and cost: economics. Individualism and self-fulfillment: culture. In asking *why* people chose to have fewer children, Landry and Notestein were getting close to advancing a theory that would predict eventual population stabilization as prosperity and urbanization grew. Only a few years later, Gary Becker would begin to fully develop an incentives-based view.

## Gary Becker: The Economics of People

Gary Becker was not just any Chicago-school economist. He was an icon whom the younger generation of professors celebrates as their mentor. He turned the increasingly mathematical and abstract field

---

[10] Kirk (1996), p. 363.

[11] Notestein (1945).

**Figure 4.5 Gary S. Becker (1930–2014), Winner of the 1992 Nobel Prize in Economics.**

*Source*: Roger Tillberg/Alamy Stock Photo.

of economics on its ear by asking questions traditionally scorned by economists: How are decisions made within the family? What is human capital and how do you increase the supply of it? Why is the time spent in formal education increasing? These are the kinds of problems that now captivate many, if not most, young economists, in large part because of Becker's influence. Becker turned the focus of economics to *people*.

The University of Chicago named its economic research center the Becker Friedman Institute, placing Becker's name ahead of the revered Milton Friedman. And, when *The Economist* wrote up Becker's life, their lead sentence was, "Why do families in rich countries have fewer children?"[12]

Okay, why do they?

[12] *The Economist* (2017). It is noted that "this article appeared in the Economics brief section of the print edition under the headline 'The people's champion'."

## Children as Consumer Goods?

"In his 1960 paper," writes the Northwestern University professor Matthias Doepke, a Becker acolyte and expert on the economics of family decisions,

> Becker conjectured that parents derive utility from both child quantity (i.e., the number of children) and the quality of children, which can be proxied by the amount spent on each child. . . . As examples of child quality choices, Becker mentions whether the parents provide "separate bedrooms, send them to nursery school and private colleges, give them dance and music lessons, and so forth."[13]

Using some technical language that I'll explain, Doepke shows how Becker applies the concept of *income elasticity* to the "demand" or desire for children. Income elasticity is the extent to which something (anything) changes when your income changes. If a higher income means that you'll have more children, then child *quantity* has a high income elasticity; if, conversely, a higher income means you'll have fewer children but spend more on each, then child *quality* has a high income elasticity, and child quantity a low one. Doepke summarizes:

> Becker's argument for a low income elasticity of child *quantity* and a high income elasticity of child *quality* rests on an analogy with other consumer durables, such as cars or houses. As households get richer, for the most part they do not buy larger numbers of cars or houses, but instead go for higher quality (a BMW instead of a Chevy, or a house with more bedrooms and bathrooms).
>
> On this basis, Becker concludes that the income elasticity for child quality (i.e., spending per child) should be high, whereas the elasticity of quantity (i.e., number of children) should be low.

That's the theory. As incomes rise, you spend more per child, but you only need a few good ones. And you'd have to be awfully rich to afford a large brood of high-investment kids (a tendency Becker also addressed.)[14]

In 1960, calling children a consumer good and comparing them to BMWs and Chevrolets seemed a little crass, and Becker took some guff

[13] All the quotes in this section are from Doepke (2015).

[14] Doepke writes that "Becker points to evidence that the income-fertility relationship switches to positive for very high income levels" (p. 60).

for it—economists dealt in dignified concepts like airline seat-miles and ship loadings, not the decision of whether to buy a child a dance lesson. But Becker's broadening of the ambit of economics to include practically every kind of human decision revolutionized the field. Economists—most admiringly, some critically—call Becker's application of economics to everything "economic imperialism."[15]

## Responding to Incentives in Mid-century America

Earlier, I wrote—or at least strongly suggested—that the parents of the American baby boomers were responding, at least somewhat rationally, to incentives when they had large families soon after World War II, and that the freewheeling hippies in Figure 4.3 were also responding to incentives when they delayed, or completely avoided, having children a generation later. How could these assertions both be true?

### Planning a Future

Imagine yourself a returning veteran in 1945, miraculously uninjured, young enough to have most of your life ahead of you but not a kid either—you're old enough to have developed some marketable skills in the Armed Forces, and to be interested in marriage and a family.[16] You grew up in the Depression, facing hardship, but the economy is now booming and the factory system is so vibrant that all you have to do to get a job is ask for one. Children are cheap to raise, since a free high school education is all that most of them will need.

Having put personal life aside in the service of your country, you want to marry quickly. Because of the time taken up by the war, you are on the late side of what was then considered marriageable age. Due to earlier population growth and a tendency of women to seek slightly older men, potential wives are abundant. The GI Bill provides generous benefits to many returning soldiers.

How do you respond? You've been waiting (and maybe getting shot at) for some time, so you start a family almost immediately. You

[15] Becker (1960) is the best example of his own work on family size choice.

[16] This section is told from the point of view of a young man. Women faced slightly different circumstances, but couples (not individuals) make family size decisions, so the general principles herein are probably valid on both sides of the aisle.

don't need a lot of children to work on the farm that you don't have, but each extra child doesn't cost much either, housing is abundant and subsidized, and educational costs are minimal. You also remember your parents and grandparents having large families and generally deriving pleasure from them.

So . . . off to the races. (See Figure 4.6. Don't these kids look like the hippie hitchhikers 10 or 15 years earlier?) The resulting baby boom was,

**Figure 4.6 New Zealand National Party: A family affair.**

*Source:* New Zealand National Party: A family affair. [Cover. 1949]. New Zealand National Party: [Political pamphlets, fliers and election propaganda. 1949]. Ref: Eph-A-NZ-NATIONAL-1949-01-cover. Alexander Turnbull Library, Wellington, New Zealand. https://natlib.govt.nz/records/22329593

then, a natural and predictable consequence of the economic and social conditions under which it occurred.

## Living for the Moment

A generation later, the adventurous vagabonds pictured in Figure 4.3 faced very different incentives. A massive increase in the labor force had temporarily depressed wages for the young, so the opportunity cost of their time was low. At the same time, their parents were mostly prosperous, which had not previously been the case, so they had a backstop. Travel had become cheap, safe, and easy. Television, books, and magazines caused the world, not just the next county, to beckon. Finally, advances in medicine made it possible to delay childbirth until one's 30s or even 40s without dire consequences.

What a set of opportunities for postponing adult responsibilities, seeing the world, and enjoying the moment! Nothing like this particular combination of circumstances had ever happened before, and the consequences were predictable from an economic or Beckerian perspective. So was the maturation of the hippies into sober, hardworking yuppies and "grumpies" (grown-up, mature people) as the decades wore on.

This kind of analysis can be extended to any population. But it's a bit of a mystery why, in most of the world, people are choosing—seemingly all at once—to lower their fertility rates to below replacement levels. We'll cover this next.

## Responding to Incentives in a Population-stabilizing World

Consider these stories:

- The *Financial Times* reports, "Raúl Velasco remembers the time when Mexican families were big. 'There was never any question that you would have seven, eight, maybe even 10 children,' says the retired construction worker from Mexico City. 'That was what our parents did and that is what we imagined our children would do, too.' Mr. Velasco's seven children—three boys and four girls, all born in the 1960s and 1970s, and all now with families of their own . . . have only 11 children [between them] and Mr. Velasco says they do not plan to have any more."[17]

[17] Thomson (2013).

- A Kenyan father sends each of his children to private school by remitting $7 a month (that's the actual amount) per student using his mobile phone. (See Chapter 14.) The private schools are aimed at families living on less than $2 a day. How much money do you think the kids in the school will make 20 years from now?
- From Ethiopia, one of the poorest countries in the world and now one of the fastest growing, albeit from a very low level, the BBC reports, "Ayenalem Daw, a mother of six living in Weyo Rafu Hargisa village about a four-hour-drive out of Addis Ababa, is in her late thirties. She says if she had heard about family planning earlier, she would have had four children. . . . 'Things are changing now. I think my children will have only two babies each,' says Mrs. Ayenalem."[18]
- Also in Ethiopia, "Muluwork Tesfaye, a nurse in Addis Ababa . . . [is a] mother of two [who] grew up in a family of eight and her parents struggled to provide for them," reports the BBC. 'My husband is the one who took me to college,' she says. 'I wanted a better life for my children.'"[19]

These parents are not the carefree youth of America in the 1960s, deferring childbearing to some later year when their incentives might change. Nor are they limiting their fertility because they're too poor to afford even the basics for larger families. Instead, they are responding to the incentives that a more modern environment provides: children have become expensive and, for them to get ahead, each one needs a lot of parental resources. The parents in my stories are consciously creating the first generation of children in their societies to have a real chance at material success, through a strategy called *K-selection* instead of *r-selection.*

K-selection is heavy investment in a small number of offspring; r-selection is light investment in a large number. This terminology was devised by biologists for classifying species of animals—whales are K-selected, rabbits r-selected—but the concept can easily be applied to human populations.[20] Heavy investment in a small number of children

[18] "Why Ethiopian . . ." (2015).

[19] Ibid.

[20] MacArthur and Wilson ([1967] 2001). The application of r- and K-selection theory to human populations is controversial, partly because it has been used to support speculation about racial differences. A relatively sober (although academic and technical) discussion is in Heylighen, Francis, and Jan L. Bernheim, Undated. "From Quantity to Quality of Life: *r-K* selection and human development," http://pespmc1.vub.ac.be/Papers/r-KselectionQOL.pdf. Edward O. Wilson's 1975 masterpiece, *Sociobiology: The New Synthesis* (Cambridge, MA: Belknap Press, Harvard University Press), contains a broad discussion of related issues.

from new shoes and proper elementary schooling to piano lessons and engineering classes—is almost literally the definition of progress and of modern life as most of us wish to see it lived.

What we call the modern way of life will change over time, but the basic elements that make low fertility attractive are unlikely to change soon. To put it just a little too crassly, in wealthy societies and increasingly in less wealthy ones, children have become a cost center (some would even say a luxury good), not a profit center. If you have children it's because you really want them, and most people can't afford very many.

## Conclusion

As Gary Becker demonstrated, decisions within the family, such as how many children to have, are just as amenable to economic analysis as any other kind of decision. These decisions are informed by the need to survive and avoid penury, the desire to achieve comfort and maximize one's potential, and the impulse to help those one cares about.

One set of incentives faces a family struggling to find enough labor to produce as much food as can be wrung out of a small plot of land. A profoundly different set motivates a family of engineers, nurses, or teachers. Economics helps explain how each of these families decides how many children to have. Fortunately, the world is becoming more prosperous and the decisions are rapidly tilting in favor of fewer, but much better cared for, children.

# 5

# *Age Before Beauty: Life in an Aging Society*

Whoever coined the phrase "age before beauty" must have known that, someday, almost the whole world would be an aging society. Since the phrase is a little mysterious, I'll rely on the poison-pen author and joker Dorothy Parker (see Figure 5.1) to put it in context:

> Mrs. Parker and a snooty debutante were both going in to supper at a party: the debutante made elaborate way, saying sweetly, "Age before beauty, Mrs. Parker." "And pearls before swine," said Mrs. Parker, sweeping in.[1]

The ageless Parker (who was only 45) meant that the young and beautiful should respect their elders, let them pass through the doorway first, and not seize the advantages that young and beautiful people might be accustomed to. Pearls before swine . . . well, you figure it out.

Today, a 45-year-old might be counted among the young, and would be well advised to yield to his or her active 75-year-old parent or friend. We are living longer—*much* longer, for in 1893, when Parker was born, the life expectancy for an American woman who reached adulthood was around 62. Today, it's about 82.[2] This profound change has, unsurprisingly, both upsides and downsides.

[1] *The Spectator* of London, September 16, 1938.

[2] That is, expected age at death, not expected remaining life span. Source: www.infoplease.com/us/mortality/life-expectancy-age-1850-2011.

**Figure 5.1 Dorothy Rothschild (later Dorothy Parker) as a young woman.**

*Source:* https://medium.com/@penguinpress/dorothy-parker-on-why-she-hasnt-married-yet-d20dc97f5735. Public domain.

Figure 5.2 is a lighthearted summary of the up and down sides, from a talk I gave at the Q Group, a discussion group for investment executives. I made that list when rock music had not yet been almost completely replaced by rap, hip-hop, and techno. I'd now push "better music" over to the right side of the chart. (Okay, partly kidding. The best rock music, like the best jazz and much of the best classical music, was composed and performed by the young, even if our past musical heroes now look like dinosaurs. Music, like mathematics, is mostly a young person's game.)

We'll first explore what an aging society will look like and how it can be improved, then discuss some implications for pension and medical costs, the length of one's work life, and the natural tension between the generations that is documented in Richard Dawkins's masterpiece, *The Selfish Gene*.

Is an aging society bad?

| YOUNGER | OLDER |
| --- | --- |
| More fun | More skilled |
| Prettier | Wiser |
| Quick learner | Already knows quite a bit |
| Can lift heavy objects | Has more money |
| Better music | Knows some history |
| Not very savvy – makes peculiar decisions | May not feel well—has greater needs |

**Figure 5.2 Young versus old.**

*Source:* Constructed by the author.

## The Upside of Aging

Fifty years on, the appealing hitchhikers in Figure 4.3 in Chapter 4, "Having Fewer Children," probably now look weather-beaten. But they are also wiser, more accomplished, much more knowledgeable, and more deeply invested in the world's future. This is the upside of an aging society.

When the primary work of the world consisted of moving heavy objects from here to there, youthful vigor dominated. Now the primary work is solving tough problems. Although the vigor of young brains is still valuable, direction and purpose must come from those with enough life experience to make sensible decisions. Even the tech industry is dominated by entrepreneurs over 45: Bill Gates, the late Steve Jobs, Jeff Bezos, Peter Thiel, Elon Musk. They may have been young when they started their companies, but they are in their prime in middle age now. Mark Zuckerberg and his young friends are the exception, not the rule.

## . . . And the Downsides

As with any phenomenon that has an upside, there are also downsides. A society with an abundance of old people might be rich in wisdom, but be careful: when wisdom was rare because few people lived to an advanced age, it was prized. (Only an old person could tell the tribe how to avoid famine by growing crops that would survive a twice-in-a-century drought.) Now there are so many old people that wisdom may sell cheap.

Another downside is that, even more than childhood, old age is expensive. Medical costs are the most widely discussed expense, but simply staying alive when one is too old to work—that is, retirement—is a costly proposition. Many people work for 30 or 40 years and, with the proceeds from that work, try to pay for 60 or 80 years of living.[3]

No wonder it's difficult to save for retirement! We weren't meant to operate that way. Throughout most of history, few people lived long enough to experience any leisure at all after their work lives were over, and those who did received care from their numerous children. Today, we rely on government, individual savings, and the financial markets to spread the income from one's working life over one's whole life, a new and untested way of functioning.

Finally, a lot of old people, especially the very old whose numbers are increasing fastest, need help of some kind. It is not enough to have money available to purchase that help. The people who can supply the help must be available—they must have been born, and they must live close by. If there is a shortage of younger people able and willing to work as helpers for the old, wages will rise, drawing workers out of other professions; and more people will immigrate.

However, immigration is not a solution if the shortage of younger people is worldwide or nearly so. Since Africa is the only young continent, we can expect much more immigration from Africa into aging regions such as Europe and the United States to fill the needed roles.

## Ben Wattenberg's Vision of an Aging World

What will an aging world look and feel like? We can already get an idea of it by visiting Florida or Japan. In the book *Fewer*, introduced in Chapter 2, Wattenberg, an optimist by temperament (he once wrote a book called *The Good News Is the Bad News Is Wrong*), has plenty of social worries:

- "We may well see a lonelier world."
- "The . . . psychologist David Gutmann of Northwestern notes that . . . 'When the burdens of childrearing are shared out among sisters, cousins, aunts, mothers, and grandmothers, parenthood becomes more a

[3] Starting, say, at age 20 and including both their working and retired years. (The first 20 years are the parents' problem.)

joy than a burden. . . . But childrearing in the unbuffered nuclear family, relatively isolated in the city . . . is very burdensome'."

- "There is a 'grandchild gap.' People well into their 60s, 70s, or 80s are looking for grandchildren to love. Without them they can become deeply disappointed, even misanthropic."

Then there is the dependency ratio. A small number of young people supporting a large number of old ones is a recipe for generational conflict and decreased prosperity.

Finally, and this is my idea, not Wattenberg's, a society full of old people reminisces about the past, often with the selective memory that blots out the hardship and remembers only better days gone by. A young society looks forward to what can be accomplished in the future. Which would you rather live in?

The young one, you're probably thinking. But we're going to have an aging society whether you like it or not. Let's try to make it rewarding.

## Aging Gracefully

### What's the Matter with Kids These Days?

In looking for ways to improve the quality of life in an aging world, let's begin at the other end, with what's ailing youth. Right now, many young people in the developed world are looking to the future with dread. This is a serious misunderstanding of what lies ahead and can be fixed by educating young people responsibly. There's a world of difference between teaching youngsters the hard realities of life and indoctrinating them in a way that causes them to tell their parents they wish they had never been born because "we're all gonna die!"

I'm not exaggerating: this is going on in households in the United States, Europe, and, for all I know, other countries. I think it's less common in the developing world, where improvement is more obvious.

Youthful angst is not new; *The Sorrows of Young Werther* was written in 1774. The ancient Greeks chronicled it too. But this angst was usually about the self, not the world: youth has generally been forward-looking. "Bliss it was in that dawn to be alive. But to be young was very heaven," wrote William Wordsworth about the French Revolution. Widespread dread of the future by the young began with the nuclear age and was understandable under Cold War circumstances. Our prospects now are much better, but our attitudes aren't.

I was having a delicious dinner in Milan with a European business colleague who told me his children "didn't plan to have any kids because they didn't want to bring anyone into an overcrowded world that is going to end in an ecological catastrophe." I almost dropped my pasta in my lap, then recovered to sputter, "They don't know what they're talking about." I doubt that I made a friend for life that way. Of course, we face environmental challenges, but putting "world" and "end" in the same sentence is not good for children, and not true.

Books like this one might help a little. But a mental attitude widely held is a battleship, and it takes a lot of effort to change its direction. (We're also not going to fix the real problems that we face, including environmental ones, if our leaders of the future believe it's hopeless.) A lot of education, life experience, and wisdom will be needed to persuade young people that life is not only worth living but almost guaranteed to improve in both material and less tangible ways.

## Loneliness and the Folk Society

Ben Wattenberg was right to worry about an aging society being lonely. One can envision changes that will alleviate that concern. Kurt Vonnegut Jr. (see Figure 5.3) said that the young communitarians of the 1960s

**Figure 5.3 Kurt Vonnegut Jr. in 1972.**

*Source:* https://commons.wikimedia.org/wiki/File:Kurt_Vonnegut_1972.jpg.

wanted to live in a "folk society." Recounting his interaction with the sociologist Robert Redfield, who coined the term, Vonnegut wrote,

> [P]rimitive societies . . . were all so small that everybody knew everybody well, and associations lasted for life. . . . The old were treasured for their memories. There was little change. . . .
>
> [W]e are full of chemicals which require us to belong to folk societies, or failing that, to feel lousy all the time. . . . When anything happens to us which would not happen to us in a folk society, our chemicals make us feel like fish out of water.
>
> And what do many of our children attempt to do? They attempt to form folk societies, which they call "communes." They fail. The generation gap is an argument between those who believe folk societies are still possible and those who know they aren't.[4]

Why did the young communitarians fail? Not because of lack of will or dedication, although some (the more sensible ones) were only marginally committed to the experiment. They failed because, under modern conditions, at least for young and middle-aged people, *a folk society has a very high opportunity cost.* You have to give up, among many other things, the ability to pursue opportunity in Seattle or London or Shanghai, far from your family. You have to live at a standard of living your great-grandparents strove mightily to escape.

But it's old folks who might actually benefit more from attempts to recreate the conditions of a folk society, where associations are long term and the aged are treasured for their memories. For such an undertaking to work, interaction with younger people is essential.

A step in that direction is the tendency of retirees to congregate in college towns, which are cultured places where young and old can mix. The cohousing movement, in which unrelated old people choose to live together in clusters of dwellings built for that purpose, is another encouraging trend, although it could be isolating if it separates the residents from the young and middle-aged. And some of the more affluent retirees are moving back to city neighborhoods that have always been a magnet for the young. Society has much to gain if entrepreneurs, governments, and the voluntary sector put thought and energy into this vision.

The rest of this chapter deals with the financial issues that arise in an aging society: the need for an adequate and secure retirement income, and the rising cost of medical care. I conclude with a call for policies

[4] Vonnegut (1974).

that make it easier for people to work longer, easing the dependency gap about which Ben Wattenberg was concerned.

## The Pension Dilemma[5]

There are two categories of problems related to providing retirement income, and I've written about them extensively in my day job.[6] The first is that people need access to financial products and practices that better enable them to spread the income from their working lives over their whole life. This is an engineering problem that can be solved with tools we have at hand or can easily develop. Although developing the needed asset management products, including longevity insurance, is a challenge, this is the lesser of the two problems.

The bigger problem is that a great many people just don't have enough money to provide for their old age. In retirement, they will need the resources of other people, whether that means government support (which is just "other people"), their children, or other relatives, friends, and associates (such as members of a church or mutual benefit society).

## Defined Benefit versus Defined Contribution Pension Plans

A traditional or "defined benefit" pension plan, such as is enjoyed by many public-sector workers in the United States but a shrinking number of private-sector workers, provides mortality-risk sharing or pooling by its very nature. Specifically, the amount saved need only last to the statistically expected or *average* length of life, about 85 years old, rather than to the maximum possible life of about 105 years. In a pension fund, the savings of those who die young help pay for those who live a long time.

But most people, at least in the United States, now rely on "defined contribution" pension plans that are really savings plans where there is no mortality-risk sharing. The contribution isn't defined at all (that terminology is an artifact of old laws and practices); rather, it is what the employee can afford, and usually also includes an employer contribution. Figuring out how to spend the savings after retirement, when you can live to over 100 while facing a decline in cognitive ability, is the retiree's job. The evidence is that they're not very good at it.

---

[5] This section was critically reviewed by Stephen C. Sexauer, Thomas L. Totten, and M. Barton Waring (see footnotes 8 and 11).

[6] See the following sections for examples.

So much has been written on retirement and pension issues that I'll just refer the reader to a comprehensive bibliography[7] and discuss my own work, with various collaborators.

## A Riskless "Pension Promise to Oneself"

I start by assuming that defined-benefit plans are not coming back, although maybe they should, and that we have to work within the context of a defined-contribution world. A template for a very basic, self-administered retirement savings plan—a "pension promise to yourself"—is described in an article I wrote several years ago with the San Diego County pension chief investment officer Stephen Sexauer.[8] The plan is to save a predetermined percentage of one's wages or salary, with the percentage escalating over time, until enough money has been accumulated to replace (when Social Security benefits are also included) 70% of the pay rate one has been earning just before retirement.

The money is assumed to be invested in riskless, inflation-indexed US Treasury bonds, which currently yield about 1% plus the rate of inflation. The savings percentages required for this plan to work are high; for our prototype participant, a Columbus, Ohio, public school teacher, they start at 10% and climb to around 30% in the second half of the teacher's career.

Although individuals rarely can save that much on their own, employer contributions help a great deal. If the participant can save 15% and the employer matches that amount, we've achieved our goal. The savings balance is split into two pools: one (about 85% of the total) used for a spend-down from ages 65 to 85, and the rest used to buy a "deferred annuity" that begins payout at 85 and lasts as long as the person remains alive.[9]

---

[7] Collins et al. (2015).

[8] Sexauer and Siegel (2013).

[9] A deferred annuity, more properly a deferred *life* annuity, is a contract, usually bought from an insurance company, in which the money you invest at a given time is held and invested by the insurance company for a period, usually until you reach age 85, then paid back as an annual or monthly guaranteed income for as long as you live. Deferred annuities can be structured so that they pay out as long as *either of two* spouses remains alive, and other variations are available. Because many people do not live long enough to collect any income, or collect for only a short period before they die, deferred annuities are "cheap," that is, one can arrange for a large amount of income to be paid, contingent on remaining alive, for a relatively modest initial investment.

Lower-income people don't need to save as much because, for them, Social Security benefits (as presently constituted) make up a larger part of the 70% income replacement that's desired. Many financial planners agree that replacing 70% of one's preretirement income is an adequate goal, because people have fewer expenses after they retire, including the expense of saving for retirement itself.[10]

## Taking Equity Market Risk to Improve Retirement Income Expectations

Although the savings requirements for a riskless pension promise to yourself are daunting, almost nobody invests that way. They buy a mix of stocks and bonds (or, more properly, stock funds and bond funds). Although Sexauer and I briefly addressed the reduction in savings requirements that might be made possible by taking market risk, a more recent article I wrote with the actuarial-firm executive Thomas Totten studies the issue more closely.[11]

In particular, Totten and I focused on the post-retirement or spend-down part of the problem; we assumed that the retiree starts out with $500,000, having accumulated it somehow. The numbers can be scaled up or down to a given retiree's situation. The money is invested in a portfolio dominated by equities (stocks) and, as before, is designed to be spent down over 20 years, with 15% of the total used to purchase a deferred life annuity.

Figure 5.4 is a simulation showing possible spending "paths" based on 20 years' spending out of the portfolio, plus the deferred annuity payout thereafter; the paths represent market outcomes, ranging from ninety-ninth percentile (best), shown by the top line, to first percentile (worst), shown by the bottom list. The distribution of market outcomes is based on history.[12] After the twentieth year of retirement (age 85), the

[10] Some taxes are also lower after retirement, as are costs associating with commuting, dressing for work, and so forth.

[11] Totten and Siegel (2019). Accumulating the half million (or any other amount) involves having a saving strategy and sticking to it; the investment-management-firm executive Barton Waring and I have written extensively on optimal savings and spending rates (Siegel and Waring, 2017, on saving rates; Waring and Siegel, 2015 on spending rates). The work is somewhat more technical than the articles discussed earlier.

[12] More accurately, the parameters used in the simulation (expected return and standard deviation) are based on history, with the expected return adjusted for the (currently high) level of markets. The simulations assume that markets follow a random walk, which is a reasonably robust but not completely realistic assumption.

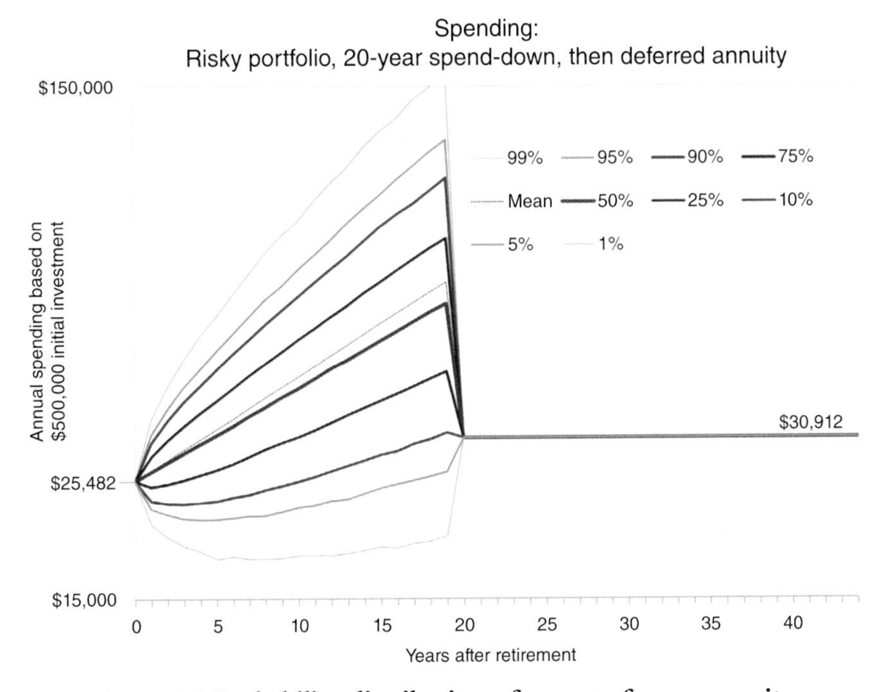


**Figure 5.4 Probability distribution of payouts from an equity-oriented retirement savings plan including deferred annuity that begins payments in twentieth year of retirement.**

*Note:* We assumed that the investment portfolio had an expected return of 6.9% and a standard deviation of returns of 11%.
*Source:* Based on data from Totten and Siegel (2019).

paths are all the same because the portfolio has been spent and all the income comes from the deferred annuity.

Note that most, but not all, of the paths yield more than 6% in nominal terms, or $30,000 per year per $500,000 invested, a big improvement over what can be achieved using riskless bonds. But, if one of the worst-case market scenarios happens, the retiree won't make 6% and would have been better off with the bonds. That's what it means to take risk; a good outcome is not in any way guaranteed. The deferred annuity at least assures that the retiree will not be destitute in his or her last years and enables her to manage the spend-down between age 65 and 85 more rationally and confidently.[13]

[13] This assumes that the annuity issuer does not default on its obligation, or that, if they do default, state annuity guarantee funds are adequate to make the retiree whole.

In realization, that is, after the fact, none of the paths experienced by an investor will look anything like the smooth lines in the figure. Instead, they reflect the bumps and potholes of the stock market. The smooth percentile lines are *averages* of the results from the many thousands of runs involved in producing the overall simulation.

Although well-engineered savings plans such as these have not been widely implemented, they're not rocket science. Employers, investment-management firms, insurance companies, trade or business associations, and legislators and regulators should work toward the streamlining of the retirement income-generating process so that our aging population has one less worry.

## Medical Costs[14]

Medical costs have risen hugely for three principal reasons:

1. The *much* greater effectiveness of the available treatments and preventive methods, compared with the past.
2. Third-party payment, which reduces the incentive to economize.
3. Bureaucracy, regulation, compliance costs, litigation and the cost of avoiding litigation, and mission creep.

In contrast to commonly held beliefs, soaring medical costs in the United States are *not* due to overutilization, gigantic paychecks for doctors, overly large numbers of specialists, or even the aging of the population. The principal real reasons are those summarized in the preceding list.[15]

### Greater Effectiveness

We could spend the same amount on medical care (adjusted for inflation) as we did 50 years ago, but most people would not be satisfied with "take two aspirin and call me in the morning." Radical advances in chemotherapy, psychoactive drugs, orthopedic devices, and cardiac and other surgeries simply cost more for the same reason a race car costs more than a quarter horse. It's better at performing the desired service.

A discussion of this sober subject could benefit from a little whimsy. In 1925, *Science and Invention* magazine, a brainchild of the science fiction

[14] This section was critically reviewed by Lee A. Kaplan, MD, a physician and retired clinical professor of medicine at the University of California, San Diego.

[15] Papanicolas et al. (2018).

publisher and futurist Hugo Gernsback, displayed a cover drawing in which a doctor is treating a patient via a two-way television connection. (The first television transmission would take place two years later.) In a fine example of life imitating art, this is now common practice, and in some cases brings medical care to those who would not otherwise get any.

But the 1925 doctor had no penicillin, no sulfa drugs, no CAT scanner or MRI machine, almost no orthopedic devices, no chemotherapies for cancer. Medical treatment was affordable, but if you were seriously ill, there was not much the doctor could do for you.

## Third-party Payment

Because of wide variation in the amount of medical care needed by any given individual, some kind of risk spreading or "insurance" is unavoidable. Food is even more of a necessity than medical care, but we don't have or want food insurance (except for the very poorest people) because everyone needs roughly the same number of calories.

But, with regard to medical costs, one person may live happily with a $250 checkup every year until she falls over and dies, while another—behaving in exactly the same way but with a different draw in the genetic lottery—may require $5 million in cancer treatments or heart surgeries over her lifetime. Lifestyle adjustments help but don't guarantee anything.

## The Problem with Medical "Insurance"

Therefore, we have medical insurance, just as we have fire and flood insurance: contracts that are supposed to pay for unusual but highly adverse outcomes. But because of the tax deductibility (for employers) of medical insurance costs—itself a relic of World War II-era wage controls in which employers were not allowed to give raises but could compete on fringe benefits—medical insurance has morphed into an almost fully prepaid health care plan for those lucky enough to get one. Regarding many medical procedures or tests, "I'll take it—it's paid for!"

That's not insurance; it's analogous to building into the price of a car all the gasoline and repairs a driver will ever need. Few drivers would buy or could afford such a "loaded" car, but the same scheme is inexplicably popular with medical insurance, probably because of the fear of crushing, life-altering expenses (that vastly exceed anything that maintaining a car could cost)—and because someone else, either the employer or the government, through Medicare and Medicaid—is paying the cost. (In the case of the employer, he is just appearing to pay the cost, because it is coming out of the total worker-compensation budget and could otherwise go to the worker in cash.)

But, as I suggested earlier, that is just the picture faced by the lucky ones, those who have permanent, full-time jobs with a large private employer or a government agency. The rest of us have to make do with overpriced medical insurance that underdelivers.

## A Modest Proposal

I don't have anything like a full solution, but here's a thought. Let government insure the "nose" and the "tail" of the beast, by paying for a very basic level of care plus expensive but rare outliers, and leave the remainder to the individual. The individual could choose to be covered by a company plan, individual commercial insurance, self-insurance (such as a health savings plan), no insurance (pay as you incur the charges), or a voluntary association (a church, an alumni association, the Raccoon Lodge, or some other granfalloon; these folks could also help with organizing pensions or retirement savings plans).[16] This arrangement would cost much less and provide better outcomes.

## Bureaucracy and Mission Creep

Only about 30% of your health-care dollar goes to actual doctors and nurses. The rest goes to equipment (mostly necessary) and an army of paper pushers, administrators, managed-care "solution providers," and paraprofessionals. The paraprofessionals are supposedly there to make doctors' and nurses' lives easier; if the system were working, we wouldn't have doctors leaving their practices in droves and nurses commanding doctor-like salaries due to their punishing hours and high dropout rates. Surely, we can do better.

I earlier mentioned mission creep. The medicalization of every problem known to man is yet another reason for high costs, with some benefits but not clearly enough to justify the costs. Faith McLellan, writing in *The Lancet*, explains:

> Once upon a time, plenty of children were unruly, some adults were shy, and bald men wore hats. Now all of these descriptions might be attributed to diseases—entities with names, diagnostic criteria, and an

[16] The granfalloon (Kurt Vonnegut-ese for "a proud and meaningless association of human beings") would, in order to secure the benefits, most likely want to reinsure with a commercial insurance company.

> increasing array of therapeutic options. . . . Medicalisation . . . refers to the process by which certain events or characteristics of everyday life . . . including sexuality, garden-variety unhappiness, childbirth, ageing, and dying, . . . become medical issues, and thus come within the purview of doctors and other health professionals to engage with, study, and treat. . . .
>
> The medicalisation of these life experiences has brought with it benefits, but at a price. And those costs, which are not just financial, are not always clear.[17]

Doctors cannot solve every human problem any more than schoolteachers can. To try is admirable, but not realistic.

## Working Longer

If you don't have enough money for something, here's a solution: go to work.

The dependency gap described by Ben Wattenberg, the small number of young people available to work for, provide services to, and pay taxes to support older people is a challenge in almost every developed country and many developing ones. In the United States, increased immigration has been proposed as a solution, and it will help: many immigrants have the needed skills, are willing to uproot their lives for opportunity, and tend to arrive young and in good health.

But immigration is not a panacea. Immigrants, like natives, get old and sick (although with a time lag). They have children who require schooling. Immigration is almost certainly a net economic positive under current conditions, but the optimal amount of immigration is not infinite. Jeremy Siegel, the famed Wharton professor who showed (not quite convincingly enough for me) that stocks are the best long-run investment, says Americans would be able continue to retire at age 62 if we imported *half a billion* immigrants. I'm not sure where we'd put them, and his proposal is not serious; he's trying to demonstrate the limits of immigration as a strategy for coping with the costs of aging.

The aging population seems to be divided into two camps: (1) those who would like to work longer, for either financial or personal reasons, and are either already doing so or need to overcome various institutional and legal barriers so they can; and (2) those who can't wait to quit their jobs and take it easy. Many in the second camp are facing physical limitations.

---

[17] McLellan (2007). McLellan, a public health scholar, was then an editor of *The Lancet*.

Let's at least try to help the first group. In a paper we are currently writing, Stephen Sexauer and I advocate policies to make it easier for older Americans to work longer. We argue:

> [M[ost people would rather be rich, healthy, and happy than poor, sick, and tired in their later years. . . . [T]he relative shortage of older workers in the economy can be traced to a set of generally well-meaning laws, regulations, and practices.
>
> [We] document the need that many older people have for more money. [There are] two common objections to working longer: poor health . . . and the desire for leisure. . . . These are longstanding and real constraints and they merit being acknowledged and discussed. Another objection is the "French Problem," the mistaken idea that there is a shortage of work or that the supply of jobs is fixed, so that employing an older person means denying a job to a younger one.
>
> [We note that] work is continually getting easier, [making it possible for older workers to perform it.] [We] present . . . evidence that working people enjoy better health . . . and also enjoy life more.[18]

I'd also add that many older workers are already trained to do the work they've done all their lives, are socially conditioned to show up on time and in the right place, are dressed properly, and have at least a tolerable attitude. Sadly, this is not universally true of the young.

What needs to be done to enable and encourage older workers to stay on the job, perhaps with reduced hours, reduced pay, and easier work? Sexauer and I write that the "challenge of getting people to work longer is largely a legal and regulatory problem." Employment law needs to be changed to provide a safe harbor for employers offering older workers the more relaxed conditions we've just described, instead of exposing the employer to the risk of litigation and punitive regulation.

Sexauer and I conclude:

> Along with the gains to the individual from working longer, there are large gains to the public. There are no losers. Let's make the needed changes, not only for the sake of older people who would like to work, but for the benefit of all of society.

---

[18] All the quotes in this section are from Laurence B. Siegel, and Stephen C. Sexauer. In progress. "Longer, Healthier, Happier." Draft available at www.pensionpromise.com.

## Conclusion

Tension between the generations is age-old and biological. As explained by Richard Dawkins in *The Selfish Gene*, parents and children compete for the same resources.[19] Each child tries to maximize the resources he or she can commandeer from the parents, while the parents must strike a balance between maximizing the likelihood of that child's survival and reserving resources for themselves, their other children, and their possible future children. This dynamic is an ongoing power struggle that, in humans, has manifested itself in the form of wars, revolutions, changing laws and tax rates, and explosive shifts in the culture.

We do not know exactly how the dynamic will play out as the population ages. We can attempt a conjecture. When children were plentiful and parents scarce, the latter held the bulk of the power. With plentiful adults and scarce children, plus a mushrooming number of adults with no direct genetic investment in the future, the arc of power will bend toward the young. Although they'll have fewer votes and fewer aggregate dollars, young people will have more influence over the behavior and priorities of their elders. What that means in practice is anyone's guess, and at the very least it will be interesting. May you live in interesting times.

---

[19] Dawkins (1976).

# PART III
# RICHER

# 6

# *Before the Great Enrichment: The Year 1 to 1750*

In the beginning was the year 1, and the Scottish economic historian Angus Maddison was already collecting data. According to Maddison, the average annual income in the world was $467.[1]

Okay, I'm kidding. Jesus of Nazareth had just been born, the dollar would not exist for another 1700 years, and Angus Maddison would not exist for another 1900 and then some. But, in the year 1 AD, the world had an economy (and the years before that did too). People worked and consumed, bartered, bought, and sold. Sometimes they bought and sold each other. They grew enough food that they could survive—our existence today is testament to that. They reproduced, fought wars, built cities, made art. An income of $467 a year is not much, but it's an average that includes many who were richer—Rome was at its zenith, Greece past its peak, China and Egypt flourishing—and many who were dirt-poor in ways that we can scarcely imagine.

[1] See Maddison (2007), and the Maddison Project, www.rug.nl/ggdc/historicaldevelopment/maddison/. The $467 estimate is actually of GDP *per capita*, not personal income (which is closely related), and is in "1990 dollars," what Maddison used as the standard "currency" in his work. To convert to 2018 dollars, add 94.7% for inflation (that is, multiply amounts quoted in 1990 dollars by 1.947). For most purposes, this book uses today's (2018–2019) dollars.

And, two millennia later, Maddison, according to Deirdre McCloskey "a bear of a man fluent in seven languages . . . a Scot living in France and working in Holland . . . was the leading authority on the history of world trade and income."[2] He was able to go back and reconstruct enough economic history, in many countries around the world, that he could come up with a defensible estimate of the productivity of the world economy in the year 1. He did the same thing, with increasing accuracy, for the years 1000, 1500, 1600, and 1700. Then, in the nineteenth century, his estimates became more frequent. The result is shown in Figure 6.1.

Looking at the data underlying Figure 6.1 (because the figure is too compressed to show every country) reveals a fascinating and turbulent story. In the year 1, Italy was the richest country, in the year 1000 Iran and Iraq (how times change!), in 1500 Italy again, in 1700 the Netherlands, in 1875 the United Kingdom, in 1925 the United States. Today, among large countries, it's still the United States, although a few small ones like Norway and Singapore have higher average incomes.[3]

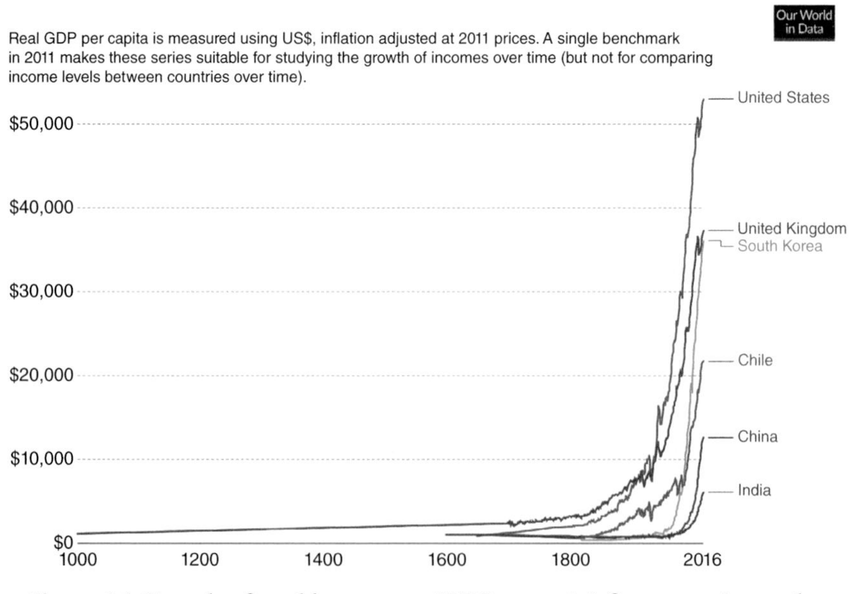


**Figure 6.1 Growth of world economy (GDP *per capita*) from year 1 to today.**

*Source:* Our World in Data, https://ourworldindata.org/economic-growth, based on data from Maddison Project Database (2018). Licensed under CC BY.

[2] McCloskey (2016).

[3] Country definitions are those that apply today. In other words, "Italy" means the part of the Roman Empire that was physically on the Italian peninsula; "Iraq" is Mesopotamia and "Iran" is Persia; and so forth.

But economic progress is not a road race. Unlike in a road race, what the best economic performer gets is not that different from what the performer in fifth or twelfth or even twentieth place gets. Economic growth is a cooperative effort, your gain is not my loss, and the goal of a decent standard of living in my country is made easier, not harder, by you achieving it in yours. We're each other's suppliers and customers.

## The Economic Hockey Stick

I've skipped past the obvious and essential message of the economic "hockey stick" in Figure 6.1, which is the central theme of this book: after about 1820, a little earlier in a few lucky countries, for the first time in human history the economy grew and grew and grew—rapidly and consistently. In the Middle Ages and the early modern period, there was some growth, but it was at a glacial pace; *per capita* consumption tripled between the year 1 and 1820.

But when growth really took off, in the countries we now call developed (western Europe and its offshoots such as the United States, Canada, and Australia, plus Japan), standards of living doubled roughly every two generations. Nothing like this had ever happened before.

Clearly, something was afoot before that, and we're all aware of brief periods of flourishing in specific locales—the Roman Empire, the era of cathedral building in the High Middle Ages, Leonardo's Florence. The English philosopher John Locke, in 1700, wrote, "a king [that is, an Indian chief] of a large and fruitful territory [in America] feeds, lodges, and is clad worse than a day laborer in England."[4]

Unless Locke's statement was just an empty boast about his homeland, this meant that significant economic growth was already taking place in England. Some economic historians have called this the First Divergence, as early modern Europe gradually outperformed Asia, Africa, and, as Locke pointed out, the Americas.[5]

Outside of a few rich cities and the villages that traded with them, however, most people, even in Europe, lived in preindustrial modern times as they had in ancient and medieval times.

---

[4] Locke (1689).

[5] The idea of a "first divergence" or "first great divergence" in this critical period in European history has spawned a rich literature. For a good recent summary, see Koot (2013). The term *first divergence* apparently originates with Van Zanden (2009). See also Moore (2015) and Karayalcin (2016).

## Why Were Premodern People Short?

I'll spare you the gory descriptions of premodern deprivation, disease, and discomfort—Matt Ridley, in his groundbreaking 2010 book, *The Rational Optimist*, does a fine job of it. It suffices to note that most people were short of stature because there was rarely enough to eat: "People in the pre-industrial past were poor—very poor," writes the economist Bradford DeLong. "If they weren't very poor," he continues,

> they would have fed their children more and better and their children would have grown taller. Pre-industrial dire poverty lasted late. Even as of 1750 people in Britain, Sweden, and Norway were four full inches shorter than people are today—consistent with an average caloric intake of only some 2000 calories per person per day, many of whom were or were attempting to be engaged in heavy physical labor.[6]

I think this assessment—that stunting and weakness were mostly due to caloric restriction—understates the role of disease, which also impairs physical growth and the ability to do work. Not everyone was poorly fed. Evidence includes the rich culinary traditions that reached well beyond the upper classes in many European countries; a starving people does not waste resources on fine cuisine or ceremonial gluttony. Moreover, the rich, who had no budget constraint on food, were also short of stature, although not nearly as short as the poor. We'll cover these questions further in Chapter 8, "Food," which introduces the work of Robert Fogel, the University of Chicago economist and cliometrician who won a Nobel Prize for work that links nutrition and productivity.[7]

Today, however, the world's biggest nutritional concern is obesity, even in many less developed countries. That is one of the most profound changes that ever occurred.

## The First Divergence

This tale of woe in premodern times obscures an important fact: not everyone was poor. In all human societies, some people are richer than others, but the moderately well off in preindustrial Europe were not a tiny minority. Everywhere you look, you find evidence of a middle class of artisans, merchants, traders, clergymen, and government officials. While

[6] DeLong (2014).

[7] A cliometrician is a person who uses data and statistical inference to study history.

medieval and early modern Europe (say, from the fall of the Roman Empire to the first stirrings of the Industrial Revolution around 1750) is sometimes portrayed as a two-class society consisting of royals and large landholders, on the one hand, and a near-starving peasantry, on the other, that is not an accurate portrayal.

Let's look at England, where history is preserved more carefully than in many other places. The handsome stone houses of the Cotswolds were built with fortunes made in the wool trade in the 1200s, deep in the Middle Ages. The poet Geoffrey Chaucer was born around 1340 to a family of vintners and merchants that would now be described as upper middle class. Shakespeare, born in 1564, grew up in a fine-looking half-timbered house that suggests a similar background, and his father was a glove-maker—an artisan, not a captain of industry.[8] His wife, Anne Hathaway, was from an even more prosperous family (see the photo of the interior of her parents' house in Figure 6.2). For some people well outside the circle of royals and nobles, then, economic activity went beyond survival to adequacy, then comfort, then, for a few, luxury.

This early breakaway of Europe—particularly northern and western Europe—from the lowly condition of the rest of the world is sometimes called by economic historians the First Divergence or Little Divergence. Based on a look at the technology of the time in various countries in, say, 1400, an observer might have guessed that China would become the dominant economy of the next few hundred years.

## In an Age with Many Paupers, a Surfeit of Geniuses

But the dominant economy turned out to be Europe—first Italy, then Spain, then France, then the Netherlands, then Great Britain. Although late medieval and early modern Europe had its horrors—the recurrent plague, the unbelievably destructive Thirty Years' War in Germany, and the seventeenth-century turmoil in England come to mind—those were the years when the foundations of the modern world were laid. Chaucer, Shakespeare, and Cervantes in literature; Leonardo, Michelangelo, and Rembrandt in art; Copernicus, Galileo, and Newton in the sciences; Erasmus, Descartes, and

[8] John Shakespeare's beautiful house, now visited by almost a million visitors a year, is cramped on the inside, as is the nearby birthplace of William Shakespeare's wife, Anne Hathaway, who was somewhat rich. Both families could have easily built their houses bigger, but they chose not to. This observation suggests that short stature was not entirely due to poverty. Stunted growth due to malnutrition is not inherited. If prosperous people also built houses to suit short inhabitants, maybe disease and other factors, rather than just poverty-related malnutrition, were major contributors to shortness.

**Figure 6.2 Anne Hathaway's cottage (interior). Despite her family's wealth, Shakespeare's wife's home was built for short people.**

*Source:* Prisma by Dukas Presseagentur GmbH/Alamy Stock photo.

Spinoza in philosophy; Machiavelli, Hobbes, and Locke in politics; Bach in music; Luther in religion: it is hard to imagine a more fertile time.[9]

Yet none of these luminaries could buy a capsule of penicillin. As noted in Chapter 1, in 1836 Nathan Mayer Rothschild, the world's richest man, with a fortune expressed in today's dollars approximating those of Jeff Bezos and Warren Buffett added together, died at the age of 58 of an infection that could have been cured, a century later, with a dollar's worth of antibiotic.

## Neo-Malthusians and Proto-cornucopians

There is a spirited debate among economic historians about the living conditions of the common people in premodern times. All agree that life was hard, but how hard? The neo-Malthusians (glass half empty)

[9] In one of history's great ironies, Martin Luther, a religious fanatic and rabid anti-Semite, and my least favorite hero of the individualist cause, was perhaps the most influential champion of the individual vis-à-vis central authority (in his case, the Roman Catholic Church) between the Middle Ages and the Enlightenment two centuries later. More than a few historians draw a direct line connecting the Lutheran Reformation, the Enlightenment, and subsequent Modernity.

cite short stature, as described earlier, and the diseases of malnutrition found in studies of old human bones, as evidence of widespread extreme poverty. When incomes and caloric intake rose, this story goes, people reproduced more, with the effect of once again lowering their level of living to mere survival.

The proto-cornucopians (my coinage) counter that most civilizations, even those early in history or thought to be primitive, have built great churches and palaces, fielded armies, and supported a leisure class that included priests, government officials, and intellectuals as well as the ordinary rich.

You can't do this if most people are almost starving. The funds for these luxuries typically come from taxation, which must be paid out of a surplus (that is, production in excess of the bare minimum level of consumption that allows people to survive). If you, the ruler, ask people to give something up—food or money—in order to get protection, roads, public administration, a system of laws and courts, or whatever you're offering in exchange, they must have something to give.

The people will not agree to being taxed to death—they'll kill you instead. The Jacquerie, Wat Tyler's rebellion, the German Peasants' War, and a thousand other revolts and revolutions are testimony to that fact. Even as late as 1789, the French Revolution, thought up by a group of well-fed lawyers in Paris, could not have succeeded without the support of the peasantry, which was experiencing widespread hunger.

But more often the people did agree to the trade, taxation in exchange for services, and the cathedrals and castles were built. The conventional explanation (the only one I know) for the explosion of cathedral building in the High Middle Ages is the power struggle between secular and religious authorities—kings and the church. The church could raise money by selling indulgences, a more attractive proposition (if you believed in their effectiveness) than paying taxes to the king. And the cathedrals, unlike the castles, were open to all, and were truly impressive—a gift *from* the Almighty as well as *to* him.

It is a little hard to visit the cathedrals of Strasbourg or Salisbury or Ulm without thinking that the proto-cornucopians were at least a little bit right, that rudiments of an affluent society could be found in that place and time. Records show that the workers, even at the lowest skill levels, were paid, unlike the slaves who grudgingly maintained the lifestyles of the noble Romans. The medieval and early modern way of life may have been to a large degree mired in misery, but its finest accomplishments are the equal of those of any period.[10]

[10] See Owen (1989).

## The Greatest Architecture in the World

How primitive could an economy have been if it produced Saint-Chappelle, the thirteenth-century Parisian masterpiece shown in Figure 6.3? "Between 1170 and 1270, 80 cathedrals and nearly 500 churches of cathedral size were constructed in Northern France. Their value in 1840 was estimated to be over one billion dollars," writes Virginia Owen, an American economist. A billion dollars in 1840 equates to about $28 billion now. And that's just northern France! Any traveler to Europe knows that cathedrals are about as common a sight all over the continent as silos are in America.

And Christianity was not the only inspiration for great architecture requiring expensive materials and highly developed skills. The Islamic world produced comparable works, as did India and China ("in Xanadu did Kublai Khan"). Xanadu is still there—it was not a fever dream of William Blake's but a real place (built in 1256), the capital of Kublai Khan's empire—and is now called Shangdu, a UNESCO World Heritage Site.

**Figure 6.3 Sainte-Chappelle, Paris (built 1248).**

*Source:* Oldmanisold, 2018. https://commons.wikimedia.org/wiki/File:Sainte_Chapelle_Interior_Stained_Glass.jpg. Licensed under CC BY-SA 4.0.

The energy, creativity, and money that went into this orgy of cathedral building present a puzzle to economic historians: In what we think of as the impoverished Middle Ages, where did the money come from?

And *what were the cathedrals for*? Commercial buildings can be beautiful, but they pay for themselves by advertising the companies that built them. Residential buildings give pleasure and comfort to their occupants. But who stands to gain from 580 cathedrals? Presumably the stonecutters, glassmakers, architects, and countless other artisans who built them—it is inconceivable to think they worked for what we sometimes call slave wages, because the demand for these skills almost certainly created a labor shortage. The cathedral era appears to have been a building boom like any other, benefiting many, but with the difference that there's no commercial motivation we can discern. The greater glory of God . . . but that virtue has been served in many ways, not all of them wildly expensive and complex, involving an army of craftsmen and engineers. The beautifully simple Protestant churches of early America celebrate the glory of God in their own, inexpensive way.

The final mystery is: Why didn't cathedral building set off a general movement of economic betterment? Or did it? Weren't these gifted artisans capable of doing other work? What else did they build, and what did they do with the money they made?

While I don't have the answers, only questions, the cathedrals remind us that, despite the economic gulf that separates us from our medieval forebears, we are not the only interesting and accomplished people who have ever lived.

Does the First Divergence show up in the data? It takes a bit of creativity to find it. Maddison and other scholars found that wages rose sharply in Britain and the Netherlands relative to other European countries. A finding of divergence between Europe as a unit and the rest of the world, however, would require more evidence than that merely local phenomenon. Two Dutch researchers, Eltjo Buringh and Jan Luiten van Zanden, find it in *book production* (Figure 6.4), which

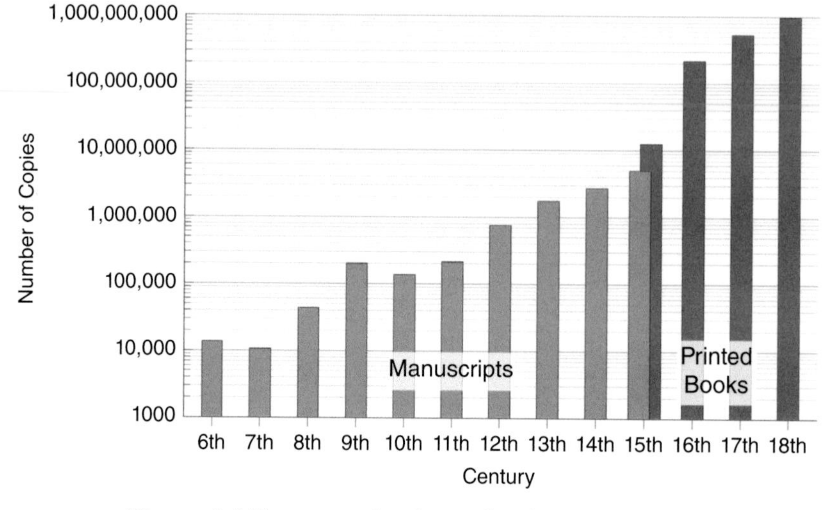


**Figure 6.4 European book production, 500–1800.***

*Without southeast Europe (Byzantine, later Ottoman realm) and Russia.

*Source:* https://en.wikipedia.org/wiki/Great_Divergence#/media/File:European_Output_of_Books_500–1800.png. Licensed under CC BY-SA 3.0.

captures one aspect of changes in the standard of living in Europe, broadly defined.[11]

Book production, obviously greatly aided by Gutenberg's introduction of the printing press in the 1450s but growing exponentially in manuscript form way before that, expanded by a factor of 100,000 between the sixth and the eighteenth centuries. All of this is before what we conventionally call the Industrial Revolution. Books, books, books—the Internet of the Middle Ages!

[11] Buringh and Luiten van Zanden (2009).

# 7

# *The Great Enrichment: 1750 to Today*

## The Industrial Revolution and the Great Divergence

What is the most important event in history? Deirdre McCloskey makes the case for the economic transformation that began in the late eighteenth century:

> Modern economic growth is the increase of income per head by a factor of 15 or 20 since the 18th century in places like Britain—and a factor of 8.5 worldwide, even including the places that have not had the luck or skill to let it happen fully. It is certainly the most important event in the history of humanity since the domestication of animals and plants, perhaps the most important since the invention of language. It bids fair to free us all, eventually.[1]

Buried within this astonishing and accurate claim are two stories. One is the Great Enrichment: the "increase of income per head by a factor of 15 or 20 since the eighteenth century." The other is the Great Divergence: that it happened in only some places, leaving others behind.

We started this Chapter 6, "Before the Great Enrichment," with bearlike Angus Maddison, the old-economic-data detective, so let's see what he had to say about the Great Divergence (not the First

[1] McCloskey (2004).

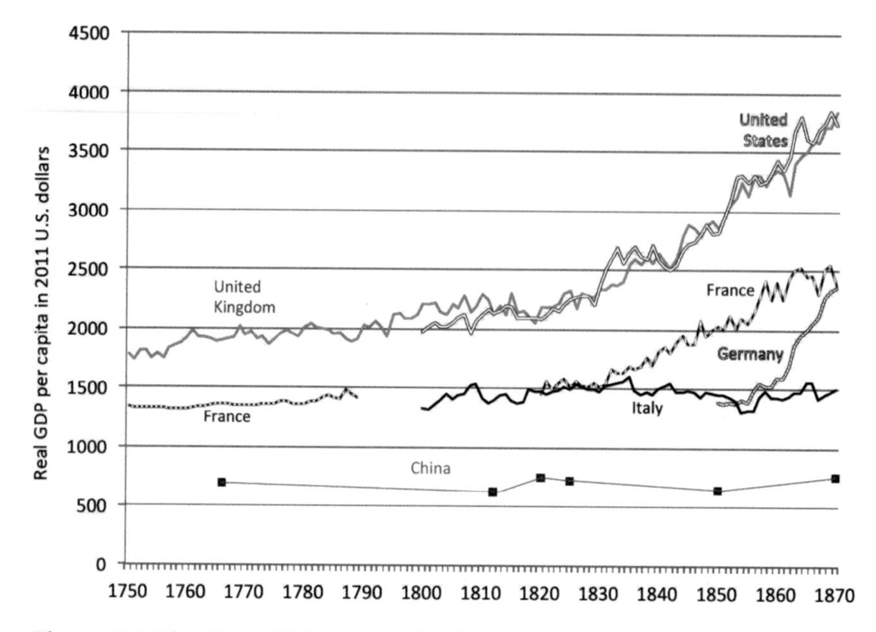


**Figure 7.1 The Great Divergence begins: the First Industrial Revolution and its aftermath, 1750–1870.**

*Note:* Real GDP is analogous to PPP GDP and is in "2011 US dollars"; to convert to 2018 dollars, multiply by 1.12.

*Source:* Constructed by the author using data from the Maddison Project (real GDP, multiple benchmarks), www.rug.nl/ggdc/historicaldevelopment/maddison/releases/maddison-project-database-2018.

Divergence, but the big one that started around 1750). Figure 7.1 shows his reconstruction of the growth of incomes in selected countries from 1750, the earliest plausible starting date for the Industrial Revolution, to 1870, when the benefits from that revolution had become widespread and a Second Industrial Revolution was starting.

The United Kingdom made considerable gains starting around 1750, as, presumably, did the colonies that would become the United States, for which we don't have data. After a period of stagnation in the early 1800s, both countries sustained high growth rates at roughly the same level. Along with the Netherlands (not shown), they were the richest countries in the world.

Italy and France had strong starts during the early modern era. *Quattrocento* (the 1400s or fifteenth-century) Italy led the world in development, but Angus Maddison would have us believe it was as poor in 1870 as in 1470 or, for that matter, at the peak of the Roman Empire. (I'll take

**Figure 7.2 Many nineteenth-century Italians could afford a trip to New York—once. That was all they needed.**

*Source:* Pictorial Press Ltd./Alamy Stock Photo.

that with a grain of salt; see Figure 7.2.[2]) France in the 1600s was the world's richest nation, but France's modern era of growth didn't take off until about 1830.

Germany and China are the real standouts in Figure 7.1. Germany's growth rate from 1850 to 1870 was like nothing that had been seen before in Europe or even in the booming United States. The perception of a "German menace" to British and American industry in that period is reminiscent of the later Japanese and Chinese "menaces" as those latter countries developed at warp speed a century or more later.

China, sadly, is a standout for remaining abjectly poor while Western Europe and North America were getting ahead. In this case, China is a proxy for the rest of the world (we don't have data for India and Africa). The Great Divergence meant that western Europe and North America got rich—although Europe had some laggards such as Italy and Spain—and the rest of the world remained as poor as it had been in the Middle Ages.

[2] Incomes may have stagnated, but the "basket" of goods for sale in the marketplace changed. Italians in 1870 could save up for a ticket to America (it was about a year's pay), something they could not do in 1470 (unless their name was Cristoforo Colombo).

This divergence would not go on forever, and the rest of the world would eventually begin to catch up. But the transition from divergence to convergence would not occur for 80 more years. The Second Industrial Revolution, which we'll define as 1870–1948, reinforced and extended the divergence between rich and poor countries.

Recall Hans Rosling's bubble charts from Chapter 1? In Figure 7.1, as well as the subsequent income charts later in this chapter (Figures 7.4 and 7.5), we're seeing some of the detail behind the movement of the bubbles to the right ("richer") as incomes grew. (Upward movement—"healthier"—represented improvements in life expectancy.) To visualize economic growth, it helps to use both bubble charts and ordinary time-series graphs.

## What the First Industrial Revolution Accomplished

What produced this once-in-human-history change in economic growth rates? Coal is a dense source of energy, so burning it—to power steam engines, heat buildings, and so forth—enabled one to do much more "work"—in the physics sense of moving an object from here to there—than had ever been possible before. The history of economic growth is, in considerable part, the history of obtaining better and cheaper sources of energy—what Buckminster Fuller called "freeing the energy slaves."[3] The original energy slaves were human slaves, but later energy slaves included animals, windmills, water mills, and trees (burned for fuel). Coal and, eventually, oil and natural gas were several orders of magnitude more efficient.

Production on a massive scale was thus made possible: the steam-powered railroad, which came into being around 1810, required laying tens of thousands of miles of steel rail that had to be mined and manufactured to close tolerances. Mechanical looms replaced the hand weaving of cloth. Because of the innovation of standardized, interchangeable parts, factories could mass-produce what previously had to be made by hand. All of a sudden, in a very wide variety of activities, machines started doing work that had been done by hand since time immemorial.

"Dark satanic mills," the poet William Blake called them, and working conditions could be brutal. But workers streamed from the farms into the cities in search of the new, plentiful factory jobs because they saw a better life for themselves there. As Matt Ridley reminds us, "the

[3] Fuller (1940).

rural pauper of 1700 was markedly worse off than the urban pauper of 1850 and there were many more of him."[4]

## The Second Industrial Revolution

The Second Industrial Revolution inaugurated a new age of freedom. The First Industrial Revolution spawned the age of steam, heated by wood or fossil fuels, and the train, which had to go where the tracks went. The Second, in contrast, was the age of electricity—also mostly produced by burning fossil fuels, but at a convenient distance—and the automobile, which could go wherever the driver wanted. These are profound differences.

The new technologies did not just liberate the individual driver or telephone caller—it brought freedom to innovate in surprising ways. A factory full of electrical devices, each independently powered using a loop of twisted copper, was a huge advance over one in which a steam engine awkwardly transmits power simultaneously to all the machines in the factory using a Rube Goldberg contraption that hardly anyone now remembers (see Figure 7.3).

**Figure 7.3 Steam powered factory. A single steam engine turned the overhead horizontal shaft, which then transmitted power to individual machines through the vertical shafts. (Pullman Industrial Complex, 111th Street and Cottage Grove Avenue vicinity, Chicago, Cook County, IL)**

*Source:* Historic American Engineering Record (Library of Congress), www.loc.gov/resource/hhh.il0377.photos/?sp=3.

[4] Ridley (2010), p. 144.

The first electrical device worth having was probably the telegraph, which became practical by the 1840s, but that was just a harbinger of a new revolution. The big changes came after the US Civil War, with the invention of the telephone by Alexander Graham Bell and others around 1876, the building of the first practical motorcars by G. W. Daimler in 1885, and the flood of inventions coming out of Thomas Edison's laboratories starting in the 1870s. (Amazingly, the airplane came pretty quickly after that: from building a primitive telephone—"Mr. Watson, come here, I want to see you"—to flying in the sky in one generation.[5])

The automobile was the most important of these innovations. It tremendously expanded opportunities for work and residence, reshaped the cities, made life in the countryside bearable, and gave autonomy to the individual. It has been observed that people down on their luck try to hold onto their cars, more so than any other possession: the car gives them the ability to recover by going where the work is.

Naturally, the economic boom sparked by these inventions was spectacular. There were, of course, interruptions including economic panics on a scale that had not been seen before; great booms often lead to great busts, but almost invariably leave society better off than before the boom started.[6] Figure 7.4 shows Angus Maddison's reconstruction of incomes and growth rates over the Second Industrial Revolution period. Note that Figure 7.4 is on a different scale (a log scale) than Figure 7.1—which shows the previous period—because the magnitude of the changes is so large.

The first part of the 1870–1948 period was part of the peaceful "long nineteenth century" (1789–1913, according to some), and economic progress was steady, even in long-suffering Italy and impoverished British India. The lines representing incomes for these now First World countries are almost exactly parallel, showing evenly distributed growth.

But then, in 1914, all hell broke loose. World War I and the Russian Revolution devastated most of the economies of Europe, and the Great

[5] With the telephone came new words and phrases. The phrase "Where are you?" was one; before communication at a distance, a speaker knew the location of the person he was speaking to. So are "hello" and "phony" ("hello" already existed but was archaic and rarely used). Aviation also modified the language: "I have landed" means something different on an airplane than on a ship; aviators' colorful language has become part of the common vernacular. Roger that.

[6] The financial historian and investment manager Peter Bernstein, in a private communication, expressed the idea that booms followed by the inevitable crashes are the price of progress. The Internet boom of the late 1990s was followed by a crash, but we got the Internet. The housing boom of the next decade was followed by an even bigger crash, but we got the houses, Bernstein argued. The same can be said of innovation-related booms and their related busts throughout history.

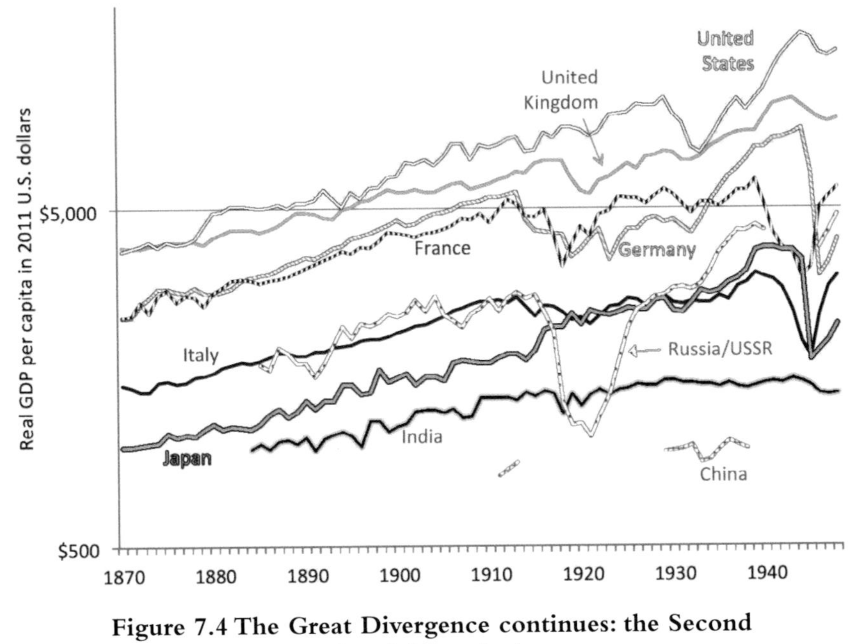


**Figure 7.4 The Great Divergence continues: the Second Industrial Revolution.**

*Note:* Real GDP is analogous to PPP GDP and is in "2011 US dollars"; to convert to 2018 dollars, multiply by 1.12.

*Source:* Constructed by the author using data from the Maddison Project.

Depression deeply affected not just the United States but also much of the world. The worldwide influenza epidemic of 1918 also sharply curtailed economic output.

The wartime recovery from the US Great Depression is legendary. Germany and Japan collapsed as a result of World War II, and France almost did. But the big surprise in Figure 7.4 is the USSR, which grew so rapidly (if we are to believe the data released by its Communist central planners) that Paul Samuelson, no communist himself, bizarrely predicted that Soviet output per capita would exceed that of the United States by the 1980s.[7] Needless to say, it did not. And 724 days after the 1980s ended, there was no more Soviet Union.

In this period, China is still a tragedy. The well-meaning republican government of Sun Yat-sen achieved little growth, and his incompetent successor, Chiang Kai-Shek, first presided over stagnation and the famines that led, just as the period shown in Figure 7.4 ended, to the loss of

[7] Tabarrok (2010).

China to one of the most murderous bunch of communists ever to rule any country. As Hans Rosling said, "1948 was a great year," and it was, unless you were Asian. China was a catastrophe, and India was also about to enter a period of prolonged difficulty.

## The Great Convergence Begins

There are a lot of "greats" in this book, but after a great divergence between the rich and poor parts of the world lasting 200 or 500 years, depending on which version of the story you believe, a catch-up by the poor is great news indeed. Today's readers might guess that it began in China or India, but it did not. The convergence, or rapid rate of economic growth in poorer countries, was most pronounced in the immediate postwar period in Southern Europe, Latin America, and the Soviet-bloc states. The Asian Tigers (South Korea, Taiwan, Hong Kong, and what would become the independent republic of Singapore) also made rapid progress, although their glory days were a little later.

Figure 7.5 shows incomes from 1948 to 2018 for a selected group of both developed and emerging countries.

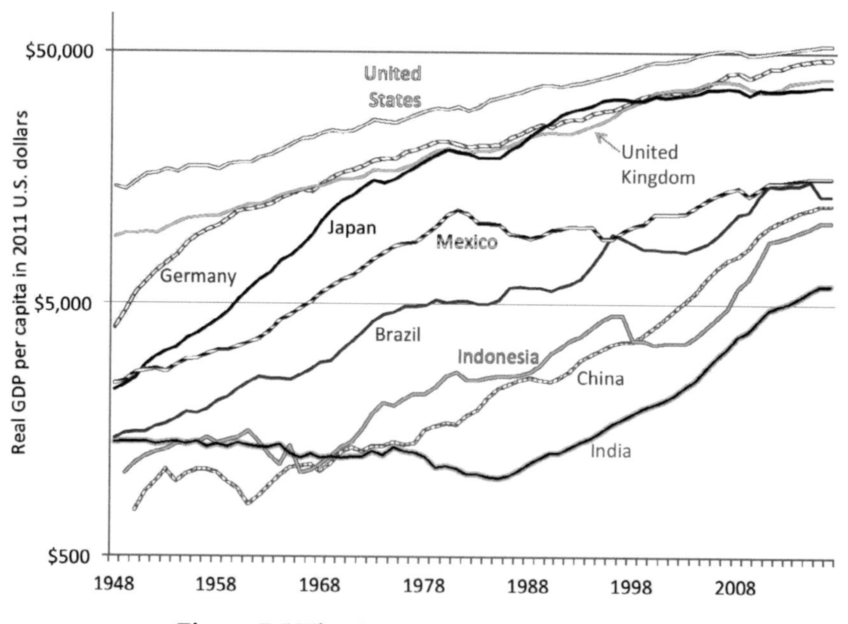


**Figure 7.5 The Great Convergence begins.**

*Note:* Real GDP is analogous to PPP GDP and is in "2011 US dollars"; to convert to 2018 dollars, multiply by 1.12.

*Source:* Constructed by the author using data from the Maddison Project.

## The Postwar Decades

Well, this is different! As Rosling showed with his bubble charts in Chapter 1, "Right Here, Right Now," economic development after 1948 is no longer a tale of two mostly separate worlds, a rich one getting richer and a poor one remaining at almost medieval levels of poverty. Instead, the poorer countries began to catch up—war-devastated Japan and Germany first, because they had been rich at one time and had the institutions and skills necessary for a quick recovery, but Brazil and Mexico also made spectacular progress in the first half of the period.

Mexico then suffered a two-decade period of stagnation caused by its failure to foresee that it would be competing with China as the "manufacturing division" of the United States; who knew? From 1948 Brazil continued its rapid growth until its recent sharp recession, and Mexico's progress began again after its turn-of-the-century political reforms. (Still, Mexico has lagged other emerging economies, mostly for ongoing political reasons, but it has all the characteristics of an economy that can be expected to grow significantly in the future if it is governed well.)

## China and India

But the big stories—not just because of their high growth rates, but because they are the countries with by far the largest populations—are China and India. (Indonesia, no slouch either, is the fourth most populous country, just behind the United States.) These Asian giants were slow to awaken. Maoism in China kept the country almost unbelievably poor until the opening of the economy by Deng Xiaoping in 1978, although public health and life expectancy did improve in the latter part of the Mao period, after the recovery from the great famine of the early 1960s.

India, despite the parallelism with China in the latter part of Figure 7.5, is a very different story. Every life matters, but in macroeconomic terms the two million dead from India's bloody civil war of 1947–1948 were a minor casualty when compared to the Chinese "Great Leap Forward" that lost 45 million lives to famine and murder. As a result, and because British India, despite its faults, had imposed Anglo-Saxon legal traditions and a semblance of competence in government, India entered the postwar period much less impoverished than China.

But then India adopted policies—Nehruvian socialism, and a "license raj" that made it almost impossible for those who were not well connected to start a business—that kept its growth agonizingly slow for almost a half-century. Some Indians blamed themselves, referring to a "Hindu rate of growth" that was slower than that of other Asian countries.

Yet, in 1993 when India's new leadership swung sharply to growth oriented, capitalist policies, the country's growth rate took off just as China's had 15 years earlier. Although China specialized in manufacturing, India specialized in services, taking advantage of the wide use of English in its population. India is now considered a lower middle-income country, a remarkable achievement for a nation that was a byword for extreme deprivation only a generation ago.

### Sub-Saharan Africa

We don't have good data for sub-Saharan Africa, the other large population center, as a unit; it's a big place, and different countries have developed—or not developed—in very different ways. But, if we had the data, it would show up as the lowest line on Figure 7.5, except right at the beginning of the period when it would have been roughly on par with China and India. The Great Convergence of Africa toward the standards of living of the rest of the world has started, but mostly just in the current century, and with occasional serious setbacks (such as the racial conflict in South Africa at present). We'd like to be able to look again in 2100.

We're going to cover these events and localities in greater detail, and their prospects for the future, in Chapter 14, "The Alleviation of Poverty," but here we just want to show the data and briefly summarize what happened.

## Where We Are Today: Toward a Middle-class World

We will have, in just a year or so (by 2020), reached a milestone: half the world's population will be middle class. This does not mean that half live at a standard that we'd identify as middle class in the United States, or Britain, or France, or Australia. But it is not a state of hardship either. The Brookings Institution scholar Homi Kharas defines middle class in this context as "people who have enough money to cover basic needs—such as food, clothing and shelter—and still have enough left over for a few luxuries, such as fancy food, a television, a motorbike, home improvements or higher education."[8]

[8] Long and Shaprio (2018).

The remarkable website Dollar Street,[9] compiled by Anna Rosling Rönnlund (Hans Rosling's daughter-in-law), displays images of families around the world in various income categories, along with their homes and possessions. Figure 7.6, from that website, shows an Indonesian family just at the cusp of entering the global middle class: they make $12,000 a year (restated in US dollars), have jobs and plenty to eat, and enjoy the newfound luxury of a motorbike big enough to carry them all (very carefully). They have aspirations well beyond this standard of living:

> Angga is 30 years old and he is a social worker. His wife Yuli is 33 years old and she is a teacher. They live with their . . . daughter, Luce, and son, Rado. . . . The house has unstable electricity, a drinking water source on the property, and an indoor toilet. About 30% of the family's income is spent on . . . food . . . [which] they cook . . . with propane-fueled appliances. Their last vacation lasted a week and was taken a

**Figure 7.6 The entry point for the global middle class: the Yanvar family of Yogyakarta, Indonesia, making $12,000 a year.**

*Source:* Photo by Gito Nirboyo for Dollar Street, www.gapminder.org/dollar-street/family?place=54b53c0138ef07015525f259. Licensed under CC BY 4.0.

[9] www.dollarstreet.org.

> month ago. They plan on buying their own house and their dream is to buy a car.

This is the future, for several billion people who never got a decent break before.

## Conclusion: Money Isn't Everything

In documenting the Great Enrichment, first in the rich countries and now spreading to the rest of the world, unevenly but rapidly, I've focused on money as a measure of wealth in this chapter. Obviously, money is just a unit of account and has no intrinsic value, but money incomes tell you a lot about human development. That's because a great deal of what people need and crave is what money can buy—food, medical care, education, culture, freedom. The next few chapters look at these indicia of human development in some detail.

# 8

# *Food*

## The Greatest Person in History?

"In one episode of the TV series *Bullshit!*," writes the author and historian Johan Norberg,

> the magicians Penn and Teller play a game of "The Greatest Person in History," with all the pretenders, religious leaders, presidents, and revolutionary leaders in one deck. Like poker, each player places bets based on how good their cards are—but they might be bluffing. Penn draws one card and immediately goes all in, because he knows he is going to win. . . . He drew Norman Borlaug.

Clearly, Norberg (see Figure 8.1) is fooling around and having fun. But a case really can be made that the Iowa agronomist was the greatest person in history. His invention of Green Revolution techniques and his work in bringing them to the world's poorest countries have given rise to a revolution in the food supply. We've gone, in a little more than a generation, from a condition where the Club of Rome, in 1972, could plausibly forecast worldwide famine by the end of that decade to one in which it is unimaginable (obesity is a bigger problem). And all of that was accomplished while the world's population doubled![1]

---

[1] This section appeared in my *Advisor Perspectives* review of Johan Norberg's *Progress,* accessed at www.advisorperspectives.com/articles/2017/03/20/middle-class-wage-stagnation-is-a-first-world-problem-the-world-is-getting-richer on September 13, 2018.

**Figure 8.1 Norman Borlaug at work.**

*Source:* Conner Flecks/Alamy Stock Photo.

The versatile author and environmentalist Stewart Brand recalls:

> Norman Borlaug, the one [biographer Charles] Mann calls "the Wizard," was a farm kid trained as a forester. In 1944 he found himself in impoverished Mexico with an impossible task—solve the ancient fungal killer of wheat, rust. First he invented high-volume crossbreeding, then shuttle breeding (between winter wheat and spring wheat), and then semi-dwarf wheat. The resulting package of hybrid seeds, synthetic fertilizer, and irrigation became the Green Revolution that ended most of hunger throughout the world for the first time in history.
>
> There were costs. The diversity of crops went down. Excess fertilizer became a pollutant. Agriculture industrialized at increasing scale, and displaced smallhold farmers fled to urban slums.[2]

Small-hold farmers were displaced because the price of food went *down*, making life easier for hungry village and city dwellers; and it also took fewer farmers to grow the food, forcing many into other

[2] Brand (2018).

occupations. This also happened in the United States a half-century earlier when farm families migrated *en masse* to factory jobs in the cities.

Farmers are doing the same thing now in China (which is far along in the process) and India and Bangladesh and Vietnam (which are not as far along). Because it takes fewer people to grow more and better food, the market sends a price signal to the surplus farmers telling them they have to find something else to do.

## "There Are Children Starving in Africa"

Of all human needs, food is the most basic—so basic that, under primitive conditions, basically all of one's effort is devoting to finding it or growing it. A persistent or serious shortage of food is the greatest hardship there is, and under such conditions people will do anything to get it.

When my children were young, "There are children starving in Africa" was my parental response to complaints about the food at the dinner table. When I myself was a child, I heard from my parents that there were children starving in China or India, and my father probably heard that children were starving in Belgium.

There is still hunger in the world—more about that later—and certainly still some children starving in Africa, but much of it is related to rulers' wartime attempts to starve out the opposition. (What an execrable practice.) Bill Gates writes,

> I run into a lot of people from rich countries who still think of Africa as a continent of starvation. The fact is, that's an outdated picture (to the extent that it was ever accurate at all). Thanks to economic growth and smart policies, the extreme hunger and starvation that once defined the continent are now rare. . . . [T]oday the issue isn't quantity of food as much as it is quality—whether kids are getting enough protein and other nutrients to fully develop.[3]

So there is a problem to be solved, just not the one most people think of when their minds turn to Africa. The problem is in supplying the proper mix of nutrients. On a caloric basis, if all the food produced in the world were distributed evenly, there would be more than enough for everybody. That is a big change from most of the world's history, when, until fairly recently, there were not enough calories to go around. This change is the result of the Green Revolution (Borlaug's baby), and of even more recent advances in technology, such as genetic engineering.

[3] Gates (2014).

## The Biomechanics of Not Enough Food

For most of human history, "not enough food" was the main problem that people faced. The French historian Fernand Braudel, whose work has revolutionized the way history is studied and written about, writes that famine "became incorporated into man's biological regime and built into his daily life."[4] The frequency and severity of famine in Europe was such that, according to Johan Norberg, "France, one of the wealthiest countries in the world, suffered twenty-six national famines in the eleventh century, two in the twelfth" and so on through "sixteen in the eighteenth. In each century there were also hundreds of local famines."[5] (Maybe the presence of "only" two widespread famines in France in the twelfth century explains the exuberant cathedral building discussed in Chapter 6, "Before the Great Enrichment.")

And that was in "advanced" Europe. According to Braudel, "things were far worse in Asia, China, and India." The "Food" chapter in Norberg's book *Progress* fills in many of the grisly details. It's tough reading, and people with a weak stomach would be well advised to skip it.

Given this history, what is remarkable about life in the world today is not that there are still hungry people, some starving, but how many fewer there are than only a couple of generations ago. Figure 8.2 shows famine deaths by decade from the 1860s to the current decade, expressed as an annual rate per 100,000 population.

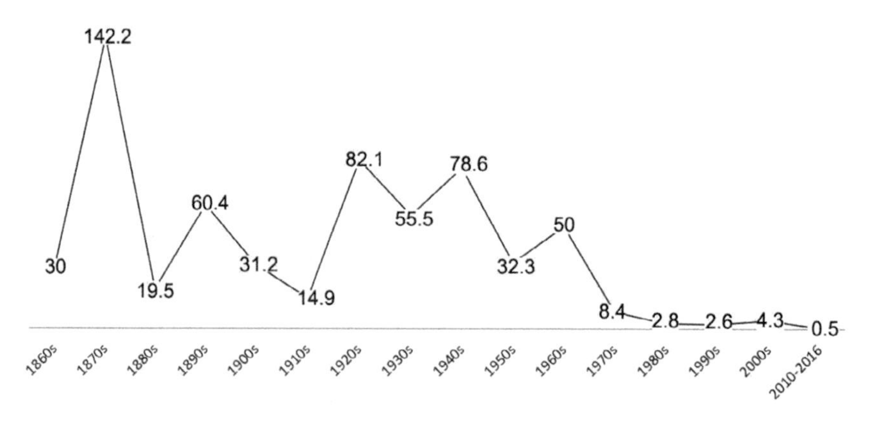


**Figure 8.2 Annual rate of people dying due to famine globally, per 100,000 people, by decade.**

*Source:* Chart re-created from an Our World in Data chart, https://ourworldindata.org/uploads/2018/03/Famine-death-rate-since-1860s-revised.png.

[4] Braudel quoted in Norberg (2017).

[5] Ibid., p. 9.

Although any deaths from famine are unacceptable in an abundant world, food has the unique characteristic (from among our basic needs) that you need it all the time. Thus, famines can come up abruptly and unexpectedly due to any interruption—climatic, logistic, military—in the food supply. As a result, deaths from famine have not been completely wiped out, but the decline in their incidence is very dramatic, and the figure doesn't even attempt to show the horrible statistics from the fourteenth or even the eighteenth century.

Death is, of course, not the only possible outcome in a famine. Most people suffer greatly and survive. But adults need nutrition to work, and children need it to grow (and require fats in their diet for their brains to develop properly). The role of famine and persistent malnutrition in retarding human progress has been a principal concern of the late, Nobel Prize–winning economist Robert Fogel of the University of Chicago.

## A New Kind of Evolution?

In a collaborative effort with several co-authors, Fogel documents the remarkable changes in the human body in the last few hundred years: an incredible increase of 50% in body mass and more than 100% in longevity. He proposes a new theory that he calls *technophysio evolution*—the co-evolution of human technologies or capabilities and of the body itself. Here are the basics (with my reservations and criticisms, meant to be inquisitive rather than argumentative, in parentheses):

1. The nutritional status of a generation—shown by the size and shape of their bodies—determines how long that generation will live and how much work its members will be able to do.[6]

(While this generalization is very important, it leaves out sanitation, penicillin, coronary artery surgery, cancer drugs, and other non-nutritional factors. But let's move on.)

2. The work of a generation, measured in both hours, days, and weeks of work and in work intensity, when combined with available technology, determines the output of that generation in terms of goods and services.

[6] Quotes 1 to 4 are from Floud et al. (2011), pp. 3–4.

(Again, what is work intensity? Intellectual work uses a lot of brain power, but your whole body doesn't need to function well as a unit to accomplish such work. *Vide* Stephen Hawking. *Vide* Chopin, who wrote some of the world's most beautiful music while battling a lifelong illness that would kill him at 39. But you need good nutrition, especially as a child, to develop that brain power.)

3. The output of a generation is partly determined by its inheritance from past generations; it also determines its standard of living and its distribution of income and wealth, together with the investment it makes in technology.

(This might explain why the twentieth century was so spectacular in its technological advances—it inherited the right mix of ingredients from the nineteenth century. On the other hand, I don't see anything inevitable about that progression; the people could have found Victorian technology and Great Power stasis to be quite satisfactory, causing the twentieth century to look much like its predecessor. In fact, there was a lucky explosion of scientific discovery in the late nineteenth and early twentieth centuries that powered the transition.)

4. The standard of living of a generation determines, through its fertility and the distribution of income and wealth, the nutritional status of the next generation.

. . . if the Four Horsemen of the Apocalypse—conquest, war, famine, and pestilence—don't happen to ride.[7]

Thus, the theory of technophysiological evolution proposes a model of social development that is "biological but not genetic, rapid, culturally transmitted, [and] not necessarily stable."[8]

The work of Fogel and his collaborators also shows that the early development of organs, as influenced by nutrition is very important later in life. This especially applies to early brain development and extends to nutrition *in utero* as well as in childhood.[9]

Thus, the most important interventions are in pregnancy and early childhood. Stunting (widespread short stature) still occurs in countries

[7] The traditional Four Horsemen are conquest, war, famine, and death, but since everybody dies, I modified the phrase to include disease, which causes death to come early and obviously affects future generations, as it did during the Black Plagues.

[8] Prinz (2015).

[9] As stated by Prinz (2015).

where overall caloric availability is adequate; it's specific nutrients—proteins, fats, and micronutrients—that appear to cause children to fail to develop to standards easily achievable in more prosperous societies. These observations provide powerful guidelines for aid and policy.

This novel approach to thinking about nutrition and physiology helps solve some old puzzles: Why were even rich people in olden times often short? (Remember Anne Hathaway's cottage?) The effects of nutrition are multigenerational. Why are so many children stunted in countries that now have adequate caloric resources? The mix of nutrients early in life is critical. Why are some populations that are smart but technologically primitive seemingly unable to get much done? They don't have the food supply to support hard work. Fogel's Nobel was unusually well deserved: he got us to think in radically new ways about old, well-known concerns.

## Our Paul Bunyanesque Ancestors

But you cannot just extrapolate body size and caloric availability back indefinitely and conclude that we are descended from starving hobbits. In *The Year 1000*, released at the turn of this most recent millennium in a stroke of marketing genius, the historians Robert Lacey and Danny Danziger write that, although "life was simple" and "for the vast majority of ordinary people life was a struggle in even the smallest respect,"

> [i]f you were to meet an Englishman in the year 1000, the first thing that would strike you would be how tall he was—very much the size of anyone alive today. It is generally believed that we are taller than our ancestors, and that is certainly true when we compare our stature to the size of more recent generations. Malnourished and overcrowded, the inhabitants of Georgian or Victorian England could not match our health or physique. . . .
>
> But the bones that have been excavated from the graves of people buried in England around 1000 tell a tale of strong and healthy folk . . . [with] sturdy limbs—and very healthy teeth.[10]

Interesting! We'd guess, just from thinking about it, that the distant past, like the nearer past, had its economic booms and busts, periods of relative prosperity and periods of deprivation. But strong and healthy bodies in Anglo-Saxon England depart substantially from our expectations.

---

[10] Lacey and Danziger (1999).

What was going on? Didn't Malthus say, backed by historical evidence, that a society with plenty of food would enlarge its numbers until the food surplus was used up?

## A Time of Plenty

The answers are necessarily speculative, but, reporting on work by the Ohio State University economist Richard Steckel, *Science Daily* observes,

> Reasons for such tall heights during the early Middle Ages may have to do with climate. Steckel points out that agriculture from 900 to 1300 benefited from a warm period—temperatures were as much as 2 to 3 degrees warmer than subsequent centuries. . . . [P]opulations had more land to choose from when producing crops and raising livestock.
>
> "The temperature difference was enough to extend the growing season by three to four weeks in many settled regions of northern Europe," Steckel said. "It also allowed for cultivation of previously unavailable land at higher elevations."[11]

## Brrr . . .

But the era of tallness in Britain came to an end. Apparently, the culprits were the Little Ice Age and rats. When the cold period began around 1300, growing seasons shortened, arable land shrank, a devastating famine struck in 1315–1317, and an increase in trade brought bubonic plague–infected rats to England. The fourteenth century has been described as the worst century ever.

Not much art survives from this period, so to illustrate it I'll use a later painting from Pieter Brueghel the Elder (who will return with better news in a moment). Figure 8.3 is a vivid depiction of the worst century ever.

So much for the possibility that the High Middle Ages and what is sometimes called the Renaissance of the twelfth century could have morphed into a great enrichment five hundred years earlier than it did. Europe had to start over.

And Europe did start over—but not exactly from scratch. Because of the plague, there was a labor shortage and wages rose. Women, less susceptible to plague and in many cases inheriting property, enjoyed a newly enhanced role. A smaller population meant more arable land per person. Most historians agree that it disrupted established social

[11] *Science Daily* (2004).

**Figure 8.3** ***The Triumph of Death*** **by Pieter Brueghel the Elder (c 1562).**

*Source:* https://commons.wikimedia.org/wiki/File:The_Triumph_of_Death_by_Pieter_Bruegel_the_Elder.jpg. Public domain.

patterns so as to pave the way for the transition from medieval to early modern times.

## Happy Days Are Here Again—for a While

Europe did eventually recover and move on to greater things, despite repeats of the plague. In 1569, the same Pieter Brueghel the Elder showed common peasants having fun again (Figure 8.4). They did not all have to wait 250 years for good times to return—there is fun in Chaucer's mid-fourteenth century poetry—although they had to wait that long for a painter as skilled as Brueghel to depict them.

Were the revelers healthy? Were they well fed? William Carlos Williams, an American doctor and gifted poet contemplating the painting 400 years later, certainly thought so:

> In Brueghel's great picture, The Kermess,
> the dancers go round, they go round and
> around, the squeal and the blare and the
> tweedle of bagpipes, a bugle and fiddles

**Figure 8.4** ***The Kermess*** **(Peasant Dance), by Pieter Brueghel the Elder (1569).**

Source: https://commons.wikimedia.org/wiki/File:Pieter_Bruegel_the_Elder_-_The_Peasant_Dance_-_WGA3499.jpg. Public domain.

> tipping their bellies (round as the thick-
> sided glasses whose wash they impound)
> their hips and their bellies off balance
> to turn them. Kicking and rolling
> about the Fair Grounds, swinging their butts, those
> shanks must be sound to bear up under such
> rollicking measures, prance as they dance
> in Brueghel's great picture, The Kermess.[12]

There was more trouble ahead—the Thirty Years' War ("a terrible time," intoned the historian Ernst Gombrich) and the plague of 1665–1666. But the First Divergence was well underway. Better times were coming.

## The Food Revolution of the Twentieth Century

Let's skip ahead a few centuries; a full recounting of the agricultural revolutions of the past would take a whole book. While farmers

[12] Williams (1962).

had been improving food production techniques ever since they unhooked oxen from the plow and switched to horses—this involved inventing a horse collar that didn't choke the horse—the biggest improvements had to wait until the twentieth century. The key to the first big advance of the century was artificial fertilizer—manure and guano being in short supply.

To produce a synthetic fertilizer, chemists sought a way to "fix" abundant atmospheric nitrogen, *fixing* meaning combining the nitrogen with other substances to make ammonia and other nitrates. Two German chemists, Fritz Haber and Carl Bosch, perfected the process in 1909 and 1910 and later won Nobel Prizes (that the Nazi regime did not allow them to claim). Some historians believe their innovation saved or enhanced even more lives than Borlaug's.[13] The reason they don't usually rank as highly in the greatest-person contest as Borlaug is that Haber also developed explosives that were used against the Allies in World War I.

Especially in the second half of the twentieth century, crop yields—and, to some extent, the amount of arable land—rocketed upward even faster than the population, turning an undernourished world into an overly nourished one. Food distribution underwent a revolution caused by refrigeration and cheap shipping, making a much wider variety of fresh foods available in all seasons.

Remember that, at midcentury, about half the world's population was underfed. Now, it's more like one-tenth. (Refer back to Figure 8.2.) We'll now explore some ways to bring the food security that is enjoyed by much of the world's population to those who are still suffering from hunger or the reasonable fear of it.

## Toward Food Security for All

According to the World Health Organization, "1.9 billion adults [in the world] are overweight or obese, while 462 million are underweight."[14]

---

[13] Pinker (2018, p. 75) says Haber and Bosch "saved" "the greatest number of lives in history, with 2.7 billion"—a third of the current world population. I put "saved" in quotes because it isn't clear what it means to save a life—allow the life to have been created in the first place, because the parents knew there would be food available? Saved a child or adult from dying young (that is probably what Pinker means)? Made someone's life worth living because it would have otherwise been a constant struggle to find enough food to survive?

[14] World Health Organization (2018).

If this is true, it's ironic that we're still concerned with hunger at all. One could make a careful accounting of the extra calories consumed by the overweight, and see if they're enough to bring the undernourished up to par, and with numbers like those, I'd bet a large sum of cash that they are.

What to do? Much has been written about inequality, but most of the literature on that topic is concerned with the unfairness (of which I'm not convinced) of a few people having so much more than they need when others have too little. If Jeff Bezos could be persuaded to give up $100 billion of his $160 billion fortune, he would still be one of the dozen richest men in the world and he would eat exactly as he is eating now. But would the hungry be fed?

## Amartya Sen and the Power of Institutions

Probably not. Amartya Sen, the Nobel Prize–winning economist who has probably studied the causes of and remedies for famine more thoroughly than anyone else, argues that neither an overall lack of food nor an overall lack of "money" (that is, resources other than food) is the cause of famine or malnutrition. There are plenty of both in the world. Nor is it inequality in the sense that ordinary people cannot compete with Jeff Bezos' buying power.

The determining factor is the presence or absence of *institutions* that make it possible for the poor to earn enough money to buy the food that is available. These institutions include property rights enabling people to profit from their own labor and land, freedom from discrimination, the opportunity to acquire marketable skills, and—most importantly in Sen's way of thinking—democratic control of the political system. He writes,

> Famines are easy to prevent if there is a serious effort to do so, and a democratic government, facing elections and criticisms from opposition parties and independent newspapers, cannot help but make such an effort. Not surprisingly, while India continued to have famines under British rule right up to independence . . . they disappeared suddenly with the establishment of a multiparty democracy and a free press . . .[15]

[15] Sen (1999), p. 4 of the PDF accessed at https://www.unicef.org/socialpolicy/files/Democracy_as_a_Universal_Value.pdf. Despite Sen's claim, no one seriously argues that hunger or even starvation has been completely eliminated in India. A famine is a widespread and serious shortage of food, and Sen presents evidence that no famines, by that stringent definition, have occurred in India since independence. Local droughts and other natural conditions in India have caused food shortages, including some leading to deaths, but relief has usually come quickly and effectively. Sen's work has also been criticized; see, for example, Massing (2003).

In another work, Sen concludes that "a free press and an active political opposition constitute the best early-warning system a country threatened by famines can have."[16]

This is a complex, long-term solution. It is achievable, and is working in many countries that formerly suffered from hunger. Look at Europe. Look at most of East Asia and much of Latin America, where hunger has mostly been eliminated. But, elsewhere, people need to eat—now. The interaction of an urgent short-term need with a set of long-term remedies presents moral and practical problems.

There is a legitimate concern that relief efforts (such as giving away food) reduce the incentive for farmers to grow food by lowering the price to a point where the farmers cannot make a profit. Relief can also delay potentially painful reforms—such as adopting democratic practices—that, in the long run, would probably eliminate the problem. But there are solutions.

## Bill Gates and the Potential of Technology

Bill Gates, who's devoting the present chapter of his life to this problem, focuses on:

- the alleviation of poverty through economic development;
- the use of technology to bring agricultural yields in poor areas closer to developed-country standards;
- the use of genetically modified crops (which Europeans have withheld from Africa for ideological reasons—see Chapter 25, "Ecomodernism.");

  and—interestingly—
- the creative use of cell phones: "The team now has two projects stemming from the initial use of satellites and smartphones: one confirming the occurrence of drought so farmers can get insurance payouts when their crops fail and the other enabling biweekly, or monthly, forecasts of climates to help pre-empt changes in food production."[17]

## A Surprising Energy Shortage

I'd add one more thing: an assault on *energy poverty*. How can you thrive when it takes more energy to obtain food than the food produces when

[16] Sen (2000), pp. 180–181.
[17] Senthilingam (2017).

metabolized? You can't. And the remedy is not to grow more food or reduce its price, but to reduce the cost and increase the availability of energy.

The consequences of energy poverty go beyond just finding food. Mercy Njima, a Kenyan doctoral student interviewed in the tech entrepreneur Peter Diamandis and his co-author Steven Kotler's book *Abundance*, says:

> Consider the women and children [in Kenya] who spent hours every day searching for increasingly scarce energy resources. . . . And once they start burning biomass, the acrid smoke causes serious lung disease and turns kitchens into death traps. . . . More people die from smoke inhalation than from malaria.[18]

Njima also points out, as Diamandis and Kotler report, that "because children have to help collect fuel during school hours, time spent on their education is severely reduced. This problem compounds at night, when students need to do their homework but have no light for studying."

Solve these problems and the food situation in Kenya will improve, because energy, effort, and intelligence devoted to meeting these minimal needs can be redirected to better agricultural practices, mechanized farming, and cheaper and more reliable transportation of food. We're going to need all the energy we can get.

We'll discuss energy poverty and its remedies further in Chapter 10, "Energy."

## Future Advances in Food Production

What is going to happen to food production in the future? Most of the arable land in the world is already under cultivation, and we are growing about enough food; but the world population is going to increase by another 3 or 4 billion before it levels off. How will we feed them?

### Are We Running Out of New Food Technologies?

Despite the success of genetically engineered organisms (GEOs or GMOs, described in greater detail in Chapter 25), there is a perception that the benefits from existing technologies are leveling off. There is a

[18] Diamandis and Kotler (2012).

"shrinking backlog of untapped technologies," says Lester Brown of the Worldwatch Institute.[19] The oceans are not only drastically overfished; "runoff from nitrogen-laden fertilizer has turned [US] coastal waters into dead zones so severe that [we] must now import 80 percent of [our] seafood," warn Diamandis and Kotler. They add that, in Japan, rice yields have "flatlined for fourteen years," and wheat production in western Europe has "similarly plateaued."[20]

So are we in trouble? Will the famines predicted by Paul Ehrlich come 50 or 75 years later than predicted?

No, argue Diamandis and Kotler. There are remarkable advances in food technology waiting in the wings. One reason they haven't yet been implemented is that we don't need them yet; we're already growing enough food for the existing population. When we do need them, the technologies will be ready. Agritechnology firms are frantically busy preparing to feed a world of 9, 10, 11 billion.

## Vertical Farming

Some of the less exotic technologies cited by Diamandis and Kotler are "vertical farming": hydroponics and aeroponics. (Hydroponics uses water instead of soil to deliver nutrients; aeroponics, a newer concept, "nourishes the plants with special LED lighting and delivers nutrients to plant roots through a liquid mist."[21]) Dickson Despommier, the premier advocate of vertical farming,[22] makes the case that "the [needed] quantity of additional arable land is simply not available," so (as with housing billions of extra people) we must build up. Despommier writes that, with such a system (see Figure 8.5),

> [N]o crops would ever fail due to severe weather events (floods, droughts, hurricanes, etc.). Produce would be available to city dwellers without the need to transport it thousands of miles from rural farms to city markets. Spoilage would be greatly reduced, since crops would be sold and consumed within moments after harvesting.
>
> . . . [O]ne anticipated long-term benefit would be the gradual repair of many of the world's damaged ecosystems through the systematic abandonment of farmland. In temperate and tropical zones, the re-growth of hardwood forests could play a significant role in

[19] Quoted in Diamandis and Kotler, (2012), p. 101.
[20] Ibid.
[21] Blackburn-Dwyer (2016).
[22] Despommier (2011).

**Figure 8.5 A vertical farm in Singapore.**

*Source:* Valcenteu, 2010. https://commons.wikimedia.org/wiki/File:VertiCrop.jpg. Licensed under CC BY-SA 3.0

> carbon sequestration and may help reverse current trends in global climate change.[23]

Vertical farming—bringing the farms (instead of the produce) to the people—has its critics. Lloyd Alter, writing in *TreeHugger*, regards it as faddish romanticism:

> [A]nyone in real estate will tell you that there is a highest and best use for land, and in the city, farming ain't it. . . . Brooklyn was once one of the most agriculturally productive regions in the United States. Manhattan was once home to innumerable factories. There's a reason that farms and factories decamped to more suitable locations. Using urban real estate in this manner is incredibly wasteful: bad for the economy and bad for the environment. Local food has its merits, but that's what [the nearby countryside] is for.[24]

I share Alter's skepticism—for now. But, as megacities grow in developing countries with gas-guzzling trucks on crowded highways, bringing the farms to the people might not be such a farfetched idea. Never say never.

---

[23] Despommier (undated).

[24] Alter (2016).

## Farming the Oceans

Oceans cover more than 70% of the Earth's surface, yet we treat the oceans more like hunting grounds than like an agricultural resource—and we're depleting fish stocks very rapidly. Why don't we farm the oceans the way we farm the land?

One reason is that farming land is just easier. Property rights in land can be clearly defined: what I grow is mine to sell, what you grow is yours, and a fence makes the difference obvious, so farmers can enjoy the fruits of their innovations. That's more difficult in the open ocean, which is both hard to survey and regarded by international law as common property.

However, the exclusive economic zones of the oceans that are claimed by specific countries and recognized as legitimate by the international community makes ocean farming more of a possibility in those zones, at least.

In fact, progress in farming the oceans is already being made. Fish are not the only seaborne product; "shellfish" and plant products are nutritionally very important, and they don't swim away as easily. But fish have a ready-made commercial demand, so fish farming has gotten the jump on other kinds of ocean farming. *National Geographic* reports:

> Eight miles off the coast of Panama on the Caribbean side (most people visit the Pacific coast) we start to see net domes peeking out of the water. They're like icebergs—most of their mass is underwater. Inside the domes are some 600,000 fish living out their days in the warm Caribbean, eating real food, drinking real water, and nudged by real currents. . . . [U]nlike conventional aquaculture farms where fish swim in their own you-know-what, his fish never see the same water twice.
>
> Below us happens to be the largest open-ocean fish farm in the world. Aquaculture isn't new. Since the days of the Chinese Shang dynasty humans have raised fish to supplement the unpredictable yield of the sea. The idea has always been to corral fish in tanks or pools. At some point, people just got tired of taking a boat out right before dinner.
>
> [Brian] O'Hanlon's farm, which is part of a company he founded called Open Blue, wants to buck 4,000 years of human innovation and farm fish back in the ocean. He says that raising an animal in its natural habitat means it will be healthier and taste better and, with the right technology, grow far more efficiently.[25]

[25] Stone (2014).

And that's just within one small country's exclusive economic zone. If the right institutional arrangements could be secure to allow fish farming in a broader area of the ocean, the fish stocks of the oceans could be replenished (this would take some careful coordination) and the world's protein supply would expand dramatically.

By the way, there may not be plenty of fish in the sea, but there's plenty of space in the ocean: the land area of all the world's continents sum to a little less than the area of the Pacific.

### Science Fiction for Dinner?

Now let's get a little more exotic. You think meat comes from animals? "We're developing cell-based meat—without the animals," brags the biotech firm Memphis Meats. This is not a joke—they're funded by Bill Gates and Richard Branson—and it's not flavored tofu. They make the meat the same way animals do, biochemically.

And Alternative Protein Corporation makes animal feed out of bugs—fly larvae, to be specific. These are companies already in existence; we don't know what's going on in the two-men-or-women-in-a-garage startups that have not publicized their efforts yet. It's all a little gross, but so is the idea of eating a cow, if you think about it, and cows take up a lot of land and energy.

We may not need any of this Buck Rogers stuff. Genetic engineering, ocean farming, and application of existing advanced agricultural methods in areas that do not now use them may be enough. But it's nice to have the exotic technologies in reserve in case we do.

## Conclusion

The more people that live on the Earth, the harder it would seem to be to feed them all. Yet, starting just about when Thomas Malthus warned of that danger, the opposite has taken place: the food supply has gotten more ample (and more varied and delicious) as the population has increased. This is due to advances in crossbreeding and hybridization, fertilization, genetic engineering, refrigeration, weather forecasting, and food transportation and storage—what can be summed up as "food technology" or "agritech."

Later, in Chapter 25, "Ecomodernism," we'll delve further into the promise of genetic engineering, ecoengineering (what Stewart Brand describes as gardening on a planetary scale), and other possible technologies of the future. For now, let's turn to health and longevity, a close relative of the betterment we've experienced and expect to continue to experience in the food supply.

# 9

# *Health and Longevity*

## The Second Most Important Event in the History of the World

In Chapter 7, "The Great Enrichment: 1750 to Today," I noted Deirdre McCloskey's comment that the Industrial Revolution was the most important event in the history of the world. There are many who agree with her. It's supremely important because it set off an era of self-sustaining economic growth, starting in northwestern Europe and spreading outward, that continues to this day.

Here's the second most important event. In March 1942, reported the *New York Times*,

> Anne Sheafe Miller . . . was near death at New Haven Hospital suffering from a streptococcal infection. . . . She had been hospitalized for a month, often delirious with her temperature spiking to nearly 107, while doctors tried everything available, including sulfa drugs, blood transfusions and surgery. All failed.
>
> As she slipped in and out of consciousness, her desperate doctors obtained a tiny amount of what was still an obscure, experimental drug and injected her with it.[1]

[1] Saxon (1999). The penicillin she received was about half of the world penicillin supply that then existed.

"Within about a day," writes Lily Rothman in *Time*, "her temperature was back to normal. Miller was cured."[2]

Thanks to this first clinical use of penicillin, Anne Miller lived for 57 more years, enjoyed a productive life as a nurse and the wife of a school headmaster, and died in 1999 at the age of 90.

Why was it only the second most important event? Because the Industrial Revolution set in motion the processes that would make the discovery of penicillin and many other life-saving and life-enhancing substances and inventions possible. But, at least to Anne Sheafe Miller and the millions of others who were saved by antibiotics, it was the most dramatic improvement in their lives that could be imagined. Without penicillin, their lives would have been over.

Penicillin was not the first antibiotic ever tried or used—arsphenamine ("Salvarsan") was in use as early as 1909, and sulfa drugs in the mid-1930s—so this is not the first "second most important event" story that could possibly be told.[3] We don't know the details of the earlier events. But penicillin was the first broad-spectrum antibiotic that could be easily manufactured in quantity, although it took a few years for the penicillin supply to expand to a useful scale. After that, events such as the tragic death in 1924 of 16-year-old Calvin Coolidge Jr. from an infected blister on his foot, despite having the best medical care in the world, would only happen again in rare circumstances instead of all the time.

## The End of Agony

Writing to Walt Whitman on his seventieth birthday in 1889, his younger friend and admirer Mark Twain wrote:

> What great births you have witnessed! The steam press, the steamship, the steel ship, the railroad . . . the telegraph, the phonograph, the photograph . . . the electric light. . . . And you have seen even greater births than these; for you have seen the application of anesthesia to

[2] Rothman (2016).

[3] I have heard apocryphally that a similar "miracle cure" story occurred in 1935, perhaps in Germany where sulfa drugs were invented. A patient, lying near death, received the drug and, within hours, was up and walking about. He, too, was completely cured. Sulfa compounds were widely used after that, and found effective—they were even used to treat President Roosevelt's son, Franklin Jr., in 1936. But I cannot verify the 1935 "first use" story.

> surgery-practice, whereby the ancient dominion of pain . . . came to an end in this earth forever.[4]

Unbelievably, until 1846, only 43 years earlier, *all surgery had been performed without anesthesia.* I cannot imagine it, and I will not try to describe it, much less show a picture of it (though some exist). There is enough horror literature in the world already.

Again, ether—the anesthetic used by William T. G. Morton, a dentist, and John Collins Warren, a surgeon, in Boston in 1846—was by no means the first painkiller ever used. Whiskey, cocaine, opium, and nitrous oxide had all been tried, the first three since ancient times. But ether actually worked. It didn't just dull the pain, it made the patient fully unconscious during the operation yet did no long-term harm. Just try to imagine life without surgical anesthetics, and you might think that what happened in Boston in 1846 was the most important event in world history.

## Life Expectancy: Just Being Alive

Although it's possible to treat life expectancy, health, and longevity (reaching old age) as separable issues, and we'll get around to that in a later discussion, simply managing to be alive is a very significant indicator of health (!) so let's begin by charting life expectancy around the world over time. We first encountered the data in Hans Rosling's bubble charts in Chapter 1, "Right Here, Right Now," but those provide snapshots of life expectancy at widely separated time intervals; a continuous time series is revealing in a different way.

Figure 9.1 shows the expected length of life at birth for residents of various countries starting with the UK in 1543. Other countries are added as data become available. The United States (not shown) had an experience very similar to that of the UK.

### Preventing Cholera

Despite the income gains and technical advances in the UK that started in the middle to late eighteenth century, as noted earlier, life expectancy did not change much (except for downward spikes caused by epidemics and famines) between 1543 and 1850—a long time for no improvement to take place. After the latter date, public health improvements were responsible for most of the gains. The best known is the discovery, by

[4] Twain (1889).

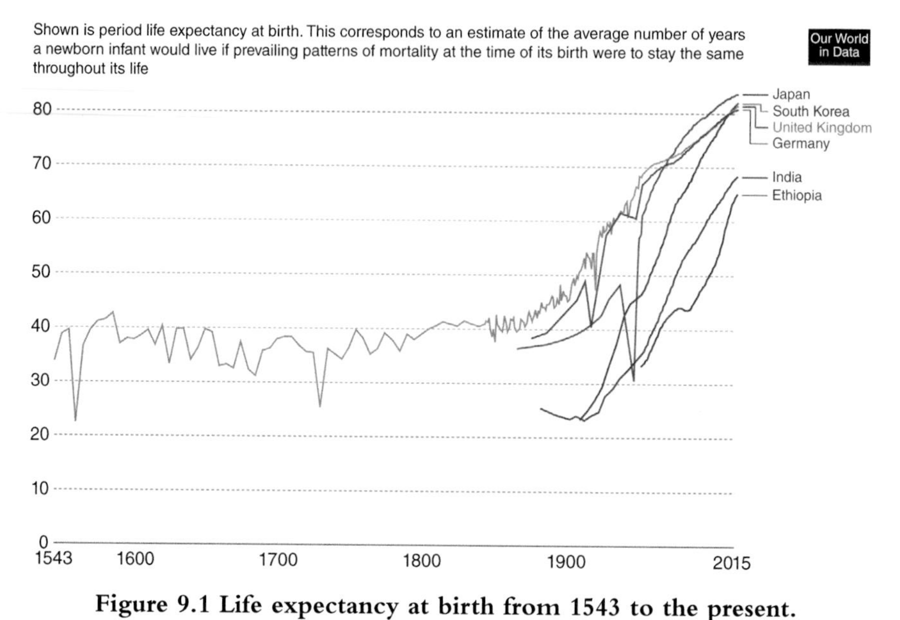


**Figure 9.1 Life expectancy at birth from 1543 to the present.**

*Source:* Max Roser (Clio-Infra estimates until 1949; UN Population Division from 1950 to 2015). Our World in Data. https://ourworldindata.org. Licensed under CC BY.

Dr. John Snow (see Figure 9.2) in London in 1854, that contaminants in drinking water caused potentially fatal diseases (cholera) and that countless lives could be saved by keeping sewage out of the water supply.

This fact must have been known at some level to the ancients, and embedded in folklore and common practice. However, even in an age and location when Faraday and Darwin were working out their complex scientific theories, Londoners were drinking filthy water and dying. (At that time, the germ theory of disease was not fully accepted; Louis Pasteur's work a decade later would establish it.)

Snow's work gradually led to public policy changes that enabled many more people to drink clean water, a process that is still ongoing today. But progress was neither quick nor smooth and illustrates the often-circuitous course of scientific knowledge and public policy. Snow's work was either dismissed or, even worse, simply ignored. The alternative theory, that cholera was transmitted through miasma (bad vapors), remained ascendant for a number of years following the publication of Snow's 1855 masterly *On the Mode of Communication of Cholera.*[5] With the 1866 cholera epidemic, scientific and political opinion began to change.

[5] Snow (1855).

**Figure 9.2 Dr. John Snow.**

*Source:* Rsabbatini, https://commons.wikimedia.org/wiki/File:John_Snow.jpg. Licensed under CC BY 4.0.

Snow is best known in the popular imagination for identifying the Broad Street pump as the source for the terrifying outbreak in London in autumn 1854. The wonderful story—that Snow removed the pump handle and stopped the epidemic in its tracks—is largely myth. The true story is more nuanced but in its own way more interesting: according to the University of Chicago professor Thomas Coleman, "a rollicking good tale—full of heroism, death, and statistics."[6]

Snow developed the water-borne theory and worked tirelessly to collect and examine a wide range of evidence. He persevered with courage and humility in both his scientific investigations and his advocacy for better health and public policy, often in the face of indifference and inertia. Snow's work serves even today as a prime example of scientific

[6] Coleman (2019).

reasoning and analysis, discussed in textbooks and used for teaching examples. Although he died before he could see his ideas vindicated, Snow is now rightly considered one the fathers of modern epidemiology.[7]

## Catching Up

The catch-up principle, the idea that good ideas and inventions only have to appear once and then they'll be copied widely and quickly, shows up vividly in Figure 9.1.[8] Life expectancy rocketed up in developing Japan (except for the World War II period) much more quickly than it had in the developed UK and Germany because the scientific principles behind the increase had already been discovered; all the Japanese had to do was implement known techniques. A similar situation prevailed in India and Ethiopia some decades later; despite being very poor, these countries' life expectancies increased rapidly because of the adoption of basic public health measures as well as the availability of antibiotics and other drugs.

A fair conclusion one could draw from Figure 9.1 is that the whole world could have a life expectancy around 80 if it were more economically developed. Long-term projections by the United Nations indicate that it will. The organization's 300-year projections have the least developed part of the world, eastern and central Africa, reaching 80 years of expected life by the middle of the next century.[9]

Remarkably, about half of the population explosion of the last two centuries is due to people living longer, rather than to excess fertility. This can be estimated by simulating what world population would have been if fertility rates had declined as they did but death rates had remained constant at the levels of 200 years ago. The world's population would be about half what it is now.[10] Although we may welcome a continued decline in fertility, we don't want people to stop living longer!

---

[7] For a good popular presentation see Johnson (2006).

[8] The catch-up principle works better in an age of instant global communication, in which people can see what they're missing out on. Before fast communication, catching up could be slow: the joint-stock corporation, a concept that is both relatively simple and extremely useful, has (according to historian William Goetzmann) its origins in ancient Rome but was not a widely used form of organization until the 1600s. That's a long wait. See Goetzmann (2016).

[9] See www.un.org/esa/population/publications/longrange2/WorldPop2300final.pdf, p. 31. I've already criticized the realism of 300-year forecasts and do not need to repeat that here.

[10] Although I haven't performed the simulation, it can be done as a back-of-the-envelope calculation. If the average person lived to 40 instead of 80 there would be roughly half as many people alive at a given time, since the birth rate would be largely unaffected (most children are born to parents under 40).

### The Long Tail of Public Health Investment

In Chapter 8, "Food," I discussed the changing heights of humans through history as a function of nutrition and disease. Robert Fogel, the Nobel Prize–winning economist, is the person most closely associated with this observation. His work has an interesting consequence for public health policy. Specifically, Roderick Floud, Robert Fogel, Bernard Harris, and Sok Chul Hong (2011) demonstrate that there can be *very* long lags between investment in public health and biomedical research and the eventual payoff. Public health policies put into effect starting around 1870, they argue, enabled life expectancy and height to grow in the United States, *even in the 1930s* when unemployment never fell below 16%—a 60-year lag between effort and reward.

## A Remarkable Reduction in Child Mortality

Much of the gain, especially the early gain, in life expectancy, is from the dramatic decline in infant and child mortality. Figure 9.3 is a vivid depiction of the decline in child mortality. In India, more than half of

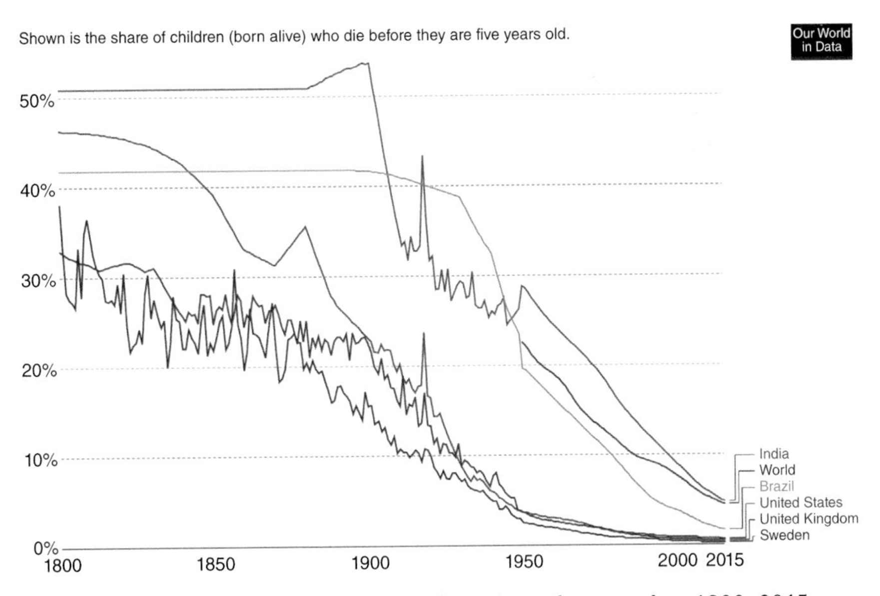


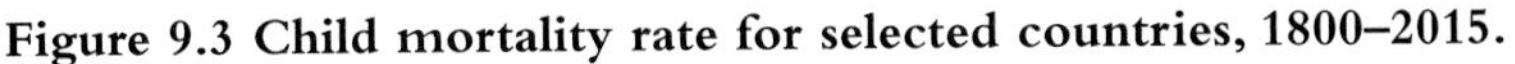

**Figure 9.3 Child mortality rate for selected countries, 1800–2015.**

*Source:* Max Roser (Gapminder estimates up until 1949 and UN Population Division from 1950 to today), Our World in Data, https://ourworldindata.org. Licensed under CC BY. Underlying sources shown in the figure.

newborns died before they were five years old as recently as 1900. The United States, with one of the lower child mortality rates at that time, is where the four Neckwinder babies died in about that year (Chapter 3, "The Demographic Transition").

By 1950, however, child mortality had declined almost to current-day levels in the most highly developed countries. The developing world, far behind, showed even more improvement in absolute terms although the resulting conditions at that time were in no way acceptable; a quarter of Indians and Brazilians died before their fifth birthday and the world average was not much different from that.

By 2015, the most recent date for which data are available, child mortality in the United States, the UK, and Sweden (as proxies for the most highly developed countries) had declined to negligible levels, with almost all such deaths due to individual tragedies such as birth defects. The improvement in the rest of the world was also remarkable. Even in sub-Saharan Africa, the region with the highest child mortality rate, the rate "has fallen from around one in four in 1960," according to Steven Pinker, "to less than one in ten in 2015."[11]

## Longevity: Getting to Know Grandma—and Great-Grandma

In gauging human progress, longevity is different from life expectancy. Life expectancy is the statistically expected length of life and is profoundly influenced by infant and child mortality. Longevity means living to a ripe old age, not just surviving childhood. Although the two concepts are closely related, we can learn much by separating them.

Even the standard by which we judge long life has changed: a century ago, a man who lived to 70 was considered to have had a full life. Today, a man who only makes it to 70 is considered to have died young.

Anecdotes may be better than data in portraying the impact of gains in longevity: a greater proportion of 20-year-old Americans now have a living grandmother than had a living *mother* in 1900.[12] In another hundred years, what will be the probability of a 20-year-old having a living *great*-grandmother? It is not negligible now, and will only grow

[11] *Enlightenment Now,* p. 56.

[12] Uhlenberg (1996). Uhlenberg describes many other remarkable changes in kin relationships that have occurred due to increasing longevity. See also Coleman et al. (2007).

(unless the trend of choosing to have children at later ages counteracts the longer life spans).

Longevity is best described, graphically, by showing the amount of remaining life one can expect at various already mature ages, typically age 65. This removes the effect of changes in child and young-adult mortality. Figure 9.4 depicts changes in longevity in the United States in this manner.

## Longevity Risk—a Financial Challenge for the Future

Note that Figure 9.4 was constructed by Social Security actuaries, who are responsible for making sure that the program they oversee is financially sound (or that they do something about it if it isn't). They are,

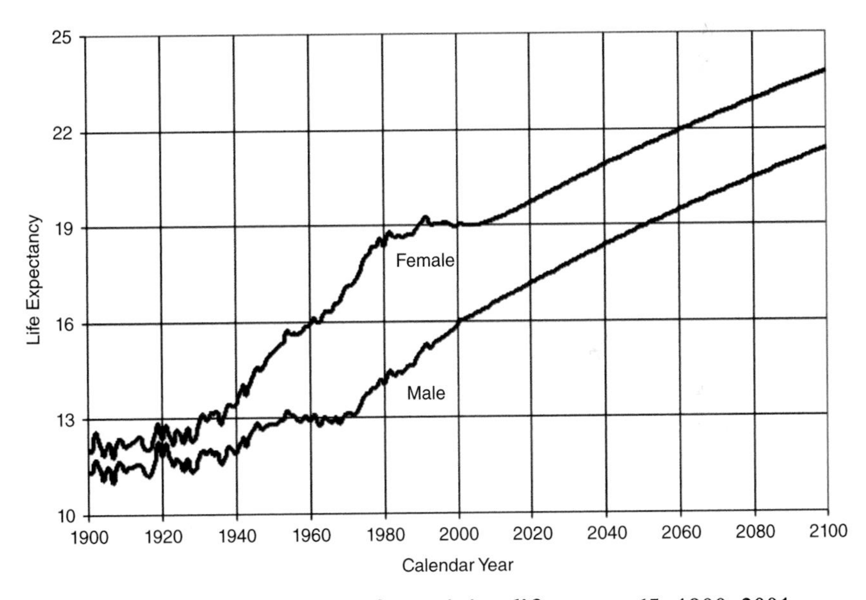


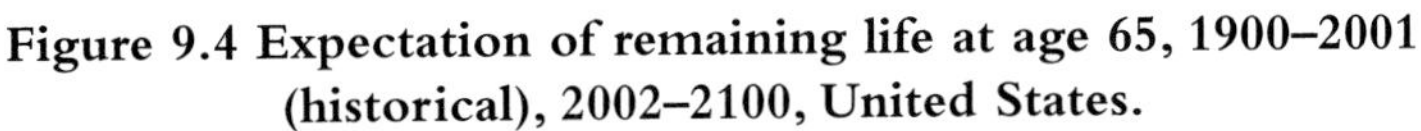

**Figure 9.4 Expectation of remaining life at age 65, 1900–2001 (historical), 2002–2100, United States.**

*Note:* Expected years of remaining life are "period expectation," estimated from mortality rates experienced in given years.[13]

*Source:* Reproduced from www.ssa.gov/oact/NOTES/as120/LifeTables_Body.html, Figure 2b.

---

[13] Period expectation and cohort expectation are different ways of estimating future average life spans. Ben Wattenberg (2004, pp. 85–86) explains: "[T]he 'cohort' measurement . . . add[s] up all the children born during a woman's childbearing years. . . . It measures what has already happened. . . . [T]he 'period' measure . . . produces a portrait of what is happening now. . . .The theory rests on the idea that women in given age cohorts today will continue to have babies at the same rate into the future.That yields lower population totals."

therefore, particularly concerned with increases in longevity that affect the payouts they are expected to make to beneficiaries.

For those paying for the old-age expenses of others (or for their own, in the case of individuals who are saving or who have saved for retirement), then, longevity becomes a *risk*, a mirror image of the better-understood risk of mortality (that is, of dying). Outliving your money is not a good outcome.

The good news from Figure 9.4, then, is that you might live a long time, an average of 20 years after retirement if you are an American female retiring now, with a wide distribution around the average. The bad news is you're going to have to pay for it. The wide distribution is the catch: if your *expected* remaining life at age 65 is 20 more years, it could be 30 or even 40. This is a very difficult risk to hedge or insure against, unless you are a participant in a traditional defined benefit pension plan (one that doesn't go bankrupt) or a life annuity (one that doesn't go bankrupt).

This risk will only get bigger over time as people live longer, as the figure suggests. If people also work longer (which they can do if they are healthy—a big if), that will mitigate the problem.[14]

## Extreme Longevity Risk: "Time's Mistress," Jeanne Calment

In what has been a fairly sober chapter, some levity is in order. Jeanne Calment (see Figure 9.5), a Frenchwoman born in 1875—a year before the Battle of Little Big Horn—lived long enough to record a rap video, recall that Vincent Van Gogh had been unkempt and smelly, and pose a real problem for the man who bought her apartment.[15] In France, it is a common practice to lock in a price for property by buying it from an elderly person and then allowing them to live there until they die, at which time the buyer can move in. (The seller gets the advantage of having the cash while she is still alive.)

The poor fellow who bought the apartment in 1965, when he was 45 years old and Calment 90, died before she did. He passed away at the age of 77, in 1995, and she lived until 1997. She was 122 years and 164

[14] See also my essay with Stephen C. Sexauer, "Longer, Healthier, Happier: How Working Longer Improves Almost Everything," https://larrysiegeldotorg.files.wordpress.com/2019/03/longer-180221-clean.pdf

[15] In an interesting sidelight, Jeanne Calment was recently accused of faking her age. The accusation was convincingly refuted in recent research by Robine and Allard (2019), which establishes the continuity of contact with her by public authorities and interviewers that is necessary to confirm that she was 122 years old when she died.

**Figure 9.5 Jean Calment.**

*Source:* Illustration by Bogatyr Khan (www.deviantart.com/bogatyrkhan). Commissioned by Laurence B. Siegel.

days old, the oldest human being who ever lived—and, unintentionally, a clever real estate operator.[16]

Moreover, she was one lucky lady. Born fairly wealthy, she did not outlive her money.

[16] https://super-mamie-du-17.skyrock.com/2572645775-MADAME-JEANNE-CALMENT.html.

## Healthspan

As the world's population ages, the medical and public health professions are increasingly focusing on maximizing not lifespan but "healthspan," the part of life during which a person feels healthy and is likely to be productive. Although some criticize this emphasis as disrespectful of the needs and well-being of the very old, resources are limited and it is rational to talk about maximizing healthspan as part of a broader discussion of what our goals are and what can realistically be achieved in terms of public and personal health.

Figure 9.6 shows, for selected regions and for the world, the World Health Organization's estimates of healthspan using a measure they designed, called HALE (healthy life expectancy), the period between birth and the average time a person contracts a serious or debilitating disease. The diseases used in the analysis are heart disease, four types of cancer, chronic obstructive pulmonary disease, stroke, lower respiratory infections, Alzheimer's, and type 2 diabetes. (I don't know

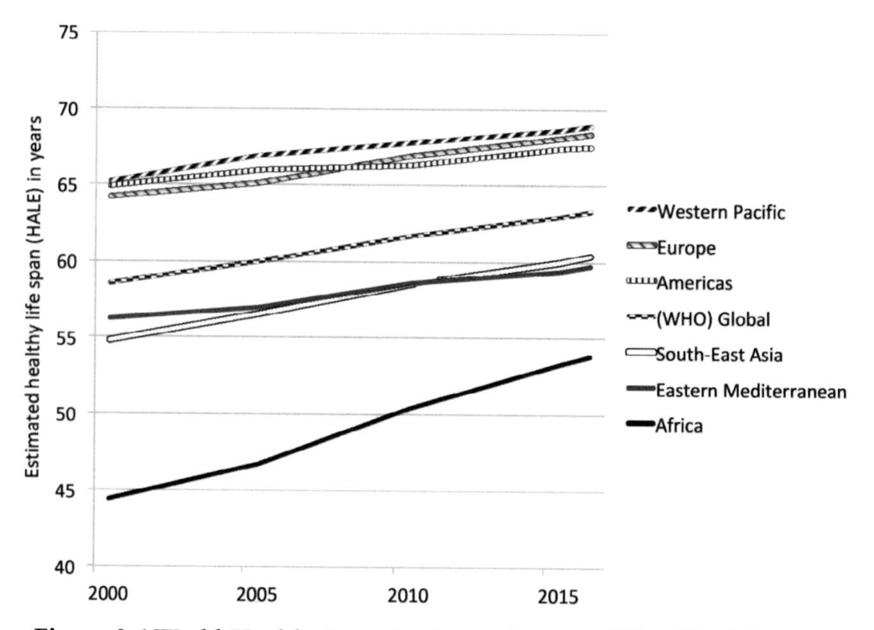


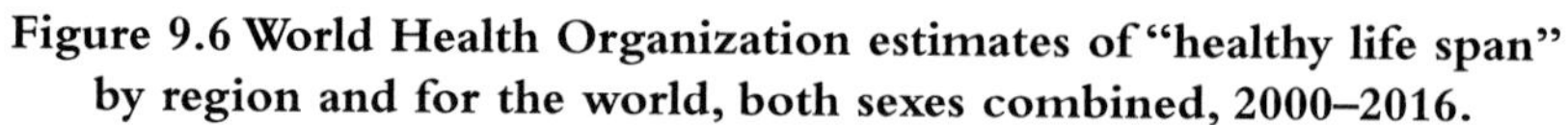

**Figure 9.6 World Health Organization estimates of "healthy life span" by region and for the world, both sexes combined, 2000–2016.**

*Source:* Constructed by the author using data from http://apps.who.int/gho/data/view.main.HALEXREGv?lang=en.

whether the analysis takes into account curable or easily treatable instances of these diseases.)

It would be useful to be able to visually compare these healthspan estimates with lifespan, to see the extent of "sick span" and whether it is increasing or decreasing overall and for each region. However, I do not have the data for the regions. An article by Dr. Tim Peterson at Washington University in St. Louis asserts that US healthspan is 63 years and life expectancy is 79 years, meaning that an American can expect to live with serious medical conditions for 16 years.[17] Whether that's a little or a lot depends on the person, the condition, and the quality, and the cost of the available treatments. It sounds like a lot to me.

## These Are Trained Professionals, Don't Try This at Home

I close on a positive note. Increasing length of life and, for some, of robust health is manifested in the remarkable accomplishments of older athletes. (Older achievers in sedentary pursuits, from Plato, 427 BC–347 BC, to the painter Titian, 1488–1576, have been a feature of human life for millennia.) Shigeyoshi Kimura, of the Associated Press, reported on August 2, 1987:[18]

> Teiichi Igarashi, [100], who was wearing heavy socks but no shoes, reached the 12,385-foot summit [of Mount Fuji] after a three-day climb that included two overnight stops in mountain huts.
>
> Igarashi's 12th conquest of Fuji came nine days after 91-year-old American Hulda Crooks from Loma Linda, Calif., became the oldest woman to reach the summit of the dormant volcano, about 62 miles southwest of Tokyo. . . . He began climbing Fuji annually in 1976 in memory of his wife, Ura, who died in 1975 at age 80. . . . In 1983, at age 96, he became the oldest climber to reach the summit.
>
> Climbing Mount Fuji has gained popularity among the elderly in recent years. Wasaburo Kato, [a climber,] . . . set up a register at the top for climbers aged over 70. Kato said the number of climbers over 70 who signed the register increased from 388 in 1974 to 947 in 1986.

[17] Peterson (2017).

[18] Kimura (1987).

**Figure 9.7 Teiichi Igarashi.**

Photo: © Gary J. Wolff. Reproduced with permission.

(See Figure 9.7.) The article is more than two decades old, so an update is in order. I can't read Japanese, but blogger Gary Wolff says that Igarashi climbed it again at age 105.[19] Bully for Teiichi!

---

[19] www.flickr.com/photos/wolffman/20784267311/in/photostream/.

# 10

# *Energy: A BTU Is a Unit of Work You Don't Have to Do*

This chapter describes how access to increasingly cheap and plentiful energy became the foundation of modern life, and begins to address prospects for the future. A later chapter, "Ecomodernism" (Chapter 25), speculates about the future of energy in greater detail.

Why do we care so much about energy? About oil and coal and natural gas, nuclear reactors, wind turbines, solar panels? Why do we revere horsepower and tigers in our tank?

Let's go back in time a few thousand years, to the very beginnings of civilization. There is much work to be done. Fields plowed, sown, and harvested, meals cooked, shelter constructed, water borne from where it's found to where it's needed. And all of the energy involved in doing this work comes from human muscle power.

In other words, it comes from food. The food is metabolized to produce energy, which is used to produce more food. It is a recipe for survival—that's what all animals, plants, and protists do—but not much more.

How do you get ahead? How do you make sure you know where your next meal is coming from? How do you get off the hamster treadmill that I just described?

One way is to harness the muscle power of other people. When someone else does that kind of work for you, instead of for themselves, we typically call them *slaves*. In fact, when there are only enough calories to survive, they more or less have to be slaves; otherwise you're expending your own energy to get the money (or food or other resources) to pay them and you're no better off than you would be if you did your own work.

Things improved (for people) when animals became enslaved. And, even in the earliest times, not everyone was a master or a slave; there were other ways to contract for labor, as indicated by recent evidence that the workers who built the Egyptian Pyramids were paid.[1] Still, enslavement was the most common way of getting other people to expend energy on your behalf from the beginning of recorded history until fairly recently.

Now, our "slaves" are machines, powered by nonhuman energy sources. How many slaves do you have?

## Freeing the Energy Slaves

In 1940, R. Buckminster Fuller, whom we met briefly in Chapter 7, "The Great Enrichment," portrayed the machines that make our lives easier (or livable) as our "energy slaves." He drew a world map with various white and red dots.[2] Each white dot represented 1% of the world's population in 1940, or 21,250,000 people, roughly the metropolitan population of New York or Mexico City today. The red dots showed the number of "inanimate energy slaves" serving the clusters of people. The total number of energy slaves amounted to 36.8 billion, or a little over 17 energy slaves per man, woman, and child living on the Earth in 1940.

Fuller's point was, of course, that mechanical slaves, powered by fossil fuels or other forms of energy, substituted for the billions and billions of real slaves that we would have needed to sustain the world's lifestyle as it existed in 1940. He regarded this substitution of mechanical power for muscle power as one of the crowning achievements of mankind and the most fundamental reason for the prosperity that the world was beginning to experience.

[1] See, for example, "Great Pyramid" (2010).

[2] Unfortunately, extant copies of the map are not in good enough condition to reproduce here.

## "They Would Need Their Own Slaves"

That was, of course, a much poorer world; how many energy slaves does each person in the world command now? Matt Ridley writes,

> Today, the average person on the planet consumes power at the rate of about 2,500 watts, or to put it a different way, uses 600 calories per second. . . . Since a reasonably fit person on an exercise bicycle can generate about fifty watts, this means that it would take 150 slaves, working eight-hour shifts each, to peddle [sic] you to your current lifestyle. (Americans would need 660 slaves, French 360 and Nigerians 16.)
>
> Next time you lament human dependence on fossil fuels, pause to imagine that for every family of four you see in the street, there should be 600 unpaid slaves back home, living in abject poverty: if they had any better lifestyle they would need their own slaves.

Ridley says that the energy-slave story can be taken two ways. One, which he calls the conventional way, is to rue the absurd wastefulness of the modern world. The other is to realize that:

> [W]ere it not for fossil fuels, 99 per cent of people would have to live in slavery for the rest to have a decent standard of living, as indeed they did in Bronze Age empires. This is not to try to make you love coal and oil, but to drive home how much your Louis Quatorze standard of living is made possible by the invention of energy-substitutes for slaves.[3]

Thus, the number of energy "slaves" per person in Nigeria, a very poor country, today is about equal to the number of energy slaves per person in the whole world in 1940. That's how much progress we've made.

By "freeing the energy slaves," of course, I'm referring to human slaves, who are mostly now freed. The mechanical slaves are not free, but they don't care.

## The History of Energy: Making Life Easier

Striving to summarize the importance of energy to economic growth and human life in general, Stephen Sexauer, game warden of San Diego County's roughly $12 billion in employee pension assets, said that a British thermal unit or BTU, a metric of energy output, measures the amount of "work you don't have to do." A machine does it instead of

---

[3] All quotes in this section are from Ridley (2010), pp. 236–237. Ridley states that he makes his living from the sale of oil and gas extracted on the land his family has owned for generations.

**Figure 10.1 Muscle power, still in use after all these years.**

*Source:* JCarillet/iStockphoto.

your ox, horse, or aching back. (Figure 10.1 shows that some people still do the work with their aching backs.) I can't think of a better way to describe what energy *is* and what it's *for.*

Figure 10.2 shows the relative importance of human labor, animal labor, biomass (mostly wood) consumption, and fossil fuel consumption in the generation of energy from 1700 to the present day. Note that the *y*-axis is a log scale in base 10, so that energy from fossil fuels has grown roughly 200-fold since 1800.

Energy from human labor has also grown, but mostly from population growth, so that the amount of physical work done by each person (not shown in the figure) has fallen drastically. Jobs have gotten so much easier (see Chapter 15, "Robots Don't Work for Free") that death from overwork, *karoshi*, has become a Japanese oddity, not a common occurrence wherever people strive to earn a living.

The figure shows that, as early as 1700, the amount of energy obtained from biomass was more than 20 times that obtained from either human or animal labor. Water wheels and windmills were also producing energy.

Before the Industrial Revolution, then, we were already experiencing the birth of the modern world, which was ongoing as early as the Middle Ages when these energy sources were first widely exploited. A

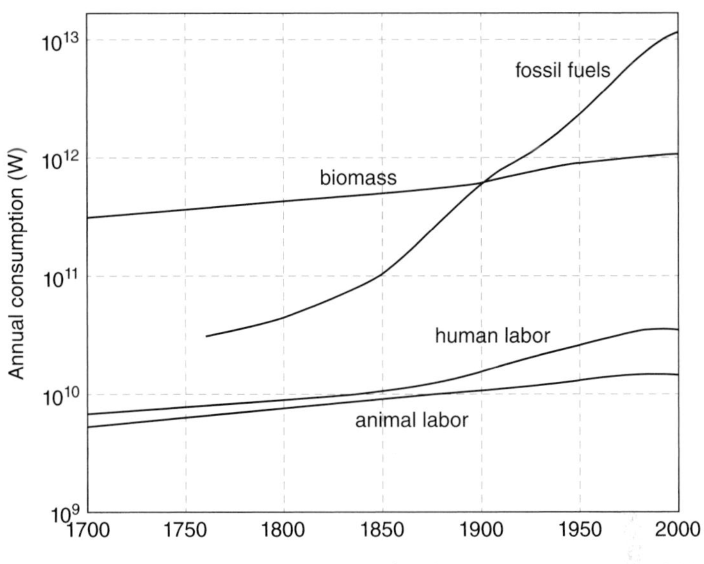


**Figure 10.2 Global consumption of primary energy, 1750–2000.**

*Source:* Smil, Vaclav. 2010. "Science, Energy, Ethics, and Civilization." In R.Y. Chiao et al., eds., *Visions of Discovery: New Light on Physics, Cosmology, and Consciousness*, Cambridge, UK: Cambridge University Press, p. 712. © Cambridge University Press 2010. Reproduced with permission of the licensor through PLSclear.

few aspects of modernity came remarkably early: by 1700 a tiny "car," powered by a primitive steam engine, had already been invented. Ferdinand Verbiest, a Jesuit missionary from Belgium who moved to China, either built one as an amusement for the Chinese emperor or tried to.[4] An artist's conception of it is in Figure 10.3.

Verbiest's toy car is an early example of Matt Ridley's lively metaphor of inventions arising from "ideas having sex." Carriage + engine + some sort of transmission (see the gears at left in Figure 10.3) = car. Make it larger, add some seats, a steering wheel, and brakes, and there's no going back: a usable car was going to be invented.[5] (It took a while.[6])

---

[4] Some question whether it was ever constructed, although it does not look that hard to build.

[5] Inventions thus build on each other: telegraph + microphone + speaker = telephone; computer + telephone = Internet. Ridley says his favorite example is the camera pill, which emerged from a conversation between a gastroenterologist and a guided missile engineer; the back story, although not known, must be fascinating.

[6] Several French inventors in the 1700s built fully functioning motor vehicle prototypes, and a reproduction of the 1769 steamer built by Nicolas Cugnot can be seen at www.youtube.com/watch?v=KP_oQHYmdRs. The 1885–1886 Benz Patent-Motorwagen was the first practical car. Good things take time, but *two centuries* is a long wait from concept to execution.

**Figure 10.3 Verbiest's 1678 "car"—sans steering wheel, brakes, or place to put a driver or passenger.**

*Source:* Public domain.

## Energy Transitions Are Gradual

The "recent" (since the Industrial Revolution) history of energy is shown in Figure 10.4. The figure shows the gradual energy-source transitions that made work easier and that made it possible to do radically more work with less human effort: BTUs doing work, so you don't have to. The first transition, from human and animal power to water, wind, and wood, was well underway by 1800; since the figure focuses only on the more modern energy sources, wood and other "traditional biofuels" represent almost the entire energy supply (as shown) in 1800.

By the mid-1800s, however, coal production was becoming significant, and "peak coal" occurred around 1910. Meanwhile, petroleum (crude oil) was beginning to grow in importance as the automobile era dawned, also around 1910, and natural gas production began to grow just a little later. Nuclear energy became a contributor in the 1960s, and modern renewables (wind, solar, and nontraditional biofuels) are just beginning to emerge in the current century. Note that "peak oil" appears to have been reached around 1970, not because we began to run out of

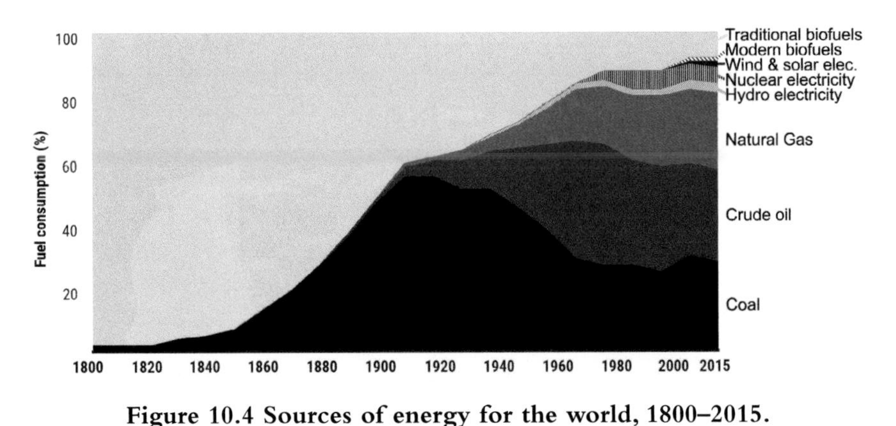


**Figure 10.4 Sources of energy for the world, 1800–2015.**

*Note:* "Traditional biofuels" are primarily wood. Animal and human muscle power are ignored.

*Source:* www.sciencemag.org/news/2018/03/meet-vaclav-smil-man-who-has-quietly-shaped-how-world-thinks-about-energy (details on page). Reproduced with permission.

the gooey stuff, as the geologist M. King Hubbert predicted, but because natural gas and nuclear power were becoming cheaper and more widely available.[7]

## Vaclav Smil and the Future of Energy

Vaclav Smil (see Figure 10.5) is one of the world's foremost experts on energy and a favorite author of Bill Gates, who says he has read nearly all of Smil's 37 books. Smil (pronounced "Smeel") uses the trends shown in Figures 10.2 and 10.4 to caution that future energy transitions will also be gradual, even if some people are crying out for the immediate replacement of fossil fuels by "renewables" for climatic or other reasons.

---

[7] I've used the expression "peak oil" without explanation because it is familiar to most readers of the news. For the curious, Hubbert (1956) predicted that oil production would peak in the United States around 1970 and globally around 2000, then decline, causing serious energy shortages. Although the United States did reach a peak around 1970, the subsequent decline was interrupted by the implementation of hydraulic fracturing ("fracking") and US production is once again approaching 1970 levels. Hubbert's forecast for peak *global* oil was way off, with world production currently at 2.5 times the forecast maximum. Still, oil production can be expected to decline someday, as other energy sources displace it.

**Figure 10.5 Vaclav Smil.**

*Source:* Drawing by Bogatyr Khan (www.deviantart.com/bogatyrkhan). Commissioned by Laurence B. Siegel.

Smil, a Czech-born scientist who emigrated to Canada after the 1968 Soviet invasion of his native country and who teaches at the University of Manitoba, describes himself as a "European pessimist," in contrast to his admirer Bill Gates, an "American optimist."

Smil writes,

> After more than 20 years of highly subsidized development, new renewables such as wind and solar and modern biofuels such as corn ethanol have claimed only 3.35 percent of the country's energy supply.
>
> The slow pace of this energy transition is not surprising. . . . In the U.S. and around the world, each widespread transition from one dominant fuel to another has taken 50 to 60 years. First came a change from wood to coal. Then from coal to oil. The U.S. is going through a

third major energy transition right now, from coal and oil to natural gas. Yet even though natural gas is abundant, clean and affordable, it will be another decade or two before gas use overwhelms coal consumption, which still generates more than a third of U.S. electricity.[8]

Figure 10.6 drills down into the trends shown in Figure 10.4, by highlighting the rate of growth of each major energy source, portrayed as a percentage of total energy supply, once that source has begun supplying 5% of global demand.

Referring to Figure 10.6, Smil explains,

> Renewables are not taking off any faster than the other new fuels once did, and there is no technical or financial reason to believe they will rise any quicker, in part because energy demand is soaring globally, making it hard for natural gas, much less renewables, to just keep up. . . . [E]nergy transitions take a long time.

The cautious tone of Smil's pronouncements is somewhat at odds with the optimistic thrust of this book. But my intent is to be realistic, not to cheerlead for change that won't happen, at least not at the hoped-for speed. Anyone who thinks that a transition to cheap, abundant, renewable, carbon-free energy will be easy is smoking a controlled substance.

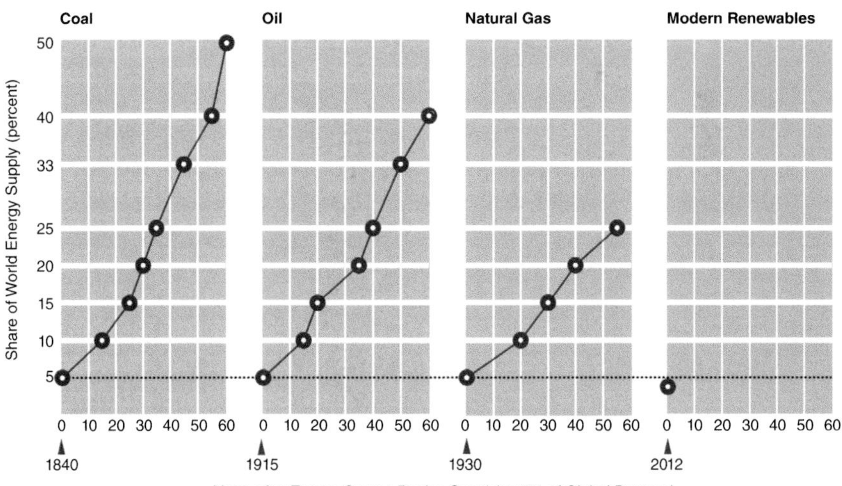


**Figure 10.6 Speed of energy transitions.**



[8] All quotes in this section are from Smil (2014).

Fossil fuels will be part of the world's energy mix for a very long time. However, they will represent a diminishing and more specialized share of it, since they will be used where their energy density and portability are most valuable instead of being used for everything. But *nuclear* energy offers a real possibility of a meaningful transition (that is nevertheless also gradual) away from fossil fuels and toward greater abundance. We address the potential of nuclear power later in this book.

## Smil's European Pessimism and the Prospects for Global Growth

Smil pushes back at the idea that technological progress will allow the whole world to live at First World standards. For Smil, energy *is* progress: its abundance is what makes us richer than our grandparents, and a hundred times richer than our sixteenth-century ancestors, to say nothing of our prehistoric ones. So, in his view, energy availability is the constraining factor on enrichment:

> With less than a sixth of all humanity enjoying the benefits of the high-energy civilization, a third of it is now engaged in a frantic race to join that minority, and more than half of the world's population has yet to begin this ascent.

(I think Smil underestimates the number currently engaged in the frantic race.) He continues,

> The potential need for more energy is thus enormous . . . [and] the probability of closing the gap during the coming one or two generations is nil.
>
> The utterly impossible option is to extend the benefits of two North American high-energy societies (. . . consuming annually some 330 gigajoules *per capita*) to the rest of the world. . . . This would require nearly 2.3 zettajoules of primary energy, or slightly more than five times the current global supply.[9]

Smil goes on to say that "neither the known resources of fossil fuels nor the available and prospective extraction and conversion techniques"

[9] All quotes in this section are from Smil (2014). For clarity, I've converted Smil's abbreviations (ZJ for zettajoules) to spelled-out words, including in the direct quotes. A zettajoule is one sextillion (1021) joules. I also refer to gigajoules, one of which equals a billion (109) joules. The definition of a joule and conversion factors for joules to other units of energy are widely available on the Internet.

could supply that much energy by 2030 or even 2050. Achieving the Japanese level of energy consumption, of about 170 gigajoules *per capita*, would require about 1.1 zettajoules, or 2.5 times the current level.

"This level is more realistic to contemplate," he says, "but its eventual achievement would . . . lead to unacceptably high levels of $CO_2$ emissions." He sees no fully acceptable solution.

What does Smil mean when he says we cannot supply the world with 1.1 zettajoules, much less 2.3 zettajoules? Does he mean the lights will suddenly go out? Of course not. Instead, energy prices will rise to levels high enough to keep the desired economic development from taking place. Following Smil's logic to its natural conclusion, the affluent countries will outbid (at considerable sacrifice of their own well-being) the less affluent for the smaller number of zettajoules that the world's energy industry *can* supply, and growth rates will slow to painfully low levels, if not slide into actual decline.

Again, this sounds woefully pessimistic. And there's no reason to sugarcoat the situation: energy is a scarce resource, we need a lot of it, and having to produce it in the face of climate risk compounds the challenge.

But there's another side to the story.

## The Case for Energy Abundance

There are two plausible responses to Smil's pessimism, one of which is supported by Smil himself in some of his other work. They are: (1) the potential for massively abundant nuclear energy, and (2) the fact that we do not need as much energy as we think—economic growth and human progress can be achieved in a more energy-efficient way.

## The Nuclear Option

I will address the potential of nuclear energy in Chapter 25, "Ecomodernism." I present the case that, outside of a few countries, nuclear power is a mostly untapped resource, waiting to be exploited. As of 2015, 10.6% of world electricity production was nuclear, and effectively none of the transportation energy was nuclear (because electric and hydrogen cars are in their infancy).[10] These numbers will and should rise.

---

[10] According to www.world-nuclear.org/information-library/current-and-future-generation/nuclear-power-in-the-world-today.aspx, "Sixteen countries depend on nuclear power for at least one-quarter of their electricity. France gets around three-quarters of its electricity from nuclear energy; Hungary, Slovakia and Ukraine get more than half from nuclear, whilst Belgium, Czech Republic, Finland, Sweden, Switzerland and Slovenia get one-third or more."

## Increasing Energy Efficiency

For now, let's focus on increased energy efficiency. The world is becoming more energy efficient at a rapid pace, and paradoxically it is doing so while *increasing* the average standard of living. This divergence is measured by *energy intensity*, defined as the amount of energy used per unit of GDP. Figure 10.7 shows the decline in energy intensity for the United States and for the world over 1995–2017. The improvement in energy intensity has been at a pace of roughly 1.5% per year for both the United States and the world. This is a huge number when compounded over long periods of time.

As with any costly resource, then, markets incentivize us to try to do more with less. Regulations encouraging or requiring energy efficiency have a role too.[11] For a discussion of doing more with less *material* input, see Chapter 23 on dematerialization; doing more with less *energy* is a parallel concept which, for lack of something better, we'll call by the old-fashioned and honorable name "conservation."

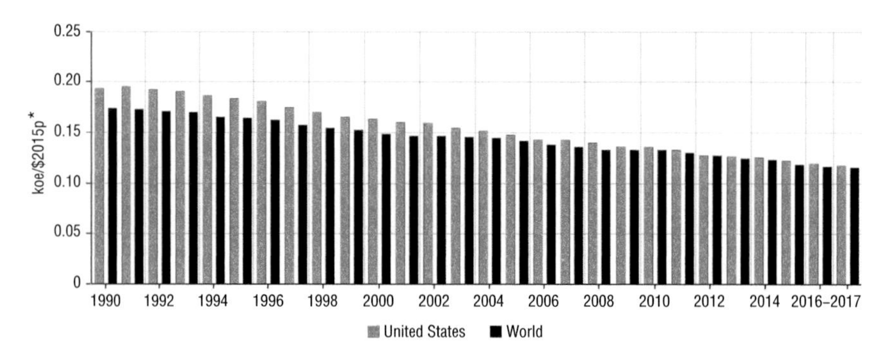


**Figure 10.7 Energy intensity (energy use per unit of GDP) for the United States and the world, 1995–2017.**

*Note:* *Kg Oil Equivalent per Unit of GDP (in 2015 USD at PPP)

*Source:* Based on data from ENERDATA, SAS (2019). Global Energy Statistics. Consulted on https://yearbook.enerdata.net.

[11] Regulations also have costs, including unforeseen consequences. Do you wonder why your refrigerator doesn't last 40 years like your grandfather's? Current regulations focus on operating efficiency but not the environmental footprint of replacing the whole unit every 8 or 9 years. Why can't you buy an old jalopy, fix it up, and drive it? Cash for Clunkers®, which took the old cars off the road. The program improved safety as well as pollution, but (as usual) imposed unforeseen costs on buyers with the least money, who now have to take the bus or walk because the cheap cars have disappeared from the market.

## From Energy Guzzlers to Energy Sippers

To accomplish this transformation, we're manufacturing more energy-efficient everything. There are large, comfortable hybrid cars that get 97 miles per gallon of gasoline, with some help from plug-in electricity sourced at a faraway power plant.[12] Light bulbs have evolved from incandescent to compact fluorescent to LED designs, providing the same amount of light using drastically less electricity. Green buildings are cheaper to heat and cool than traditional ones. And many of the energy-saving innovations are concealed within manufacturing processes and logistical techniques (think Amazon versus retail megastores).

Automobiles have become radically more energy-efficient as gasoline prices have risen, environmental regulations have taken effect, and consumers have sought ways to help cut carbon emissions. Figure 10.8 shows various options for powering a 1,000-mile drive in an automobile. Although most of the options are currently impractical (there's just not that much solar or wind energy yet available), the figure shows what can be accomplished when technology and economics work together to produce solutions to difficult energy and environmental problems.

The increasing efficiency of automobiles casts doubt on Vaclav Smil's claim that conservation just encourages more energy use. (His claim is based on the assumption of a highly elastic demand curve, where a small price decline causes a large increase in the quantity demanded. This is

| *Vehicle* | *Fuel* | *$CO_2$ produced* |
|---|---|---|
| Gasoline car | 40 gallons | 350 kilograms |
| Coal-powered EV* | 130 kilograms | 310 kilograms |
| Natural gas-powered EV | 2.5K cubic feet of gas | 170 kilograms |
| Solar rooftops-powered EV | 7 rooftops charging for 1 day | 0 |
| Wind-powered EV | 33 minutes on a typical 80-meter tall wind turbine | 0 |
| Large-scale solar array-powered EV | 3.2 minutes | 0 |

**Figure 10.8 Gas guzzlers to electron eaters: fuel consumption and CO2 emissions for a 1,000-mile drive.**

*Note:* *EV = electric vehicle.
*Source:* Constructed by the author based on data from Murtaugh, Dan. 2018. "You'll Need 286 Pounds of Coal to Fuel That Electric Road Trip." Bloomberg.com (October 31).

[12] The full-sized 2018 Ford Fusion Energi Plug-in Hybrid, four-cylinder, 2.0-liter automatic (variable gear ratio) model, is rated at 97 MPGe (miles per gallon equivalent) when gasoline and electricity are both used. When gasoline only is used, it is rated at 42 MPG. Ratings are for combined city and highway. MPGe is defined at www.edmunds.com/fuel-economy/decoding-electric-car-mpg.html.

sometimes observed in practice but is far from a universal rule.) Energy use can go down! We'll see that in more detail in the last section of this chapter, on lighting.

## Why the Energy Curve and Growth Curve Will Continue to Diverge

The problem with Smil's analysis, then—and I say this cautiously because he is almost universally respected—is that he glosses over the nonenergy contributors to economic progress. He does this even though his own work implicitly recognizes those contributors by specifying relatively modest energy requirements for a decent life:

> [A] society . . . determined to extend a good quality of life to the largest possible number of its citizens and hence willing to channel its resources into the provision of adequate diets, good health care, and basic schooling could guarantee decent physical well-being with an annual per capita use (converted with today's prevailing efficiencies) of as little as 50 gigajoules. A more satisfactory combination of infant mortalities below 20, female life expectancies above 75 years, and HDI above 0.8 requires annually about 60 gigajoules.

An annual energy budget of 60 gigajoules per person is about what is now consumed in Brazil, Mexico, and Turkey. Smil continues,

> Surprisingly, even [a much higher standard of living] is achievable without exorbitant energy consumption. Physical conditions that now prevail in affluent Western societies—infant mortalities below 10, female life expectancies above 80 years, and, needless to say, a surfeit of food—can be combined with high rates of house ownership (more than half of households), good access to post-secondary education, and [a Human Development Index] above 0.9 at energy consumption levels as low as 110 gigajoules per capita.[13]

That is a little more than is used in Spain and Poland and a little less than in the United Kingdom. None of these countries are poor.

And the 110 gigajoules *per capita* benchmark is at "today's prevailing efficiencies," which are going to improve dramatically, albeit slowly. If the rate of improvement continues to be 1.5% per year, energy intensity will fall in half in the next 46 years.

---

[13] Smil (2010), pp. 724–725.

Thus, while Smil sometimes sounds as though he believes growth is synonymous with more energy consumption, his own writing reflects a different view. And that alternative view is correct: growth is multidimensional. Building a more beautiful and comfortable city, replacing vinyl sound recordings with weightless digital ones, perfecting technology for communicating with co-workers without driving or flying to visit them—all of these innovations cause the energy consumption curve to diverge from the overall growth curve. *We can become better off without using more energy.*

### Let's Not Forget the Poorest

Of course, a great many people still suffer from energy poverty, as we saw in the poignant tale reported by the Kenyan student Mercy Njima in Chapter 8, "Food." The energy-starved do not need to conserve energy, they need to use more—but only up to a point, an idea to which Smil alludes when he says that energy use of 110 gigajoules per capita, or possibly less in the future, might be enough.

## A Deep Dive into the Light

To conclude this chapter, we take a detailed look at lighting so we can better understand the interplay of changing energy demand, increases in efficiency, and decreases in cost. Lighting is ideal for this examination because it's a universal need, and because data on lighting are abundant and go back a long time in history.

In *Enlightenment Now*, Steven Pinker's brilliant bestseller (that's two lighting metaphors in this sentence already), he writes:

> Light is so empowering that is serves as the metaphor of choice for a superior intellectual and spiritual state: *enlightenment*. . . . The economist William Nordhaus [who shared the 2018 Nobel in economics] has cited [its] plunging price (and hence . . . soaring availability) . . . as an emblem of progress.[14]

Figure 10.9 shows the cost of light in England in inflation-adjusted terms from 1300 to today: the cost per million lumen-hours fell *twelve-thousand*-fold. And that's just the money cost; the cost expressed in hours

[14] Pinker (2017), p. 253.

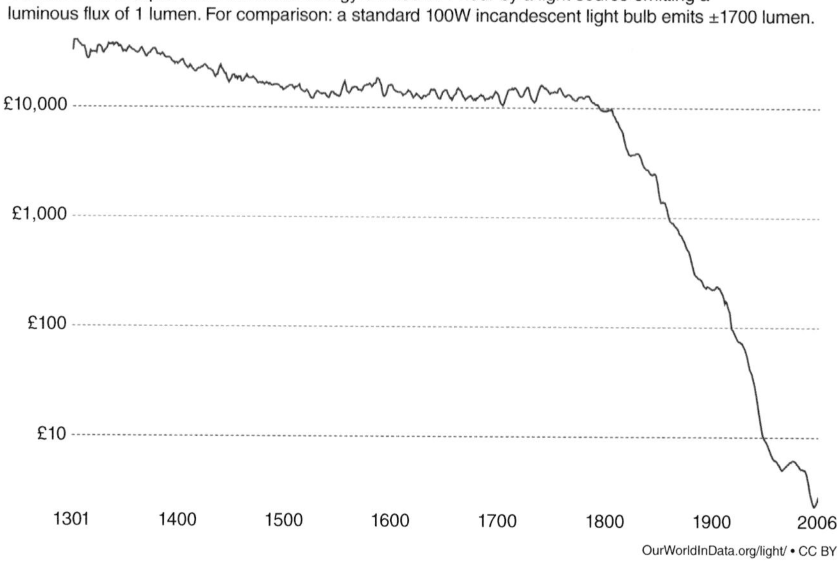


**Figure 10.9 Real (inflation-adjusted) price of lighting from 1300 to the present in the UK.**

*Note:* The price is adjusted for inflation and expressed in prices for the year 2000. Shown is a five-year moving average.

*Source:* ourworldindata.org/light, based on data from Fouquet, Roger, and Pearson, Peter J.G. 2012. "The Long Run Demand for Lighting: Elasticities and Rebound Effects in Different Phases of Economic Development." *Economics of Energy and Environmental Policy* 1: 83–100.

of labor fell much more, because of rising wages. The figure uses a log scale, emphasizing the speed of the decline once the use of coal gas in lamps, including streetlights, became widespread around 1810.

Max Roser writes,

> In the long period up to the 19th century nothing much changed and lighting stayed prohibitively expensive for most. The period of stagnation came to an end with . . . "town gas," . . . supplied via piped distribution systems [and] produced from coal, and [thus] also referred to as "coal gas." The transition from candles to town gas began in the 1810s and [happened] so rapidly that it was concluded by 1850.[15]

Yet another energy transition at exactly the rate Vaclav Smil documented! Although the 40-year transformation seems almost instantaneous

[15] Roser, "Light," Our World in Data, https://ourworldindata.org/light.

when compared to the long period of lighting stagnation that preceded it, it is consistent with the speed of the transitions in Figure 10.6.[16]

Steven Pinker emphasizes the reduction in the amount of labor involved in obtaining light, rather than the money price. "Nordhaus," Pinker writes, "estimated how many hours a person would have to work to earn an hour to read by at various times in history. A Babylonian in 1750 BCE would have had to labor fifty hours to spend one hour reading . . . by a sesame-oil lamp." (It's amazing the Babylonian Talmud ever got written, although maybe rabbis didn't have day jobs and could pursue their religious studies at leisure.[17]) "In 1880," Pinker continues, "you'd need to work 15 minutes, and in 1994, *a half-second* for the same hour [of illumination]."[18] That's a 360,000-fold improvement in 3,750 years![19]

Figure 10.10 contrasts the fall in the price of lighting itself with the fall in the price of the energy used to power the light source. After about 1800 the price of lighting fell many times faster than the price of energy, showing the rapidly improving *energy efficiency* of the lighting technologies being used. (In other words, if the cost of a given amount of lighting fell by a factor of 5,000 between the year 1800 and today, but the cost of the energy used to produce that amount of light fell by a factor of only 10, then lighting became 500 times more efficient.[20])

Naturally, it's electricity that did the heavy lifting after Sir Humphry Davy's carbon-arc lamp (1802–1809), shown in Figure 10.11, and Edison's carbon-filament bulb (1879) were invented. However, early electric lighting was very inefficient—one of Davy's lamps required a 2,000-cell

---

[16] The decline in the price of lighting between about 1375 and 1520 is also worth commenting on. In the diagram used by Pinker [2017], Figure 17-4, page 253, which shows the same data as our Figure 10.9 but on a linear scale, the improvement in the late Middle Ages is very dramatic, with the price declining from £34,000 to about £12,000 per million lumen-hours. (A million lumen-hours is the amount of light generated by operating a 1000-lumen, or a little over 60-watt, incandescent bulb for 1000 hours, giving an idea of how insanely expensive late-medieval lighting was.) Tallow candles were the principal lighting source, so the price decline was probably due to rising consumption of meat as incomes in England rose in the late Middle Ages, causing meat by-products such as tallow to be more abundant and thus cheaper.

[17] I'm cheating a little, for effect. The Babylonian Talmud was written much later than 1750 BCE —about 2,200 years later. Lighting was still expensive.

[18] Pinker (2017), p. 254.

[19] That's still only a 0.34% compound annual growth rate, but growth at that pace adds up over the millennia. I wonder what the labor price of lighting will be in the year 5769 (3750 years from now). I can't see how it can get much cheaper, but neither did the Babylonians.

[20] These numbers are rough estimates obtained by reading the graph, without having access to the underlying data, but they are correct in an order-of-magnitude sense.

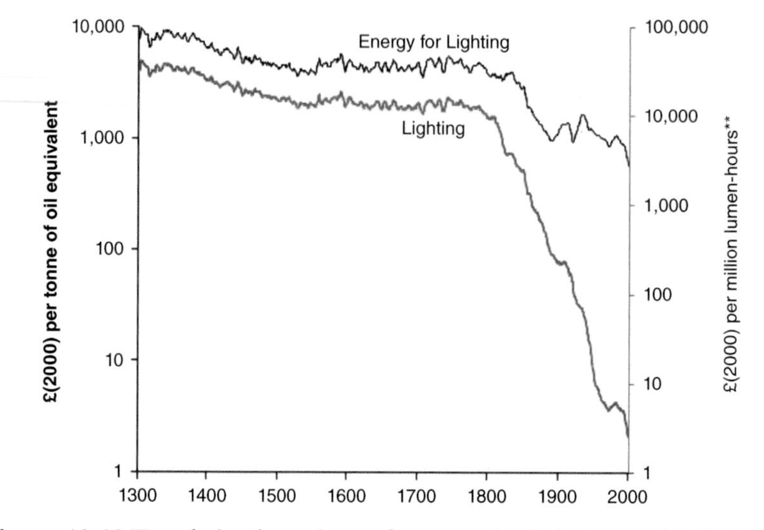


**Figure 10.10 Trends in the prices of energy for lighting and of lighting services, 1300–2000.**

*Source:* Fouquet, Roger, and Pearson, Peter J.G. 2012. "The Long Run Demand for Lighting: Elasticities and Rebound Effects in Different Phases of Economic Development." *Economics of Energy and Environmental Policy* 1: 83–100. Reproduced with permission.

**Figure 10.11 Carbon arc lamp (1802), invented by Sir Humphry Davy, the forgotten hero of the Industrial Revolution (if only, as Davy joked, because his assistant Michael Faraday was his greatest discovery).**

*Source:* https://en.wikipedia.org/wiki/Arc_lamp#/media/File:Arc_light_and_battery.jpg. Drawing created December 31, 1872. Retrieved September 17, 2013 from *Augustin Privat Deschanel 1878 Elementary Treatise on Natural Philosophy, Part 3: Electricity and Magnetism,* D. Appleton and Co., New York, translated by J. D. Everett, p. 702, fig. 509 on Google Books.

battery!—and only became practical on a mass scale when tungsten was substituted for carbon in incandescent bulbs around 1904. Roser writes that "the dominance of electricity was . . . not established overnight and only in the 1920s did electric lighting became cheaper than gas lighting."

Thus, gradual technology transitions worked in tandem with gradual energy-source transitions to produce the massive 150- or 200-year decline in lighting cost and corresponding improvement in energy efficiency shown in Figure 10.10. It was a miracle, all right, but it took a century and a half to mature. Incremental innovations, such as the fluorescent bulb and the light-emitting diode (LED), produced the slower but still significant cost and efficiency improvements since 1950, and can be expected to continue.

So what happened to lighting *consumption* over this period of enlightenment? Figure 10.12 shows that, as prices fell, consumption rose much more—an incredible 100,000-fold between 1700 and the present—so that the total amount of energy used for lighting increased even as energy cost per unit of lighting was plunging. This observation validates Vaclav Smil's concern that increases in energy efficiency just mean more consumption and no actual reduction in the use of energy.

However, this one instance does not represent a universal principle. Many energy uses *saturate*—you don't want to heat your house beyond about 72 degrees Fahrenheit, nor can you use a computer or a smartphone more than 24 hours a day (unless you're a teenager). We can see saturation at work for lighting in Figure 10.12, after about 1960; in a rich

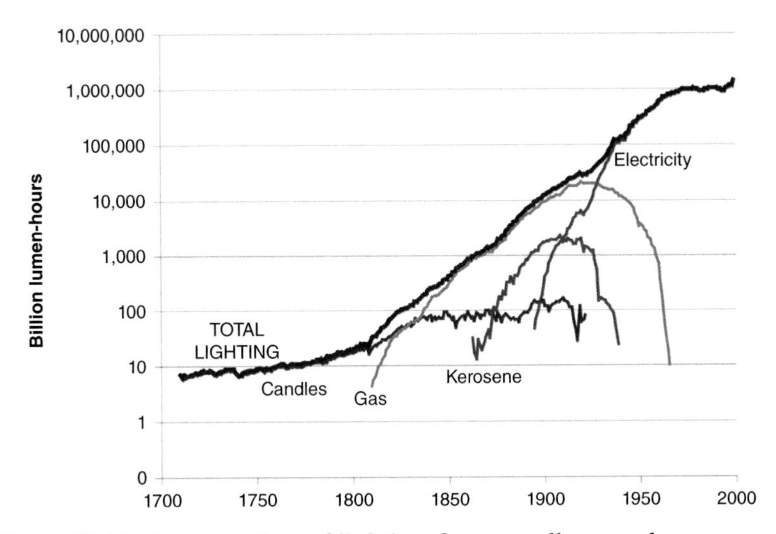


**Figure 10.12 Consumption of lighting from candles, gas, kerosene and electricity in the United Kingdom (in billion lumen-hours), 1700–2000.**

*Source:* Fouquet, Roger, and Pearson, Peter J. G. (2007). "Seven Centuries of Energy Services." *The Energy Journal*. 27 (1). Reproduced with permission.

country such as the UK, the lights are already on whenever somebody wants them on. So, as lighting becomes more energy-efficient, we'll actually use less energy.

That is what the supply-demand-price system, the linchpin of a market economy, is *for*. Energy prices—it's still not free—are nature's way of telling us to use less of it per unit of output, while encouraging energy producers of all kinds to produce more so that the Bangladeshi will someday live at the level of the Chinese, the Chinese at the level of the French, and so forth.

We cannot know which force, greater efficiency or more total want or need, will "win" in any given period. We just have to wait and see. Here's a defensible forecast: in the intermediate run—a few decades—total energy use will grow, but in the longer run it will decline.

We'll return to energy, in particular the future of energy, in Part V, "Greener."

# 11

# *Cities*

What is a city? Isn't it just a bunch of buildings that are close together?

No. Blow up or burn down a city and it springs right back where it was. Even Hiroshima couldn't be wiped out: today it has a thriving population of 1,196,274. Chicago, Dresden, New Orleans, you name it—whether destroyed by fire in 1871, war in 1945, or flood in 2005, a city regrows itself like a patch of injured skin, almost exactly in the pattern that existed before the destruction.

The reason is that a city is not a collection of buildings (those are the symptom, not the cause) but a *network of human connections*. *Knowledge* connections. In the more or less urban neighborhood where I live, there is a shoe repairman named Mike. (See Figure 11.1.)

What makes him a good shoemaker? Not just his personal skill, but the network of connections he's developed: he knows where to buy leather, get the tools of his trade repaired, and acquire the paraphernalia that all trades have in common—a credit card acceptance system, a telephone line, and so forth. He can (and did) locate on a crowded block with lots of foot traffic. And, because he's in a city, he can comparison shop among competing suppliers without traveling immense distances, so the supplies are affordable.[1]

[1] The Internet changes this dynamic a little, but personal, local connections are still important and build trust between parties to a transaction, a factor that has been widely shown to improve economic efficiency.

**Figure 11.1 The locus of a network of knowledge connections.**

*Source:* J. Needham.

The other network Mike needs is on the demand side: customers and potential customers. Just as Mike knows how to find his suppliers, *I* know where to find *him*. Enough of my neighbors need custom shoes or shoe repairs that my recommendation helps his business. We have all heard of the supply chain, which includes, say, the leather supplier's connections and so on down the line to the farmer who grows food for us all. But there's also a *demand chain*. I can afford Mike's services because I have my own customers, who have their own customers. A city, or even a village, is a trading economy-in-miniature that has arisen in search of, and in reaction to, low transportation and transaction costs.

## Why We Should Care About Cities

In a book on economic growth, why a chapter on cities? It's because that is where most growth takes place. As with Mike and his suppliers and customers, cities foster easy, cheap, productive connections as well as true innovations.

Cities are where the brains are. I don't mean to disparage rural people in any way, but the gifted and ambitious have always gravitated to cities—if only because they need someone to talk to. That's where the other gifted and ambitious people are.

There is a long thread in economic and philosophical thinking that is appreciative of the immense contribution made by cities, running from Aristotle through Locke and Dr. Johnson ("when a man is tired of London he is tired of life") to the grandiloquent Lewis Mumford; the beloved urbanist, Jane Jacobs; and that contemporary champion of cities, Edward Glaeser.[2] This book is not the right place to lay out that whole intellectual history, or to counter the fear and loathing of cities by counter-Enlightenment figures such as Rousseau.

I'll just note that the words *city, citizen, civic, civil,* and *civilization* all come from the same Latin root, *civitas*, which means not the physical city (*urbs*)[3] but something more like "the people acting together." Cicero used the word to refer to Rome as an *idea*, and we should perhaps use the word *city* to help us think about our modern urban agglomerations in the same way.

*Why* do cities foster growth and innovation? Because these benefits require specialization and trade, which in turn rely on population density, diversity, complexity, and concentrations of intelligence. The physicist Geoffrey West, in a wide-ranging book of popular science called *Scale,* demonstrates that, as I said in my review of that book,

> [e]ach doubling [that is, 100% increase] in the size of a city results in a 115% increase in each of the key economic variables such as wages, patents and industries. Because commerce is transacted more rapidly, per capita GDP is higher in larger cities: Los Angeles is richer than Tulsa, which is richer than a typical rural town. The relationship is causal, not coincidental—the connections made possible by the larger city's size created the additional wealth.[4]

---

[2] Mumford (1961); Jacobs (1961); Glaeser (2011).
Jacobs was a combative and effective advocate of a philosophy of urbanism that marries preservation and progress. She appreciated market outcomes and cities formed through organic growth. Those achievements and views are what is beloved; as a person, she may be best remembered for New York housing administrator Robert Starr's wisecrack, "What a dear, sweet character she isn't."

[3] Pronounced, oddly, "erps"; *civitas* is KEE-wi-tahs, with the Latin hard *c*, not the English soft *c* or the Italian *ch*. Latin *v* is usually pronounced like the English *w*.

[4] West (2017). The quote is from Siegel (2017).

Creativity, hence progress, thus takes place when—to recycle Matt Ridley's memorable phrase, "ideas have sex." For ideas to have sex, you need more than one idea. You need random—or intentional—encounters between people who *have* ideas, not just "great" ideas, but little ones. Little ideas improve processes in manufacturing, or artisanship, or logistics, or sales, or customer service. Such serendipitous encounters are more likely to take place in a coffeehouse, pub, corporate office, bank, government agency, university, or city park than on an endless dusty road between two subsistence farms.

## Momentum Cities

Why are cities located where they are, and why do they spring back in the same place when destroyed? As schoolchildren many of us learned that Chicago became a major city because it was the lowest point of portage between the Great Lakes waterway and the rivers that flow into the Mississippi; it's an easy walk from one watershed to the other. It might not be that easy if you're carrying a canoe over your head, but it's doable. (Hence the names Portage Park, a neighborhood of Chicago, and nearby Portage, Indiana.)

But we no longer travel by canoe. Even if you take into account big ships, a point of easy transshipment between the Great Lakes and the Mississippi is not as important as it used to be. But Chicago then evolved into a railroad capital and now one of the world's great airline hubs. It is still a focal point for transportation. Why? Because it *was* the focal point for transportation a long time ago. There's a momentum effect.

Seattle, likewise, was the Boeing city until it became the Microsoft city and then the Amazon city. All these companies use a lot of engineers. *Cities reinvent themselves, then, along the lines that caused them to be created in the first place*, not identically (because times and needs change) but using the same general skill sets, labor types, and managerial talents.[5]

The rule is not foolproof. Sometimes momentum doesn't work. Cleveland was the technological capital of the United States in 1890, Detroit in 1925. Both cities are now suffering. I blame bad policies, not the failure of the momentum model, because Pittsburgh, which could have suffered the same fate, is thriving. But there is no guarantee that a great city will remain great—it's just a tendency, not an ironclad law.

---

[5] Cities can and do change their character. Before Seattle was a city of engineers, it was a shipping, fishing, and forestry capital. But that was a long time ago, and Seattle was much smaller then (it was the sixty-third largest metropolitan area in 1900 and is now fourteenth).

## All Roads Lead to—and from—Rome

Even Rome, the grandest of all ancient cities, fell in population from one million at its peak to fewer than 20,000 in the early Middle Ages. It could have been ruined as utterly as the Cahokia Mounds in Illinois.[6] But there was something about the location of Rome, people's knowledge of what Rome represented and what might be found there, and (as we'll soon see) the legacy of Roman engineering that caused it to be resettled once its darkest days were over. Today it is a metropolis four times larger than in Caesar's time.[7]

One factor that led to Rome's revival, and Europe's emergence as the world center of civilization in the early modern era, is the network of roads that the Romans built. (They were built to last. Some are still in pretty good shape; see Figure 11.2). The roads formed a dense network

**Figure 11.2 Roman road in Seville, Spain, built about 200 BC, 1450 miles from Rome.**

*Source:* ©Vitantonio Caporusso/Alamy.

[6] The Cakohia Mounds are a pre-Columbian Native American ruin in Illinois that may have housed 15,000 to 30,000 people at its peak around 1275. No living Native American residents remain (Hodges, 2011).

[7] That is still not a high growth rate: one million to four million in 2,000 years is 0.07% per year. Relative to the world's overall population at a given time, ancient Rome was probably the largest city that has ever existed.

over a vast area that ranged from England to Portugal to Romania to Libya—"massive infrastructure projects even by modern standards," writes the *Washington Post* reporter Christopher Ingraham. Although the roads were built mostly for military purposes, they had the side effect of making trade much easier and cheaper.

A group led by Carl-Johan Dalgaard, an economist at the University of Copenhagen, studied the links between the Roman routes and current patterns of prosperity. Reporting on Dalgaard's study, Ingraham writes,

> Prosperity begets prosperity: On a global level, economists and historians have shown that places that prospered 100, 500, even 1,000 years ago tend to be more economically developed today.

The evidence is in the remarkable Figure 11.3, an overlay of modern nighttime lighting patterns in Europe over a map of the Roman roads.

**Figure 11.3 Map of Roman roads overlaid with modern urban lights at night.**

*Source:* Klazina Stanwick.[8] Commissioned by Laurence B. Siegel.

*Source of map:* NASA's 2016 Black Marble photograph, https://earthobservatory.nasa.gov/features/NightLights/page3.php, combined with the work of Mapping Roman Roads by the Pelagios Project http://pelagios.org/maps/greco-roman/about.html.

[8] Inspired by artwork in Ingraham (2018).

The lighting pattern is a proxy for the current intensity of economic activity in a location.

The correspondence between Roman road intersections and modern, brightly lit cities is uncanny. It is especially visible in France, where "you can clearly see the paths of ancient roadways connecting not just major modern cities, like Paris and Lyon, but also many minor ones, too. Across inland France, nearly every junction of ancient roads is marked by a splash of light in the modern era."[9]

The researchers' conclusion is that large infrastructure investments can have payoffs far beyond the short term. They compared Roman roads in Europe with those in North Africa and the Near East. The European roads were heavily used for trade by wheeled ox-cart long after the Roman empire had fallen; on the roads in North Africa and the Near East, camels had become the main mode of transportation. As the Middle Ages proceeded, Europe thrived relative to the North African and Near Eastern areas. Because good roads are much more important for transportation by wheeled ox-cart than by camel, the authors concluded that the roads were an important *causal* factor for subsequent prosperity in Europe relative to the other areas.

## A Magnet for the Ambitious, an Escape for the Oppressed

São Paulo (Figure 11.4) is nobody's idea of a beautiful city. The gem of Brazil, Rio de Janeiro, a couple of hundred miles away, is far more fetching: it looks like a fantasyland if you cast your eye in the right direction. So why do 10 million more people live in São Paulo than in Rio? Why has São Paulo been richer since the 1940s, "the locomotive 'pulling the rest of Brazil' and . . . the hub of an immense megametropolis"?[10]

It's because Rio is for romantics and nostalgics, and São Paulo is for business. Once a city establishes itself as a magnet for the hardworking and the ambitious, it tends to stay that way and become more so, leaving other locales behind. Momentum. Matt Ridley writes:

[9] Ingraham (2018). A close-up of the French part of the map is in the *Washington Post* article.

[10] Leite et al. (undated).

**Figure 11.4 São Paulo: Eyesore or magnet?**

*Source:* Andrea Dal Max/Shutterstock.

> Trade draws people to cities and swells the slums. Is this not a bad thing? No. Satanic the mills of the industrial revolution may have looked to romantic poets, but they were also beacons of opportunity to young people facing the squalor and crowding of a country cottage on too small a plot of land. As Ford Madox Ford celebrated in his Edwardian novel *The Soul of London*, the city may have seemed dirty and squalid to the rich but it was seen by the working class as a place of liberation and enterprise.[11]

That was London in 1905; how about Mumbai in 2018? Isn't Mumbai a much tougher environment? Ridley explains:

> Ask a modern Indian woman why she wants to leave her rural village for a Mumbai slum. Because the city, for all its dangers and squalor, represents opportunity, the chance to escape from the village of her birth, where there is drudgery without wages, suffocating family control and where work happens in the merciless heat of the sun or the drenching downpour of the monsoon.

[11] Ridley (2010), pp. 188–189.

> "[F]or the young person in an Indian village, the call of Mumbai isn't just about money. It's also about freedom," says Suketa Mehta. . . . Rural self-sufficiency is a romantic mirage. Urban opportunity is what people want.

Noting that, for the first time in the history of the world, more than half of people now live in cities—a number that some say is an underestimate—Ridley concludes, "That is not a bad thing. It is a measure of economic progress that more than half the population can leave subsistence and seek the possibilities of a *life based on the collective brain* instead. Two-thirds of economic growth happens in cities."[12]

## Cities and Loneliness

Some even claim that urbanization in China has cut the suicide rate by alleviating loneliness. In a Cato Institute online publication, Chelsea Follett writes:

> As more Chinese have left farms in the countryside to work in factory cities, the suicide rate has plummeted. This may be shocking to many people in rich countries. That is because many people who enjoy post-industrial prosperity worry about "sweatshop" conditions and exploitation in factories [and] may also have an idealised opinion of rural peasant life.
>
> In reality, factory work is typically an improvement compared to poverty in the countryside. Factory conditions can be harsh and no one is claiming they should not improve. But far worse back-breaking labour and grinding poverty often define rural existence. The option of migrating to a city to take up factory work . . . [offering] higher wages and a better standard of living . . . can be a lifeline to those contemplating suicide.[13]

Echoing Ridley, Follett notes, "it also can mean freedom from the more restrictive social norms of the countryside—particularly for women . . . since gender roles are less limiting in cities." She concludes, "Globalisation has, quite literally, saved many of their lives."[14]

---

[12] Ridley (2010), p. 189. The datum regarding more than half of the world's population living in cities is at UN News (2014). The percentage in 2014 was 54% and is higher now.
[13] Follett (2018).
[14] Ibid.

## Squatter Cities: The Creativity of the Desperate

Pursuing a similar theme, Stewart Brand, the pioneering environmentalist and author whose work we'll study closely in a later chapter, also waxes enthusiastic about that least aesthetically appealing kind of urban settlement, the squatter city. *Wired* magazine writes:

> Some people see a squatter city in Nigeria or India and the desperation overwhelms them: rickety shelters, little kids working or begging, filthy water and air. Stewart Brand sees the same places and he's encouraged. . . . Brand . . . makes a counterintuitive case that the booming slums and squatter cities in and around Mumbai, Nairobi, and Rio de Janeiro [see Figure 11.5] are net positives for poor people and the environment. *Wired* asked him to elaborate.
>
> *Stewart Brand:* That's where vast numbers of humans—slum dwellers—are doing urban stuff in new and amazing ways. . . . People are trying desperately to get out of poverty, so there's a lot of creativity; they collaborate in ways that we've completely forgotten how to do in regular cities. . . .
>
> People come in from the countryside, enter the rickshaw economy, and work for almost nothing. But after a while, they move uptown, into the formal economy. The United Nations did extensive field research

**Figure 11.5 Inside the Rocinha favela in Rio de Janeiro, 2016.**
*Source:* © Ranimiro Lotufo Neto/123rf.com. Reproduced with permission.

> and flipped from seeing squatter cities as the world's great problem to realizing these slums are actually the world's great solution to poverty.

Brand looks at squatter cities, like almost everything else, from an environmental perspective. *Wired* asks him how they could possibly be good for the environment:

> *Brand:* Cities draw people away from subsistence farming, which is ecologically devastating, and they defuse the population bomb. In the villages, women spend their time doing agricultural stuff, for no pay, or having lots and lots of kids. When women move to town, it's better to have fewer kids, bear down, and get them some education, some economic opportunity. Women become important, powerful creatures in the slums.

Finally, the *Wired* interviewer asks how governments can help nurture these positives:

> *Brand:* The . . . main thing is not to bulldoze the slums. Treat the people as pioneers. Get them some grid electricity, water, sanitation, crime prevention. All that makes a huge difference. [15]

## Why Cities Are Green

Cities don't look particularly green. It's possible to stand in the heart of some cities and not see a tree or a blade of grass. Yet economists and many environmentalists agree that the division of population into a heartland of specialized producers and a hinterland with a small number of very efficient food growers is the greenest—that is, the most ecologically efficient—system ever devised.

Actually, it was not devised. It just happened. Producing one thing extremely well and consuming a lot of different things from all over is just what people want to do, when they have the choice. The result is city living for most, with farming becoming its own technologically advanced specialty.

The least efficient system, of course, is for everyone to do everything for themselves. That's how primitive man lived (very poorly). Self-sufficiency is still a recipe for poverty: Matt Ridley has noted that all

[15] *Wired* (2009).

Dark Ages were retreats from specialization into self-sufficiency. If you have to do everything yourself, you're not going to be very good at it, so a population of self-sufficient farmers will mostly just grow enough food to survive and never produce much else.

If something like 11 billion people are going to live on the surface of the earth, it's much better to pack most of them into small areas. Such a design allows for two kinds of benefits: (1) people can interact with each other and create things, and (2) it leaves more space for nature. We need a lot of land for food production, and a lot more for forests, which suck out the carbon emitted by the burning of fossil fuels: that's called having a *negative carbon footprint*, and it's a good thing.

## Deforestation … or Reforestation?

Thus the ongoing concern about deforestation. The destruction of forests continues to take place in countries where the still-growing need for agricultural land is resulting in forest cutting. But in the United States, there are more trees today than there were in 1950.

This dramatic *reforestation* began when the opening of the Erie Canal, in the 1820s, made it practical to farm in the rich lands of Ohio and beyond. The previously valuable Northeastern farmland became marginal. Thus, much of the Northeastern land reverted to forest. As the Case Western Reserve University law professor and market environmentalist Jonathan H. Adler noted,

> [Since about 1950], timberland east of the Mississippi has expanded by 3.8 million acres, in addition to the nearly three million acres in the eastern United States that have been declared wilderness in the past two decades. By 1980, New England contained more forested acres than in the mid-19th century; Vermont is now twice as forested as then. Fifty-nine percent of the northeastern United States is covered by forest. [16]

This is not simply a matter of land returning to its natural state. Forests are a positive good, for which people are willing to pay if they can live close enough to one to visit it.

---

[16] Adler (1993).

## England's Green and Pleasant Land

England, not the whole of the British Isles but just England, is one of the most crowded countries in the world, with 55,619,400 people in 50,301 square miles: 4.3 times the population density of California and about one-third the density of Bangladesh. It nevertheless has a superabundance of green space and a lot of poetry devoted to that space. Blake: "England's green and pleasant land." Housman: *A Shropshire Lad*. Shakespeare: "this sceptred isle . . ./This other Eden, demi-paradise . . ./This blessed plot, this earth, this realm, this England." How on earth did the English do it?

The countryside is green *because* the cities are densely packed (Figure 11.6). The wonderful livability of crowded England is also

**Figure 11.6 Royal Crescent, Bath, UK, aerial view.**

*Source:* Skyscan Photolibrary/Alamy Stock Photo.

partly due to government protection of open spaces, and partly to its natural endowment. But it's mostly traceable to simple economics—when England was being built, transport costs were high: dragging around all their possessions and merchandise by horse or perhaps floating them up a canal, villagers needed to live near other villagers

*(continued)*

*(continued)*

and city dwellers near other city dwellers. England is also green because of the selfish motives of kings and landowners who wanted to preserve its natural beauty for themselves—an act with highly beneficial unintended consequences.

The United States, developing later along different principles, reflects cheap transportation: fast trains and an early commitment to highway building. So it is not as ecologically efficient. Nevertheless, US cities and their suburbs take up only 4% of the United States' vast land area, up from 3% a generation ago.

How the world will balance urbanism and nature in the future remains to be seen, but if rapidly developing East Asia is any indication, the future will be intensely urban. And I'll argue that that result is efficient and helps nature, rather than hurting it as some might surmise.

The return of the forests to the Northeast and elsewhere are an outcome of changes in economic demand. The use of land for forest, at least in that part of the country, has become more valuable than use of the same land for agriculture or urban development. Ellen Stroud, a Bryn Mawr College environmentalist and urbanologist, has written, "It is no coincidence that the most heavily urbanized part of the country has experienced the most dramatic return of woodlands. . . . *The desires and physical needs of city people encouraged and required the return of the forest.*"[17]

As countries develop, and the world becomes increasingly urbanized, the deforestation of the world can be expected to reverse course; reforestation has already started in developed countries and it's only a matter of time before it does elsewhere.

## Conclusion

Cities are the very soul of economic development, of betterment, of the Great Enrichment. They allow people to connect in ways that produce new ideas, new products, new books and music, new cures for diseases—in short, better modes of living.

[17] Stroud (2012).

There is a downside to cities. Disease spreads more quickly in close quarters. Families may not look after their more vulnerable members as closely in an impersonal city as in a smaller community. But the market has spoken—people have voted with their feet—and a vast number of them would rather take their chances in cities than settle for stasis in their places of origin. Cities are the future.

One might guess that the Internet and cheap, safe air travel will make it less important for talented people to locate physically close to one another. We've gotten a peek at that trend with telecommuting, distributed businesses, and work from home. But, each year, the biggest cities get bigger and more important, and millions migrate to them from rural areas in search of the opportunity and stimulation they so desperately crave. The need for proximity and interaction seems to be much greater than the need for privacy and simplicity.

As more people move to cities, we'll see an improvement in energy use per capita, an increase in opportunity for the individual, an intensified exchange of skills and ideas, and the continued acceleration of innovation. And, thanks to cities, the forests can return.

Geologists may have recently decided to call the current era the Anthropocene, but it's really the age of cities, of the interconnected mind.

# 12

# *Education: The Third Democratization*

We could be on the cusp of a worldwide golden age of education.

Not schooling. We've pretty much ruined that. Bueller? Bueller? Bueller?

But education is much more than schooling. It's learning by doing. Learning by reading and listening and talking and writing. Learning by having to face real problems and come up with potential solutions that are empirically testable with regard to whether they work, producing a feedback loop for the learner. That is what education is—not sitting in a classroom listening to a lecture and being tested on it, although that's a legitimate part of it.

Fifty years ago Milton Friedman wrote "The Higher Schooling in America."[1] The article was about how to pay for it, not how to structure or deliver it, but it's telling that he called it schooling, reserving the word *education* for a higher calling.

But maybe Friedman protested too much. Despite the sublime silliness of some of the course offerings in US universities, the US university system is the envy of the world. It attracts full-fee-paying students from countries that can afford to send their best young people anywhere.

[1] Friedman (1968).

## "Surely You're Joking, Mr. Feynman!"

At its best, the university system employs some truly extraordinary teachers. Although research is supremely important for scientific progress, teaching is what counts for the *democratization* of education. This is true whether the teaching is provided in person or via the Internet. At Caltech in the early 1960s, the physicist Richard Feynman (see Figure 12.1) recorded a series of lectures that are almost universally considered classics. They are all available free of charge as videos on the Internet, as are transcripts of them. Watch them.[2]

I have no special interest in learning college-level physics, especially as it was understood 50 years ago, since much has changed, but I just can't stop listening to the guy. So I've heard the whole series of Feynman lectures. To what end? Just for fun, I guess, and because I wouldn't be writing this book if a love of learning wasn't one of my basic personality traits.

**Figure 12.1 Richard Feynman in early 1986.**

*Source:* https://commons.wikimedia.org/wiki/File:Richard_Feynman_1988.png.

[2]Videos are at www.youtube.com/user/FeynmanVideoLectures/videos, and transcripts are at www.feynmanlectures.caltech.edu/info/. *Surely You're Joking, Mr. Feynman! Adventures of a Curious Character* is the title of Feynman's 1985 memoir (New York: W. W. Norton).

My reason for mentioning the Feynman lectures goes beyond just saying how good they are. If I can watch these instructional videos, free of charge and at my leisure, then so can everyone. We have thus reached the third inflection point in the democratization of knowledge in modern history. Johann Gutenberg, who invented the printing press around 1450, brought about the first. The second was the spread of public education in America and elsewhere in the late nineteenth and early twentieth centuries, including the establishment of "land grant" colleges (what we now call state universities).

The third democratization of education is being made possible by the Internet and is just getting started.

Listening to the Feynman lectures is not the same as watching *Hamlet*, or a dramatization of Dickens's *Christmas Carol*, at the movie theater or on TV. These great plays or stories were intended as entertainment and only incidentally have an educational purpose; but the Feynman lectures, and 10,000 (mostly not as good) others like them, are explicitly instructional. They are intended to *substitute* for going to the lectures in person and having discussion groups or tutorials afterward.

## First Out of the Gate: MOOCs

A MOOC is a massive open online course. MOOCs made a big splash around 2006 to 2012 and were expected—by some—to catch on quickly and act as competitors to conventional university educations. Among the best-known providers are Khan Academy and Coursera. The two organizations follow different models. Khan is a nonprofit, focuses on precollegiate as well as collegiate learning, offers courses in many languages, and does not grant degrees. Khan has also started a bricks-and-mortar lab school (private school) in Silicon Valley. Coursera is for-profit, college-level, and offers bachelor's and master's degrees. Both claim millions of registrants.

In fact, Khan Academy videos are now used in many public schools. In a method called "flipping the classroom," teachers give students the Khan videos as homework (and have mechanisms for gauging students' progress) and then use them as the basis for class discussions. It's an encouraging trend.

Universities themselves have also gotten into the act. MIT has put essentially all its courses online.[3] They encourage (but do not actually provide)

[3] https://ocw.mit.edu.

integration of their online materials with in-person teaching. A number of top universities have formed consortiums for putting their best professors' lectures online free of charge, or to offer courses for a fee. Tyler Cowen and Alex Tabarrok of George Mason University in Virginia have founded Marginal Revolution University, a free online resource focused just on economics.[4]

How well have these efforts worked out so far? The results are mixed. For "pure" MOOCs (those with no in-person teacher), as few as 5% to 10% of students who register for the courses finish them, but 5% or 10% of a population of millions is still a lot of uptake. At any rate, pure MOOCs are a flawed model, "an Internet version of a book" in the words of WizIQ CEO Harman Singh, founder of a company that blends online lectures with live instructor-led learning.[5] A lot of people start books and don't finish them, and we don't regard that as a failure of the publishing industry!

But, as both Benjamin Franklin and Michael Faraday responded when queried about their silly inventions, "Of what use is a newborn baby?" Recall our earlier discussion of cars. Did an 1895 car work well? It worked horribly, and not just because cars were newborn babies. There was also no supporting infrastructure—no good roads (see Figure 16.2 in Chapter 16), no good maps, no trained mechanics to speak of, no warehouses full of spare parts, *no gas stations*. You had to buy motor fuel in drug stores.

## MOOCs Are Like Primitive Cars—They're Primitive

MOOCs are, today, where cars were in 1895. To start with, MOOCs are very poorly named, inviting bovine humor.[6] (Inventors are the worst at naming their inventions: the inventor of the crossword called it the word-cross, and Monopoly started out life as The Landlord's Game—really!)

Moreover, channeling Voltaire, who said the Holy Roman Empire was neither holy, nor Roman, nor an empire, a massive open online course doesn't have to be massive, or open, or a course. Massive? It can be narrowcast, to speakers of Lakota or Zulu or to students of an arcane aspect of nuclear engineering. Open? There are already ways of imposing prerequisites, enforcing admissions requirements, and charging fees. A course? A TED talk can be educational and worth hearing without stringing a bunch of them together in course format. There is no limit to the permutations that some future entrepreneur can offer.

---

[4] www.mruniversity.com.

[5] Lynch (2016).

[6] In an example of life imitating art, Moo U in Jane Smiley's academic novel *Moo* (1995) predates the first MOOC.

The manufacturers of an 1895 car didn't think about doors, a roof, a windshield, a speedometer, a gas gauge, or a starter, yet cars "evolved" all of these pretty quickly. Online education can evolve quickly too; there are plenty of innovators who are chomping to have a go at it.

And, as with cars in 1895, there is not much supporting infrastructure. For a curious and able student, the best part of a university education is stopping by the professor's office to chat, being invited to his or her home for a discussion of advanced material, or being hired as a research or teaching assistant. It's hard to do that online. The enrichments of college life—sports, parties, dance classes, road trips—are also hard to replicate online. Is all that really necessary for obtaining an education? No, but well-roundedness is important and I'd hate to see the pursuit of it abandoned in favor of the single-minded delivery of an affordable credential.

One reason that so many attempts to merge Internet education with traditional schooling have failed or floundered is that the effort is just too new. To continue my comparison with cars, between 1900 and 1950 about *two thousand* auto manufacturers were established in the United States. Three of them survived to the end of the century.[7] Another reason is teachers' unions, which resist any form of competition. And the government subsidization of public universities causes them to be hard to compete with, although the best public universities are asking some pretty fancy prices, making them an easier target.

## MOOC-ish Models That Might Work

My recommendation is to keep trying different things: let a hundred flowers bloom, let a hundred schools of thought contend. (Mao should have put down his gun and stuck to writing epigrams.) The top-university consortium idea is promising, since they not only have some of the best professors but also recognize the value of a prestigious credential, which is what many would-be online students seek.[8]

Another model is trade schools; they recognize the need to put a dollar in the student's pocket. But teaching people how to work with their hands involves a lot of contact with the physical world, so the blend of online and in-person education has to be different for trade school than for more intellectual pursuits.

---

[7] GM, Ford, and Chrysler all survived to 2000, but only Ford survived a fifth of the way into the next century without either a bankruptcy (GM) or acquisition by a foreign company (Chrysler).

[8] For a serious exploration of these ideas in greater detail, see an essay by the serial entrepreneur, business historian, and author Gary Hoover (2016).

If the "trade" is computer programming, now mysteriously called *coding*, a trade school that delivers its teaching by computer can actually work. This model is already booming and disrupting university computer science programs. For example, Codecademy is a website that lets people learn coding for free.[9]

But in-person *programming bootcamps* are the real disruptors chipping away at four-year university computer science degree programs.[10] Effectively 15-week-long courses, these boot camps involve 10 hours a day of very intense programming instruction with mentors and coaches. Obviously they blend technology with aspects of traditional instruction. At the end of the 15 weeks, most students are at a level of proficiency and on a trajectory of self-education such that large corporations will employ them at middle-class wages. Not bad.

A tightly compressed variation on the programming boot camp is called a hackathon (see Figure 12.2). With 24 to 48 hours from start

**Figure 12.2 Junction Hackathon in Helsinki, Finland.**

*Source:* Vmuru, 2015, https://commons.wikimedia.org/wiki/File:Junction_2015.jpg. Licensed under CC BY-SA 4.0.

[9] www.codecademy.com.

[10] See, among other resources, www.switchup.org/rankings/best-coding-bootcamps.

to finish (and free of charge), participants gather to build out new and interesting ideas from scratch, then present them to their peers to win prizes. Under such time constraint and social pressure, sleep-deprived participants are forced to eliminate the nonessential elements of their ideas, and often enter "flow" states where they lose track of time and of their surroundings.[11] Outcomes range from the fun of learning-by-doing to starting up a new company.

Thus, we shouldn't give up on in-person education for cost reasons. According to a website that provides information on bootcamps, "the average full-time programming bootcamp in the United States costs $11,906, [and] can range from $9,000 to $21,000."[12] Peanuts when compared to a four-year college education. (We touched on the downside of missing out on a traditional college education earlier; if you can afford one, it's almost certainly a more attractive alternative.)

Recall that Matt Ridley said most innovation comes from ideas having sex. In actual sex, complex combinations are rarely productive, but in idea sex, they are. Waze, the navigation app, is a threesome: digitized street maps; location and speed information about the user's car, broadcast by the app; and social-network sourcing of traffic and road hazard information. A fully developed *hybrid* (online and in-person) university is a virtual orgy of general-purpose technologies, ranging from streaming video to the steam plant used to heat the gym—it is a city, a civilization in miniature. It may be the low-cost educational institution of the future.

## Are We Getting Stupider?

Looking back on the past accomplishments of high school and college students, it's hard to avoid the conclusion that we used to be smarter. However, the Flynn effect, a well-known phenomenon in psychological testing wherein average IQ scores *rise* significantly over time, suggests the opposite: we're getting smarter.[13]

---

[11] See Csikszentmihalyi (1990).

[12] Crispe (2019).

[13] Explanations for the Flynn effect vary. One possibility is that the largest factor has been an improvement in nutrition, eliminating much of the left tail of the roughly normal distribution of cognitive abilities and thereby increasing the average. Supporting this is the controversial but valid observation that, on average, tall people are smarter. The reasoning is that among the tall there are fewer who were nutritionally stunted and thereby cognitively stunted. So it's possible (I'd say likely) that we are achieving higher IQ scores by fulfilling a greater portion of our innate ability. See, among other works, Case and Paxson (2008).

I'd venture that neither is correct. As high school and college education has become radically more democratized in the last century or so (see Figure 12.3 for US data), *more and more people are getting a less and less rigorous education.* This produces both effects! The average college student today knows less than the average college student 100 or even 50 years ago, but these are averages across extremely different populations: we used to educate only the elite. (In 1900, the US *high school* graduation rate was 6%.) Yet the average person (not the average student) knows more, because the average person is now much more likely to have been a student.

And IQ tests, which purport to measure innate ability, don't really—or the Flynn effect would not be observed, because the truly innate part of intelligence is biological and evolves very slowly, not on a scale that we could observe within a century. Mostly IQ tests measure the ability to take IQ tests, which is correlated with one's knowledge base and life experience as well as innate intelligence.

Still, it's fun to look back on the educational accomplishments and standards of the past and see how we compare. It's not always pretty. Three examples follow.

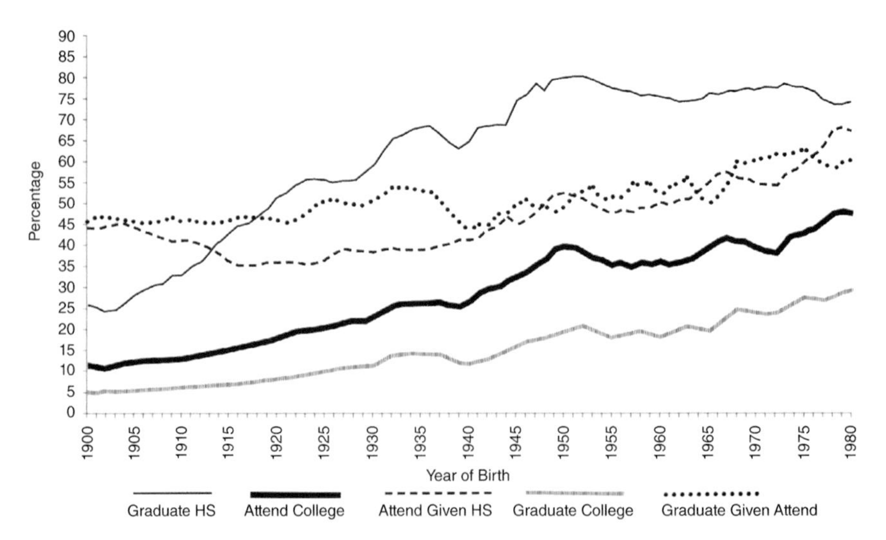


**Figure 12.3 US high school and college attendance and graduation rates for persons born between 1900 and 1980.**

*Source:* Adapted from Heckman and LaFontaine (2010), figure XIII.[14]

[14] Heckman, J. J. and LaFontaine, P. A. 2010. "The American High School Graduation Rate: Trends and Levels." *Review of Economics and Statistics* 92 (2, May): 244–262.

## Buddy Holly's Homework

It is the *lower* schooling, through the end of high school, that's the real problem: we waste the best years of our kids' lives. It was not always thus. An article in a long-past issue of *The New Yorker* described Buddy Holly's high school homework. He was a genius at writing deceptively simple melodies and love-struck lyrics, but was academically unexceptional—at an undistinguished high school in a forgettable town in western Texas.

Yet the work required of him would challenge many of today's college graduates; the star students could handle it nicely, but average ones would struggle. Sample question: "State briefly how the following contributed to the discovery of America. (a) the Renaissance; (b) the Crusades; (c) Marco Polo and other travelers; (d) the invention of printing; (e) the commercial revolution."[15] Imagine today's kids on *Watters' World*, who don't know who Franklin Roosevelt was or whether Africa is a country or a continent, tackling that! We have not advanced in every dimension.

## *Evangeline* versus Harrison High

Or compare a current-day middle school reading assignment with one from 1908.[16] Here's the current one: "Coach Jamison saw me in the hall and said he wanted to make sure I'm trying out for the track team!!!! Said my middle school gym teacher told him I was really good!!!! Then he said that with me on the Harrison High team we have a real shot at being county champs. Fantastic!!!!!!"[17]

All right, I'll stop. It's just too embarrassing. Let's see what kids of a comparable age were being asked to read in 1908:

> This is the forest primeval. The murmuring pines and the hemlocks,
> Bearded with moss, and in garments green, indistinct in the twilight,
> Stand like Druids of eld, with voices sad and prophetic,

---

[15] www.historyforsale.com/buddy-holly-autograph-manuscript-unsigned/dc157592. In addition, a blogger writes, "Buddy neglected his schoolwork through his junior year. [Nevertheless,] years later, when his tests and term themes were auctioned at Sotheby's, the *New Yorker* magazine reviewed his book report on Robert Frost and pronounced it—by homework standards—'a masterpiece.'" https://jake-weird.blogspot.com/2015/11/buddy-holly-rock-roll-soldier-influence.html.

[16] Holmquist (2016).

[17] Wortis (1991).

> Stand like harpers hoar, with beards that rest on their bosoms.
> Loud from its rocky caverns, the deep-voiced neighboring ocean
> Speaks, and in accents disconsolate answers the wail of the forest.[18]

After reading the opening lines of Longfellow's *Evangeline*, I feel like I've been on a wilderness adventure, in awe of glorious Nature. (I still don't know what "harpers hoar" is.) After reading the opening lines of *Nothing But the Truth*, by "Avi," I don't feel anything . . . except pity for the students who have to read it.

Neither choice is right for middle schoolers. As author Ann McMan says, Longfellow's *Evangeline* is "hot . . . a page turner," full of tragedy, lost love, adventure travel, coincidence, and reconciliation. It was a bestseller. It is also magnificently written, and a student who "gets" it absorbs a new way of hearing the English language.

But even McMan, a Longfellow admirer who teaches *Evangeline* to freshmen at Davidson College, remembers that in eighth grade, when she was required to read it, "I had to claw my way through [it]." She finishes her recollection with some expletives that I won't repeat. Why this hostility? She didn't have the maturity to understand it, and neither did I at the same age. But we both needed something much better than "my middle school gym teacher told him I was really good!!!!" Doesn't everyone?

## What a Young Woman and Man Should Know—1933 Edition

I could go on and on, citing as examples the recently publicized Harvard entrance exams of a century and a half ago,[19] and Willa Cather's fantastically difficult University of Nebraska (not Harvard!) education in the 1890s. My favorite is a 1933 take by Margaret Culkin Banning in *Harper's* on "What a Young Girl Should Know": "Write a Latin poem . . . Play golf and tennis . . . Dance well . . . Face the brutality of stag lines at parties" (love it!), "read intelligently such books as *The Imitation of Christ*, Emily Dickinson's *Poems* . . . and William Faulkner's *Light in August*." And change a tire. The author finally admits, "It sounds as if I got my ideas of parenthood from Simon Legree."[20]

---

[18] Longfellow (1856).

[19] http://bibletranslation.ws/down/Harvard_Entrance_Exam_1869.pdf.

[20] Banning (1933).

Wait, who's Simon Legree?

The companion article to Banning's, an essay by Robert Littell titled "What the Young Man Should Know," also published in *Harper's* in 1933, is even more frightening. It assumes that extensive academic knowledge will be acquired and focuses on life skills I can't even begin to imagine having. Whoever masters them, I want with me if I am ever stranded on a desert island—or faced with situations that only James Bond would enjoy.[21]

At any rate, all this ambition for one's kids was presumably aimed at the elite—maybe defined as those who could afford a subscription to *Harper's* in the depths of the Great Depression. That's why I focused on Buddy Holly's more modest, yet still creditable, academic achievements. But enough about the glories of the past; I recount them only to remind ourselves of what we are capable of, and that we should set our sights high. We need to think about what kinds of education people, worldwide, need in the future and how we can provide it.

## What Went Wrong?

What happened in the 60 years since Buddy Holly did his homework that makes his modest achievements so impressive by today's lights? One problem is that, with the greater prevalence of two-career couples and single parents, schools have been forced into *in loco parentis* roles that they're not prepared for and that overwhelm their resources, causing instruction to take a back seat to caregiving and discipline. Some blame can be laid at the feet of parents (both "uninvolved" and "helicopter"), rigid teachers' unions, and officious administrators; some, on low teacher salaries. (Great teachers are worth three times what they're paid; bad ones are worse than worthless and should pay reparations to the kids whose natural curiosity they've destroyed.)

But something also went right. As jobs have become more complex, educating the common man and woman at a more than basic level has become imperative. But an educational model designed for the intellectual elite probably needs to be modified for students of more limited ability or more earthy interests. Just giving them C's and telling them to go to community college isn't doing them any favors.

---

[21] Littell (1933).

Although the democratization of learning is an admirable goal, it has not been done well. What we did was design a program to teach a larger and larger number of young people how to be professors, but they don't want to be professors, aren't qualified to be professors, and there are no jobs for professors.

Why is *The Odyssey* taught in ninth grade? Because that is where a chronological study of literature starts, and is where you'd better start if you're preparing to be an English teacher. This is not a great reason. Ninth graders don't have the maturity to understand *The Odyssey*, and most of them will never teach English. We destroy their curiosity and natural desire, ability, and need to learn before they turn 14. The best time to "get" them is in middle school. Let's put some extra effort, and creative thinking, into that.

## Surprising Reservoirs of Knowledge

So far, we've reflected on the fact that we took an educational structure designed for the very capable and democratized it with the purest of intentions but not the best results. How can we do it better?

Let's start with a moral principle: that everyone, regardless of ability, deserves exposure to the best teachers, mentors, and guides available and appropriate to his or her level of talent. The Internet makes this possible through means we've already discussed. But there are surprising reservoirs of knowledge all over the place, many of whom are available and eager to serve in roles that fill the hunger for individual, in-person guidance as well as the need for best-in-class teaching delivered via the Internet.

### Expertise in Unexpected Places

Some of this expertise is hiding in unexpected places: branches of non-flagship state universities, religious colleges, high schools, technical schools, engineering departments of corporations, research institutes, trade associations, government agencies. And let's not forget the trades and crafts, where much of the world's work is done; the economist Robert Lerman has made a career out of promoting apprenticeship and explaining its value.[22] It's a crime to waste all these reservoirs of talent.

[22] Lerman (2016).

To access this widely scattered expertise, it will take professional guidance, dedicated effort, and the Internet. Young students cannot be expected to find it; that is what schools, teachers, and librarians should be for: ideally, they're *guides* to the knowledge that is out there. Adult learners might do better on their own, and we'll have more to say about them in a moment.

Let's continue inventorying the resources that are already available:

- The cost of most classics is zero: www.bartleby.com.
- Project Gutenberg, at www.gutenberg.org, "offers 57,000 free eBooks."
- Printed books from all over the world, in many languages, are available on Amazon and eBay at deeply discounted prices.
- Don't forget museums, historical societies, and independent research libraries.
- The cost of listening to a TED talk is zero. Some TED talkers are intellectual hucksters, but many are real experts on their fields, and almost all have been selected for their ability to communicate.
- If you're intellectually *really* ambitious, you might be excited that, as noted earlier, almost the entire MIT curriculum has been put online at https://ocw.mit.edu/index.htm.

## Asset Pricing for Wizards

In a parallel to the MIT effort, the celebrated University of Chicago finance professor John Cochrane engaged in what was then regarded as a quixotic deed: in 2014, he put his PhD-level Asset Pricing course on the Internet, using Coursera as the platform. On its face, the course sounds appealing. Wouldn't you like to know how to determine the value of assets? Maybe you could make some money in the stock market or in real estate.

But, no, Cochrane's course uses not just ordinary, college-level math such as calculus but also "stochastic calculus," which involves "Ito's Lemma" and related math from the physics of Brownian motion. The reason for all this higher math, much more difficult than simple calculus, is to model financial assets in continuous time. This daunting content shows that a Web-based course can be quite narrowcast and still get traction: Asset Pricing attracted 37,000 registrants in the first year; 4,000 of them finished watching the videos "and about 250 completed all the homeworks [sic] and final exam."[23]

---

[23] Nguyen (2014). At the time the course was first put online, Cochrane was at the University of Chicago, but he is now at Stanford.

Is this a terrible outcome? No. The number of people worldwide who are both interested in a PhD-level mathematical course on asset pricing and qualified to complete it might not be a large multiple of 250. Getting them all to the University of Chicago would be a logistical nightmare, and the school charges a bundle. Cochrane's online course is free. The students missed the experience of questioning him after class or helping him do research, but he answers his e-mails (I have several from him), so the whole of the personal aspect is not lost.

## Self-education: Chautauquas Then and Now

Self-education has always been a force. And autodidacts, those who are primarily self-educated, are among the most interesting and creative people (although the gaps in their formal knowledge are often amusing). Books on how to learn, from Mortimer Adler and Charles Van Doren's *How to Read a Book* (1940)[24] to Gary Hoover's less well-known but extremely valuable (and up-to-date) *Lifetime Learner's Guide to Reading and Learning* (2017)[25] are essential resources for both autodidacts and conventionally educated readers.

From the 1870s to the 1920s, adult self-education was a massive craze, called the Chautauqua movement after the summer camp in western New York state where the first and largest sessions were held. It was a mix of a traveling university and a carnival. One writer notes, "During this time, hundreds of touring Chautauquas presented lectures, dance, music, drama, and other forms of 'cultural enrichment.' In rural America, big tents served as temporary theaters for these productions. Lectures by author Mark Twain, suffragette Susan B. Anthony, or a production of *A Tale of Two Cities* are the kinds of entertainment one could expect. . . . Teddy Roosevelt once called the traveling Chautauquas 'the most American thing in America.'"[26] Figure 12.4 shows the enthusiasm that attended such meetings.

Note that these were *rural* Americans flocking to the gatherings. Rural life has become an object of pity or derision among some people, but an hour of listening to classic country, bluegrass, or old-time music might convince you that it shouldn't be. People from all walks of life are not only capable of learning but interested in it.

[24] Adler (1940).
[25] Hoover (2017).
[26] CredeCalhoun (2018). Some spelling errors in the original are corrected here.

**Figure 12.4 Telling the Red Cross Health Story from the Chautauqua Platform, late nineteenth century.**

*Source:* American National Red Cross photograph collection (Library of Congress). https://lccn.loc.gov/2017670244.

Then why don't we have Chautauquas anymore? One reason is that public higher education has spread far and wide, so one does not have to travel far (by horse) to get some intellectual stimulation. Another is that we have television, radio, and the Internet. Finally, almost everyone goes to high school, a big change from the 1870s.

But maybe we could use a new Chautauqua movement. If you don't mind doing without the social aspect, TED talks and similar online services are the new Chautauquas. But the social aspect is important. It is where the free exchange of ideas takes place most fluidly (have you ever gotten frustrated arguing with a stranger on the Internet?), but more importantly it is fun. And our hard-working, wealthy society is short on fun.

What we do have is meetups. They are forums, organized online but conducted in person, for discussing every topic imaginable. Combined with online interaction, MOOCs, digitized free books, and all the other resources we've mentioned, we could be on the path to a modern version of a Chautauqua movement. It will take some social (and business) entrepreneurs to get it right, but it's a promising avenue for those who would further democratize education and make it as much fun as it appeared to be in the 1870s.

## "Terence, This Is Stupid Stuff . . . ": A Personal Recollection

The history of schooling includes much that is terrible (the cruel treatment of boys in British private schools) and much that is admirable (the remarkable stock of knowledge the boys got). Let's make the most of this past.

I didn't get the abuse, thank God, but I got the drilling—I can recite A. E. Housman's "Terence, this is stupid stuff" from memory. I'm not sure why. Maybe, as a kid, I just identified with its theme: to experience the best in life, prepare for the worst. "Mithridates, he died old," the poem concludes, because the ancient king steeled himself with small doses of poison in preparation for his enemies trying to kill him.[27] (The rest of the poem is not that grim—it's wryly humorous.)

Housman's late-Victorian rhythm has been drummed into my head just as surely as that of Beethoven or the Rolling Stones. To what use? Through a very indirect path, it helps me communicate, just as knowledge of *Alice in Wonderland* or the poetry of Bob Dylan or the rhetoric of Lincoln or Churchill does. You can't have a civilization without some kind of collective memory.

More importantly, my study of Housman forces me to underlie the seemingly boundless optimism of this book with an appreciation, if not full acceptance, of the tragic vision. "Terence" (Housman):

> Therefore, since the world has still
> Much good, but much less good than ill . . .
> I'd face it as a wise man would,
> And train for ill and not for good.

If you prepare for the worst, every surprise will be pleasant. It really will. Try it.

## "A Higher Plane of Life"

I learned about "Terence" at Hawken, then a boys' school, now co-ed, in Cleveland. It was not designed for geniuses, but for motivated learners who were well above average—the typical IQ of a Hawken student was

[27] Housman (1896). Full text online at http://ceadserv1.nku.edu/longa/poems/housman9.html. Interestingly, Housman, a genius at his avocation (poetry), was not one at his vocation (classical scholarship). He was ordinary–a rank-and-file academic. By rights he should have just been a poet. But poetry does not pay the bills.

probably that of an average student at a good state university. Hawken aimed high. It emphasized character building—"the whole boy"—as well as outdoor life, individualized attention, and academic rigor and seriousness. In each classroom were two mottoes: "Fair Play"; and a favorite motto of the school's founder, James Hawken, who was 27 years old (see Figure 12.5) when he established the institution in 1915, "That the better self shall prevail, and each generation introduce its successor to a higher plane of life."[28]

**Figure 12.5 James A. Hawken, about 1915, with his school's two bywords.**

*Source: Hawken Review*, Winter 2015. https://issuu.com/hawkenschool/docs/review_w15_nonotes_web/24. Reproduced with permission.

---

[28] The quote itself is from Spalding, John Lancaster. 1890. *Education and the Higher Life*. Chicago: A. C. McClurg & Co. Spalding, a Catholic priest, was the founder of the Catholic University of America in Washington, DC. Spalding's book is not very good, but I'm glad that Hawken, a lapsed Catholic and former seminarian who had been excommunicated for questioning the infallibility of the pope, reached back and found inspiration in it.

In the century since Hawken was founded, it has, of course, adopted technology wholeheartedly, as one might expect of a forward-thinking school rooted in tradition. But it's revealing that the school's website, while acknowledging the need to be "forward-focused" and "prepare ... students to navigate a complex and dynamic world," reaches back to place special emphasis on its founding values. "We pay purposeful attention," it says, "to the development of character through the cultivation of virtue, helping students understand Hawken's motto of 'fair play,' and its founder's call to find their 'better selves.'"[29]

Character, virtue, fair play, better selves. Just as at any other time in history, young people who absorb these values inherited from the past will face the challenges of the future with a preparedness that few have—and everyone needs.

## Back to the Future[30]

To rebuild the human capital base in the United States, at least among young people left out of the professional rat race, we may need to reinstitute programs like Franklin Roosevelt's Works Progress Administration (WPA) and Civilian Conservation Corps (CCC). These agencies provided desperately needed paychecks and sometimes (not always) accomplished valuable tasks, but they did something else too. They gave the participants the gift of responsibility.

Instead of teaching people that their "job" is to figure out how to qualify for a handout, these programs taught the unemployed to *show up*, roughly on time and in the right place, to dress appropriately for the job, and to treat the boss with a modicum of deference and respect—even if the work itself was menial and poorly paid. So, when employment demand finally returned around 1940, WPA and CCC "graduates" were prepared for the world of private-sector work—or for the military's sudden need for manpower.[31] Today's unemployed usually are not. This is an easy fix.

---

[29] www.hawken.edu/about-hawken/purpose-promise-principles.

[30] I thank John O'Brien of the Berkeley Haas School of Business for putting these thoughts, which I've had independently, into a clear and compelling form.

[31] The discipline and skills training instilled by military service itself was another factor supporting the postwar boom.

## Conclusion

- Use technology to bring the best teachers to anyone who wants them.
- Combine online lecturing with in-person instruction; high touch.
- Keep costs low.
- Personalize, but put great effort into finding common ground and building a common culture.
- Be innovative; there is no clearly winning model yet.
- People have differing abilities, but set your standards high.
- Respect the past; there is much to admire in old models of education.
- In terms of delivering education, people in other countries who do not speak English are just as important as people in the United States who do.
- Teach *knowledge*, not just how to find it, so that adults will have a common intellectual language and not rely on Google and Wikipedia as "hive minds." The individual *human mind* is what you are trying to develop.

So, "introduce your successors to a higher plane of life." And tell them to leave their computers and smartphones on.

### Can the Way We Teach the Brightest Students Be a Model for Everyone?

*By David L. Stanwick*[32]

Of course it can. If there exists a teacher with the best-structured fourth-grade geography lesson, such that the student learning and absorption rates are measurably greater than in the average course, it is irrational, given the alternative, to require the bored and stressed teacher of ordinary ability to teach that course. Video lecture courses, Khan Academy, and Duolingo are improving their layering of mini-lessons so as to maximize the absorption rate; why not use resources like these maximally? In the future, students may have individual accounts across multiple learning tools with different tracks so their

*(continued)*

[32] David Stanwick is an entrepreneur, writer, and chairman of the American Business History Center (https://americanbusinesshistory.org). He was the research assistant for this book.

*(continued)*

lesson plans can be tailored for their learning styles—visual learners, auditory learners, and so forth. The object is to get every student drinking from the deepest waters of knowledge.

There's an interesting school that began in Austin, Texas, called Acton Academy. Referring to themselves as one-room schoolhouses for the twenty-first century, Acton Academy schools serve first grade through high school, borrowing from Montessori techniques while also incorporating the Socratic method and latest technologies. With an annual tuition ($10,230) just under the overall US national average for private schools and one-third the typical elite-school tuition, Acton is intent on developing the curious, independent, lifelong learner; the teachers are "guides" in that they are trained to answer questions with questions, guide students toward resources that might provide answers, and help the student to think on her own. The students teach one another; each has an accountability partner, whom they meet with and at regular intervals set goals, hold the other accountable for said goals, and encourage each other.

There's a strong entrepreneurial side to Acton, with emphasis on communication and making the most of the real world and its resources. By the time its students are 16, they are often competitive with graduate students in real world understanding and experience. From what I hear, the students have difficulty interacting with typical public school students, having little understanding of sitting at a desk 8 hours a day for 12 years while being tack-hammered with the word *no*. These students are more interested in interacting with adults and experiencing the world.[33] Many of the brighter students move directly into the business milieu, through an internship, employment, or involvement with a startup, short-circuiting the struggle that so many young people experience in transitioning to adult life.

To sum up, there are many outstanding educational models being developed. Although schools like Acton are targeted to higher-ability students and are thus not for everyone, a shift in how the best students learn will have nudging effects on the rest. Democratization should never mean dumbing down.

[33] www.actonacademy.org/ and www.childrensbusinessfair.org/what_is_acton.

# 13

# *Conflict, Safety, and Freedom*

In his landmark book, *The Better Angels of Our Nature*, the Harvard psychologist and linguist Steven Pinker argues that we are living in the most peaceful and safest time in human history. The claim is contrary to intuition but he's right, and I'll summarize the data that he and others have used to demonstrate that fact.[1]

We also live in a time of unprecedented economic freedom, a proposition that may be easier to accept than the decline in violence. Most of human history has been characterized by tyranny, widespread desperate poverty, and slavery. Thankfully, we now think of these evils as aberrations, not the normal state of things, although a slim majority of the world's people still live in countries that are politically unfree or only partly free.

This chapter links these concepts, wrapping them into a package that I'll call *autonomy*. The word connotes the uniqueness and dignity of the individual and the notion of self-ownership. It is what we strive for when trying to ensure the absence of conflict, the safety of the person, and the ability to live and transact freely.

---

[1] Since this book is also a reader's guide, I'll venture the opinion that *The Better Angels of Our Nature* is better than Pinker's more recent [2017] *Enlightenment Now* because it covers less material in greater detail. The *Better Angels* title, which evolved from Shakespeare through Dickens and William Seward to its full flower in Lincoln's Second Inaugural, is also immeasurably better. Pinker's 2011 *Enlightenment Now* was referenced in Chapter 10.

Necessarily, our treatment of each major subtopic has to be brief. To make the effort manageable, I'll separate the issues (although they are interconnected) as follows:

- *Conflict:* both war (state-sponsored conflict) and enmity between individuals (best revealed by crime statistics);
- *Safety:* Avoiding natural disasters and preventable injury and death, such as by fire, workplace injury, and transportation accidents;
- *Freedom:* individual rights—both the political kind, as represented by electoral democracy and freedom of speech and the press; and the economic kind, the ability to choose or quit one's job, or start, operate, and sell a business without undue interference.

**The Brotherhood of All Races of Man Is a More Recent Discovery Than You'd Think**

The work of the British science-fiction author H. G. Wells reminds us of that fact. He's hard to admire, other than for his long-lasting dalliance with the luminous writer Rebecca West.

In *The God Delusion*, Richard Dawkins quotes Wells's *Anticipations*: "And how will the New Republic treat the inferior races? How will it deal with the black? . . . the yellow man? . . . the Jew? . . . those swarms of black, and brown, and dirty-white, and yellow people, who do not come into the new needs of efficiency? Well, the world is a world, and not a charitable institution, and I take it they will have to go."

Go where? His wish sounds like a call for genocide.

I've always regarded Wells as a sort of proto-Nazi. My daughter, an astute reader but by no means a scholar of English literature, points out that, at least in the movie version of Wells's *Time Machine*, the Eloi resemble a degenerate version of the Master Race and the Morlocks look like a horror-movie caricature of the people the Nazis sought to enslave (see Figure 13.1). (In the story, the two races have been evolving apart for 800,000 years, hence the exaggerated differences.)

Wells wrote *The Time Machine* in 1895, long before the beginning of the Nazi movement. If a progressive gentleman like Wells, widely admired, could express genocidal thoughts openly, then maybe we shouldn't be so surprised that Nazism caught on. Today, we look back on those attitudes with horror. In another century, people may regard some of our own practices as barbaric. Moral progress is an enigma: every society thinks it is basically a good one, yet its successors often regard the past as benighted. We will just have to wait and see which attitudes change profoundly over the next generations.

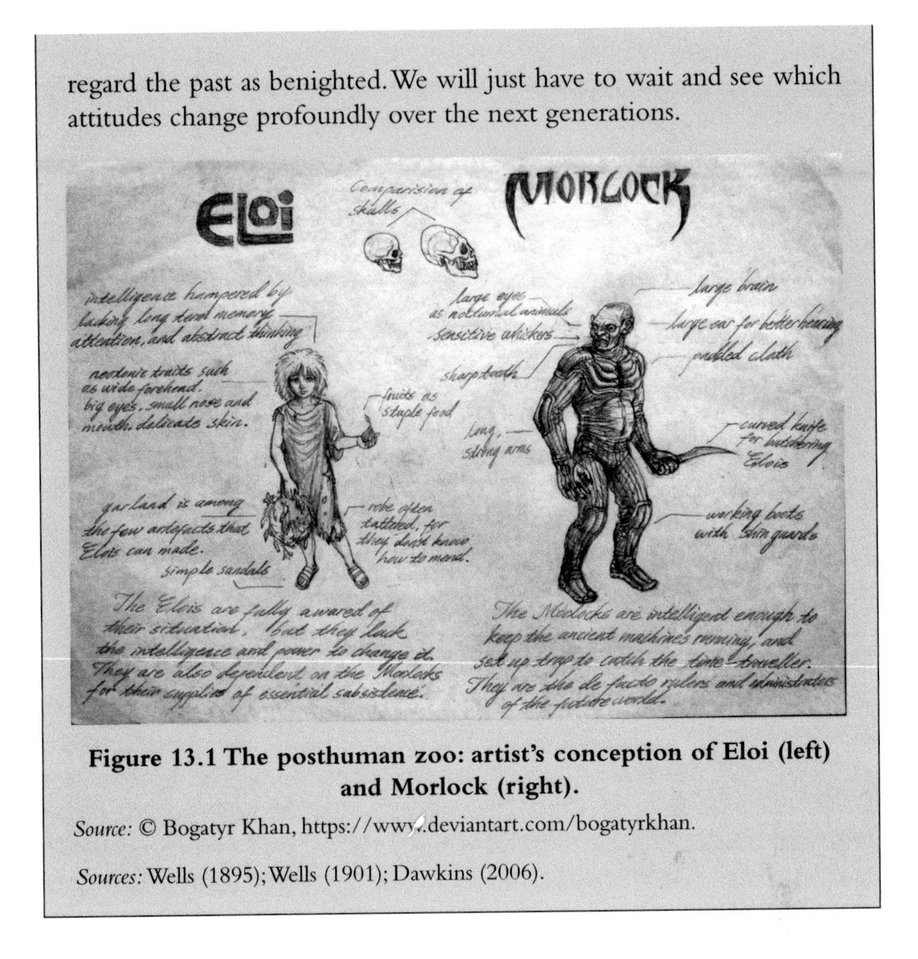


**Figure 13.1 The posthuman zoo: artist's conception of Eloi (left) and Morlock (right).**

*Source:* © Bogatyr Khan, https://www.deviantart.com/bogatyrkhan.

*Sources:* Wells (1895); Wells (1901); Dawkins (2006).

## Conflict: The Decline of State-sponsored Violence

We begin with the phenomenon that Steven Pinker has so effectively brought to our attention: the decline in war over the centuries, from what Thomas Hobbes called "the war of all against all" to a much more restrained use of state-sponsored violence in pursuit of policy aims.

### The History of War in Europe

The history of Europe, where most of the accurate data on war deaths originate, was absurdly bloody in ancient, medieval, and early modern times. Figure 13.2 is the great artist Francisco Goya's rendition of the battle of Madrid in the Peninsular War, which formed part of the larger Napoleonic Wars.

Note that Goya's spectacular battle scene takes place in one of the world's most sophisticated cities, not all that long ago. It was 1808, the age of Jefferson in the United States and Goethe in Germany.

**Figure 13.2** ***The Second of May 1808*****, by Francisco Goya.**
*Source:* https://upload.wikimedia.org/wikipedia/commons/7/76/El_dos_de_mayo_de_1808_en_Madrid.jpg. Public domain.

Despite a relatively peaceful eighteenth century, then, war did not consistently yield to peace until the Napoleonic Wars ended about 200 years ago. The first Long Peace in Europe, that of the Congress System established at the Congress of Vienna, lasted from 1815 to 1914, nearly a century, and was only punctuated by the brief Franco-Prussian war in 1870. The period from 1870 to 1914 includes of some of the best years in Europe's long history.

What was to come next could not be foreseen by anyone. It took what Pinker calls a *hemoclysm*, an orgy of bloodshed, between 1914 and 1945 to usher in a second Long Peace. The Hemoclysm includes both World Wars, the Holocaust, and the Russian Revolution and its aftermath. The second Long Peace persists to this day,[2] but of course we don't know how long it will last, since wars arise at irregular and unpredictable intervals. The modern history of Europe makes one wonder about the claim that the world has consistently become more peaceable over the centuries.

[2] At least in Europe, the Americas, and Japan. Some of India's and China's bloodiest years occurred after World War II had ended.

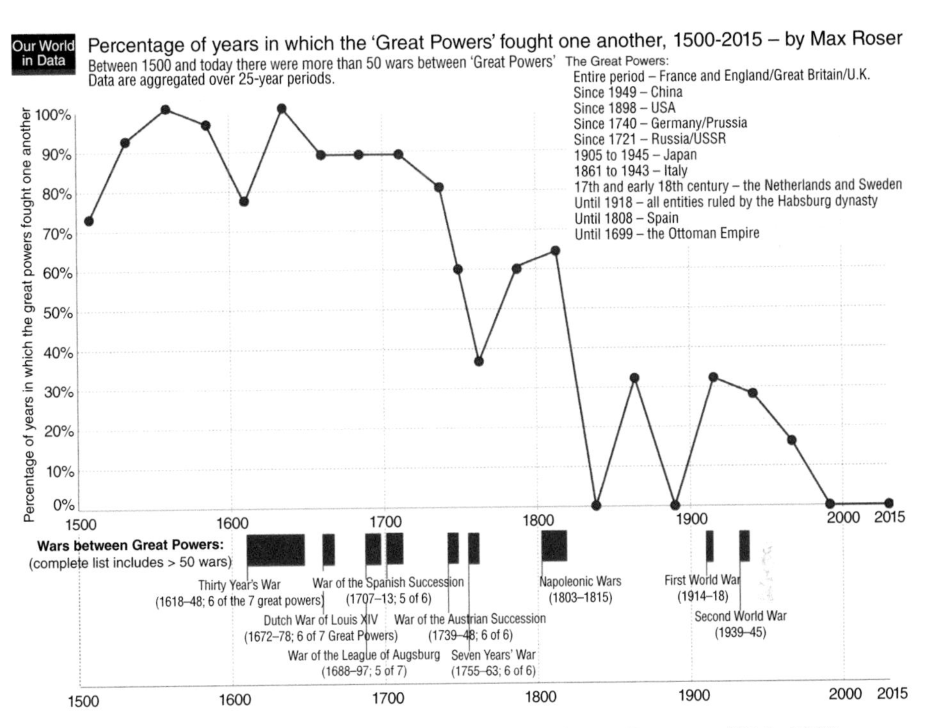


**Figure 13.3 Frequency of wars between Great Powers, 1500–2000.**

*Source:* https://ourworldindata.org/wp-content/uploads/2013/08/ourworldindata_percentage-of-years-in-which-the-great-powers-fought-one-another-1500–2000.png, based on Pinker (2011), figure 5-12 on p. 224. Licensed under CC BY-SA 3.0.

Pinker has studied this question in depth and, oddly, has come up with two conflicting answers. That Great-Power wars have become rarer is beyond dispute. Figure 13.3, from *Better Angels*, shows the frequency of wars between the Great Powers over a 500-year time span.

## Has State-sponsored Violence Really Declined?

But, while Pinker's broader thesis about the reduction in violence is well-supported, extremely bloody wars have continued almost up to the present (1945, or a little later if you include the Chinese and Indian civil wars). This fact gives the impression that the world is becoming more violent when it isn't, at least when war-related deaths are scaled relative to the (growing) world population. Figure 13.4 shows death rates per 100,000 people from selected major conflicts before the year 1400, and from all known conflicts thereafter. The Chinese War of the Three Kingdoms (220–280 AD) was gigantic, as were two of the other pre-1400 conflicts; the log scale suppresses their size, but it suffices to say that the Three Kingdoms war killed an unbelievable one-fifth of the world's population.

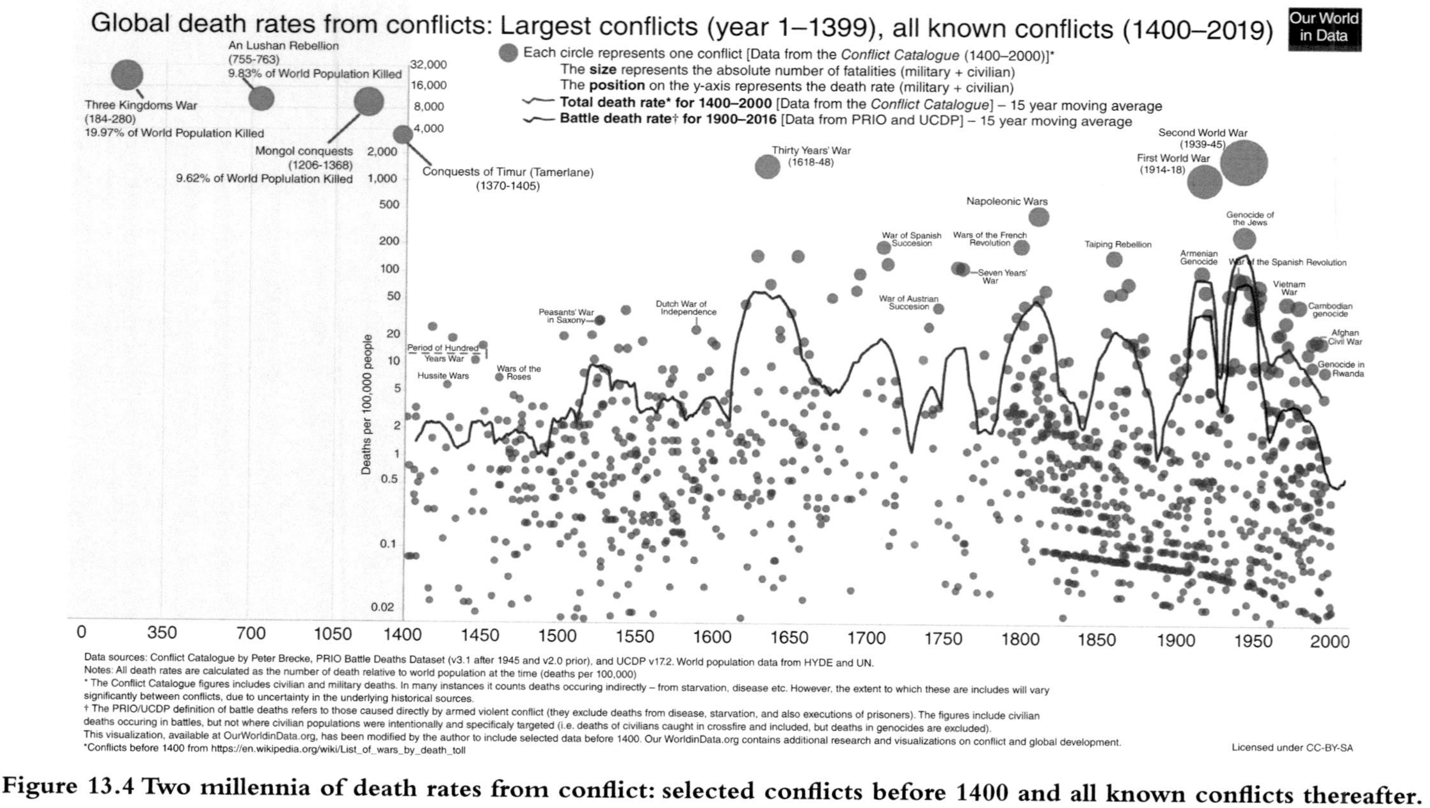


**Figure 13.4 Two millennia of death rates from conflict: selected conflicts before 1400 and all known conflicts thereafter.**

*Source:* Our World in Data, https://ourworldindata.org/uploads/2018/09/Bubble-and-lines-FINAL-03.png (data and graph after 1400); Wikipedia, https://en.wikipedia.org/wiki/List_of_wars_by_death_toll (data before 1400 for four selected conflicts, graph constructed by the author). Licensed under CC BY-SA 3.0.

After 1400, the death rate held fairly steady at a very high level until the European Long Peace began in 1815. But the Long Peace that we're in now, ushered in by the international agreements crafted after World War II and by nuclear deterrence, is more peaceful and more widely spread across the globe than the last one. I hope it turns out also to be longer in duration; we won't know in our lifetimes.

While it's still not a safe world, Figure 13.4 vividly illustrates just how terrible the distant past was.

I'll admit that it's hard to tease out a large and enduring decline in violence from the visual impression given by Figure 13.4. You'll see it more clearly if you consider World Wars I and II—the Hemoclysm—to be outliers, and block them from view with your finger. That's cheating, but you'll see that the underlying trend is downward and that the trend after 1945 is very dramatically downward.

## Understanding the Hemoclysm

The Hemoclysm was so awful that we cannot simply gloss over it. It must be addressed, either as an extreme aberration, a statistical outlier punctuating a long gradual decline in state-sponsored violence, or as evidence that Pinker's theory is wrong. John Gray (the British religious scholar, not the American preacher or the relationship counselor) is one of the more eloquent objectors to Pinker's optimism. In a lengthy column in *The Guardian*, Gray writes:

> While knowledge and invention may grow cumulatively and at an accelerating rate, advances in ethics and politics are erratic, discontinuous and easily lost. Amid the general drift, cycles can be discerned: peace and freedom alternate with war and tyranny, eras of increasing wealth with periods of economic collapse. Instead of becoming ever stronger and more widely spread, civilisation remains inherently fragile and regularly succumbs to barbarism.[3]

I do not think that Pinker's theory is wrong: today we abhor practices that even our relatively recent forebears accepted as normal (or a necessary evil). Slavery is only one of many examples. The reality that slavery has not been completely abolished everywhere in the world does not negate the fact that it is now almost universally considered evil. That was not always the case: even in the United States, less than two centuries ago, it was widely regarded as a force for good (because, some people

[3] Gray (2015).

thought, Africans in America could not take care of themselves). There is such a thing as moral progress.

Yet I agree with John Gray that we must always be vigilant lest we retreat into barbarism in pursuit of some goal that, at the time, a large group of people thinks is worthy. Communism, Nazism, and other forms of totalitarian control are too recent in human history for us to regard them as discarded relics.

The revival of politicized religious passions is another concern. And, closer to home, in the words of the investment manager and writer Michael Edesess,[4] "An extreme, utopian political and economic philosophy has been embraced in America by a network of highly influential and well-placed apostles. . . . [T]his new extreme philosophy threatens to bring about the collapse of American capitalism," the system that has brought us much of the betterment discussed in this book. The collapse of capitalism is unlikely, but we should be on guard against it.

## Why Has Conflict Abated?

The traditional answer to the question of why state-sponsored conflict has declined, at least after World War II, is that nations have consciously created a complex web of interconnections that strive to prevent war. This web includes treaties, a body of international law, and an alphabet soup of supranational organizations that promote peace and understanding and take positive steps to diffuse conflict.

A second reason is trade. It is not good business to blow up your customers—or suppliers. The first Long Peace in Europe reflected the tremendous growth of international trade over that period. The postwar *Pax Americana*, from 1945 to the present, has also been mostly peaceful (a few devastating local wars notwithstanding) for the same reason. Remarkably, the Cold War did not become a hot war, except in Korea and Vietnam—two relatively small proxies for the main belligerents—and the underlying conflict between the United States and the Soviet Union ended without a shot.

## Capitalism and the Decline of Conquest

I'd like to propose a third reason. In traditional or noncapitalist societies, the easiest and sometimes the only way a leader could raise the standard of living of his people was by conquest—to force subject peoples to pay

[4] Edesess (2019).

tribute to the conqueror. (Until a couple of centuries ago, and more recently in a few places, the aim of war was also to capture slaves.) In modern times, tribute evolved into taxation, which enriched not just the rulers of conquering powers but also their subjects; the British Empire, which survived until 1945, is an example.[5]

With capitalism, however, internal or "organic" growth became an easier way to raise the standard of living. Why amass armies and seek plunder—and risk defeat—if you can build factories, railroads, automobiles, farm equipment, and public utilities, and trade for goods you can't manufacture at home?

Maybe because, if you were a country's leader, war was easier than the grand civilian construction project I've just described. But, in a democracy, unpopular wars result in unpopular, and eventually unseated, leaders. Thus, we find a connection between rising democracy and declining conflict.

Moreover, with massive international trade and travel, and a new age of migration, perceived differences between peoples have diminished. Thankfully, it is more difficult than it used to be to rally a country to go to war against a supposed enemy whose people are familiar, who work with us on a daily basis, who are our suppliers or loyal customers in trade,[6] and whose countries we have visited with enjoyment and enthusiasm.

## Living in a "Peace of Westphalia Prison"

Many of those who are left behind, to endure lives continually threatened by state-sponsored violence and deprivation, reside in what the investment manager (and my frequent co-author) Stephen Sexauer calls a "Peace of Westphalia prison." The 1648 agreement, ending three decades of brutal conflict between Protestants and Catholics in Germany, set a precedent of nonintervention in other countries' affairs and established the permanence of internationally agreed-upon borders, unless a rogue state becomes a threat to other countries.

This precedent, which still holds and is the basis for much of international law, is the reason Venezuela is (as of this writing) ruled by a butcher who starves his people, while its neighbors, Colombia

[5]Tax revenue was not the only benefit that Britain received from its colonies, and (in Britain's case but not in all empires) the colonies also received many benefits including public works, British law and administration, and trade. But it is fair to say that taxation of the colonized was a major reason for empire building until very recent times.

[6]The UK and Germany were each other's largest trading partners before World War I (see https://unstats.un.org/unsd/trade/imts/Historical%20data%201900-1960.pdf) and their rulers, George V and Kaiser Wilhelm, were first cousins (both grandsons of Queen Victoria)!

and Brazil, mostly prosper. Venezuela poses no military threat to any country, so no one can invade it and end the humanitarian catastrophe. George W. Bush had to find a dubious reason—"weapons of mass destruction"[7]—for justifying regime change in Iraq; his administration's intervention was still widely condemned. The Rwandan genocide of 1994, not so much a civil war as a local hemoclysm, could have easily been stopped by an international military effort but was not, because it was regarded as a domestic matter, the Rwandans' own problem.[8] The Chinese could end the misery of North Korea almost instantly.

This can be changed. The decline of violence will accelerate if international agreements are modified to allow policing by organizations with earned legitimacy. The United Nations played this role for a while; a contemplated League of Democracies might do so in the future. If I fall in a lake or drive off the road into a canyon, I expect someone to try to rescue me. If I have the rotten luck to be born into an autocracy so malign that normal life is not possible, I, too, want to be rescued. No one should live in a Peace of Westphalia prison.

## Remarkably, Homicide Is a Bigger Killer Than War

Surprisingly, Pinker tells us, homicide is a much bigger problem than war, and has been through most of history. As an extreme example, 63,880 lives were lost to homicide in Brazil in 2017, and none in international wars involving that country.

Despite Brazil's ongoing violence, the worldwide decline in crime over the long run has been much more dramatic and clear-cut than the decline in state-sponsored violence. The idea that violent crime is in a long-term decline makes for a difficult conversation, especially since we've just been through a 30-year crime wave in the United States that is finally abating but that dominates most people's memories. The crime wave produced a decline in social trust that is costly and disruptive to living as one pleases. Crime has also eased in Europe from the peak levels of late in the last century, despite what you may be reading in the paper.

So let's look at the very long-term data favored by Pinker (and me). Figure 13.5 shows homicide rates in England from medieval times to

[7] I still chuckle at William F. Buckley's comment that a regime containing characters called Chemical Ali and Dr. Germ probably had weapons of mass destruction. I don't know whether it's true.

[8] An exception is that NATO, including the United States, did intervene in the violent disassembly of Yugoslavia into its constituent states. Maybe the bloodshed was just a little too close to home for its peace-loving neighbors, which include Italy and Austria.

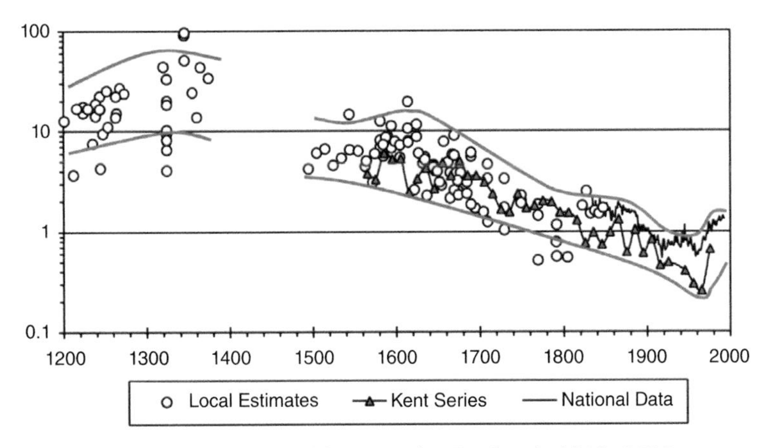


**Figure 13.5 Homicide rates in England, 1200–2000.**

*Source:* Eisner, Manuel. 2003. *Long-Term Historical Trends in Violent Crime.* www.researchgate.net/publication/279936907_Long-Term_Historical_Trends_in_Violent_Crime/figures, figure 3. Reproduced with permission.

the year 2000. Note that the *y*-axis is a base 10 log scale, so the highest homicide rate, a "local estimate" from the 1300s, is about 400 times the lowest rate, recorded around 1960. No wonder Shakespeare wrote so many murders into his plays!

Manuel Eisner, the author of the study from which Figure 13.5 is taken, found similar results for Belgium, the Netherlands, and Sweden. The sharp spike up, near the end of the time line, is the crime wave we just mentioned, now abating.

In the United States, where, as noted earlier, we are just coming off the peak of a seemingly unprecedented crime wave, violent death used to be quite commonplace, although not on the scale of medieval England, except right at the beginning of our own colonial period. Figures 13.6 and 13.7 show the homicide rate in various parts of the United States from colonial times to the present. (Homicide is the best indicator of the general level of violent crime because reporting rates are very high and the definition of homicide does not change much over time.)

Homicide rates were as sky-high in early colonial New England as in England of the 1300s. (I do not believe the numbers include casualties from settler-native conflict.) The situation improved dramatically over the 1600s and rates stayed very low, except in the biggest cities, through 1900, after which Figure 13.7 takes over. (The two figures are not directly comparable: New England, especially rural Vermont and New Hampshire, was probably more peaceable than either the big cities or the Wild West.)

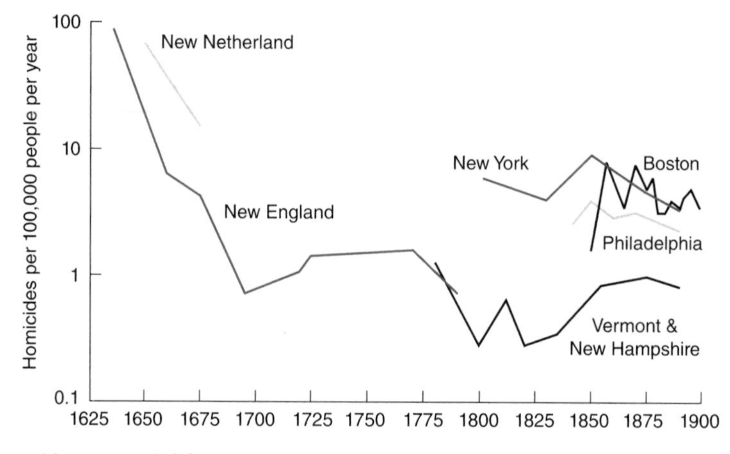


**Figure 13.6 Homicide rates in various parts of the eastern United States from colonial times to 1900.**

*Source:* Our World in Data cites Pinker (2011). The most important source of Pinker and the source for this is Roth (2009). https://ourworldindata.org/wp-content/uploads/2013/03/homicide-rates-in-the-northeastern-united-states-1636-1900-pinker-2011-jpg.jpg. Licensed under CC BY.

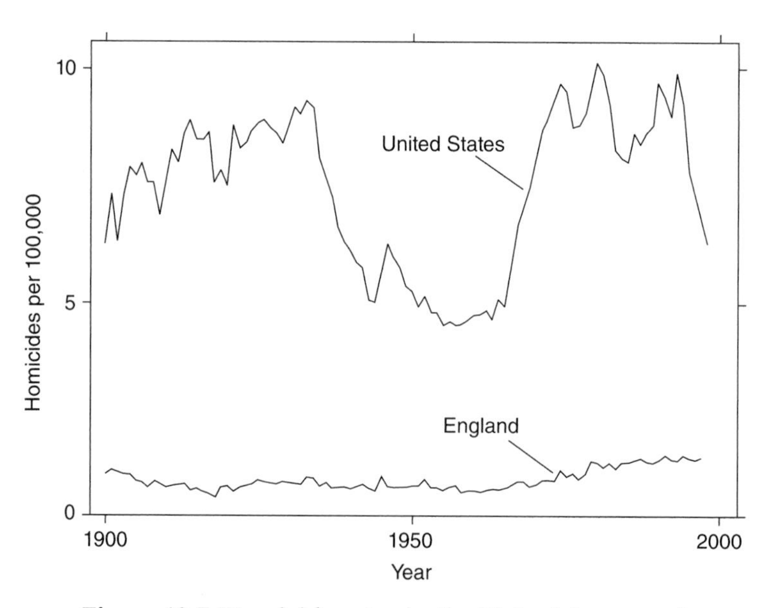


**Figure 13.7 Homicide rates in the United States and England, 1900–2000.**

*Source:* Our World in Data specifies Pinker (2011). Pinker's source is Monkkonen (2001). https://ourworldindata.org/wp-content/uploads/2013/03/homicide-rates-in-the-united-states-and-england-1900-2000-pinker-2011-jpg.jpg. Licensed under CC BY.

We don't have the data needed to properly link the two graphs, but violent crime in the early 1900s in the whole of the United States, shown in Figure 13.7, was as bad as in the biggest eastern cities in the decades immediately prior. Moreover, homicide rates continued to rise until the mid-1930s, then fell drastically to what we sometimes recall as the country's golden age from the 1940s through the early 1960s. (This perspective could not be more wrong in terms of economic prosperity and race relations but it has some validity when referring to crime rates.)

Given the sharp declines in homicide in the developed world over the centuries, and in the United States in the middle twentieth century, the crime wave from the mid-1960s to the mid-1990s was just startlingly awful. In the mid-1980s, black males, many of them not men but boys, were being killed, mostly by other black males, at the horrific rate of 72 per 100,000 per year, almost 70 times the murder rate in England (the lower line in Figure 13.7) and more than double Brazil's current rate. The reasons for the wave and its eventual abatement are still widely debated and are beyond the scope of this book; see Pinker's fascinating subsection, "Decivilization in the 1960s," in his *Better Angels* for a discussion of the various explanations.

Violent crime has certainly not been eliminated (665 people were murdered in Chicago in 2017, almost all of them in two areas of the city), but it's dramatically better. Crime rates have continued to decline after the end of the period covered in Figure 13.7 and, in 2010 to 2014, were comparable to rates in 1950 and 1960. Homicides have started to creep up again, but not at the alarming speed of two generations ago.

## Safety

The biggest danger most people face is from ill health. In this book, however, we separate that out, having covered it in Chapter 9, "Health and Longevity," and now we will focus on dangers from nature, work, transportation, and other "safety hazards." Life has become immeasurably safer over time, with wealth and technology protecting people from hazards that were common causes of death, disability, and discomfort in ages past.

### Nature

Depending on one's culture and location, we don't have to go back too far to find people worrying about being *eaten by animals*. The Brothers Grimm (born in 1785 and 1786 in comfortable circumstances) found an

**Figure 13.8 The personal war with nature: Little Red Riding Hood in peril.**

*Source:* "Red Riding Hood Meets Old Father Wolf." From Gustave Dore's 1864 edition of *Histoires ou contes du temps passé,* a collection of literary fairy tales written by Charles Perrault, originally published in 1697. https://en.wikipedia.org/wiki/File:Dore_ridinghood.jpg. Public domain.

enthusiastic audience by telling unbelievably violent tales of encounters with wolves, bears, witches, and human murderers. The medieval folktales from which the Grimms drew their storylines resonated with the people of Germany and (in translation) with many other Europeans and Americans, who were not all that far removed from a state of personal war with nature.

Today we revere nature, a development that coincided with the move to cities. Separated from nature, we crave some connection with it, albeit a sanitized one. John Muir is the man we associate most with restoring that connection in the United States: the National Park Service, called by filmmaker Ken Burns "America's best idea," is a Muir brainchild that brings people and nature together under relatively controlled, and thus safe, circumstances: our national parks are a beautifully designed, half-natural, half-built environment.[9]

---

[9] Although Muir is best known to Americans, German nature lovers got there first. Both the German Enlightenment writer Johann Wolfgang von Goethe (1749-1832) and his Romantic, perhaps counter-Enlightenment counterpart, the philosopher Johann Gottfried Herder (1744–1803), advocated a connection to nature and experiences in natural surroundings.

**Figure 13.9 A truce with nature: John Muir (1838–1914), circa 1902.**
*Source:* Library of Congress.

The war between man and nature lives on in the justified concern with natural disasters. We are still vulnerable to hurricanes, tornadoes, earthquakes, forest fires, and volcanic eruptions, as well as new risks (that are actually old) such as climate change. (The first recorded year-without-a-summer, apparently due to a volcanic eruption in Iceland, produced a continent-wide European famine in 536.[10])

Like many of the stories in this book, the story of increased protection from natural disasters deserves its own book, and some have been written. It suffices to say that the California (1999) and Alaska (2018) earthquakes that measured 7.1 and 7.0, respectively, on the Richter scale produced *no* fatalities; earthquakes of comparable magnitude in poor countries and long-ago times have cost hundreds of thousands of lives—despite 7.0 being only a moderately severe tremor. The difference is due to wealth, and is reflected in building codes, the ability to move supplies quickly, and the treatment available for victims.[11]

[10] Gibbons (2018).

[11] This comparison is a little unfair: many of the fatal earthquakes were in high population-density areas, whereas the California and Alaska quakes were not. If the same earthquakes had occurred near major cities in a wealthy country, however, fatalities would have been very few in number. We saw this in the massive Chilean earthquake of 2010, which measured 8.8 on the Richter scale and cost 525 lives, a remarkably low death toll considering the earthquake's severity. The more recent 8.3 Chilean earthquake, still a very strong tremor, only cost 11 lives. Chile is only a moderately rich country, but has sound building codes. See Bonnefoy and Lyons (2015).

Today, using big data and advanced knowledge of earth science, we are better able to identify and classify earthquake zones as well as other hazards such as flood zones. This information enables us to enact building codes and craft insurance practices that discourage living in such zones, greatly reducing injury, property damage, and death. Using satellites, we can now see hurricanes coming from hundreds of miles away, and spot tornadoes and predict their paths as they form, saving countless lives. We will never win the war with nature, since we are a part of it (and, as the playwright David Mamet wryly observed, "Anyone ever lost in the wild knows that nature wants you dead").[12] But we are winning many battles and, better yet, avoiding many others.

## Agriculture

"Back to the land" is one of the worst ideas ever (apologies to Stewart Brand; his accomplishments are stellar despite having championed the back-to-the-land movement in his youth—see Chapter 24). Agriculture is traditionally one of the most dangerous occupations, competing with mining and fishing for fatalities on the job. There's still risk in farm work, but now that agriculture directly employs only 0.64% of the US population, compared with 15.3% in 1900, far fewer people die or are seriously injured while farming.[13] And we produce a lot more food than we did in 1900!

## Industry, Mining, and Energy

In 1892 President Benjamin Harrison said, "American workmen are subjected to peril of life and limb as great as a soldier in time of war."[14] Accidents in industry, construction, mining, and allied fields such as oil and gas exploration were a source of everyday worry for workers and their families. Thanks to both changes in the laws and technological advancement, this situation has improved dramatically.

We don't have the space for detailed data, but one anecdote related by Pinker suffices to give an impression: "In the 1890s, the annual death rate for trainmen was an astonishing 852 per 100,000, almost one percent a year. The carnage was reduced when an 1893 law mandated the use of air brakes and automatic couplers in all freight trains, the first federal law intended to improve workplace safety."[15]

---

[12] Mamet (2011).

[13] By "directly employed" I mean field workers, and do not include other members of farm families, who of course are much more numerous.

[14] Pinker (2017).

[15] Pinker (2017), p. 186.

And, in Australia, as we see in Chapter 23, the traditionally very hazardous job of coal mining is being done by computer-controlled robots. It's an understatement to say this is an improvement over my coal-miner father-in-law, in the late 1940s, wondering every day whether he would live to see his family that night. He was only trapped in a burning coal mine once. Luckily, he escaped.

## Transportation

We all know that air travel has gotten safer—there were no fatalities caused by US commercial jet airline accidents on US soil from February 13, 2009, to April 17, 2018, when there was *one* death—but we may not appreciate the more general improvement in transportation safety over the years. Long-distance travel by horse was unbelievably dangerous.

The Pony Express was a case in point. Tammy Real-McKeighan of the Fremont *Tribune* recalls: "Riders faced many dangers such as buffalo stampedes. They faced robbers who wanted to take the money and jewels sent in the mail, said Megan Roach, museum store and admissions manager for the Pony Express Museum in St. Joseph, Mo. Some riders died. . . . Roach estimates there were 80 to 100 riders. The youngest was 11-year-old Bronco Charlie Miller."[16] (Child labor is not that distant a practice in our history.) Figure 13.10 provides an impression of the hardships the riders faced.

Pinker presents graphs of worldwide plane crash, pedestrian, and automobile fatality rates over time[17]; we won't bore you with them—they all look exactly the same, starting very high and ending low. In the early 1920s, automotive deaths in the United States exceeded 23 per hundred million vehicle miles; by the 1950s the number was down to five; now it's about one. Although US automotive accidents still claim an unacceptable 40,000 lives per year, our population has grown tremendously and we drive a *lot* more miles than we did a half-century ago, much less almost a century ago, and at much higher speeds.

Automobile and bus fatalities are higher outside the United States, mostly due to the poorer condition of roads and vehicles. As other countries become wealthier, the road systems will improve, and newer, safer cars will be on the road—if we do not encounter a disruptive technological revolution on the way.

---

[16] Real-McKeighan (2010).

[17] Pinker (2017), pp. 177–180.

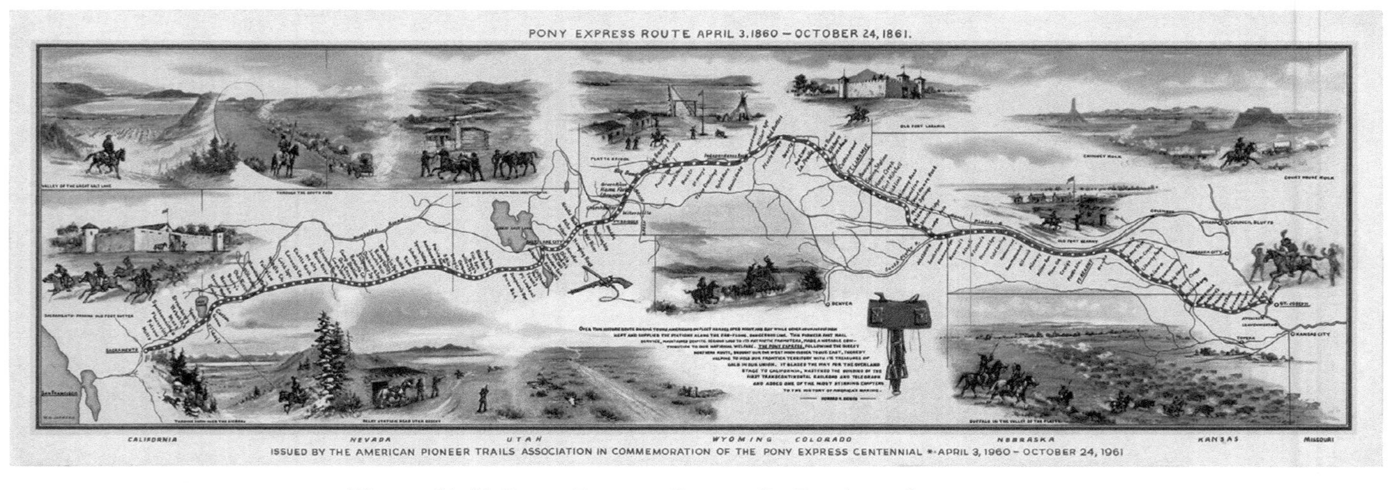


**Figure 13.10 Pony Express Route, St. Louis to Sacramento.**

*Source:* Library of Congress.

And you all know what that is. Sooner or later, self-driving cars are coming.[18] The biggest risk from them, after a period when robots, human drivers, and pedestrians all have to adjust to one another's quirks, is that we're expected to have a shortage of organ donors.

## Freedom, Political and Economic

These are tough times for lovers of economic and political freedom. Yet the long-term trend for both is so positive that it's not farfetched to call the current populist moment in many parts of the world, which I regard as a retrenchment, a blip in a long worldwide journey from tyranny to liberty. At least that's what I fervently hope it is.

"Freedom," over much of human history, meant the state of not being imprisoned or enslaved. Thanks to the struggles of our ancestors over the centuries, we have much higher expectations. The Declaration of Independence, regarded by some thinkers as surpassing even the Constitution as the foundational document describing American values and ideals, holds that all "men" (male and female) are "endowed by their Creator with certain unalienable Rights, that among these are Life, Liberty and the pursuit of Happiness" (see Figure 13.11). The Constitution expanded on this list, enumerating procedural rights such as freedom of speech and trial by jury.

Lists of enumerated rights are, of course, not an American invention. The Magna Carta (1215) and the English Bill of Rights (1689) are precursors. Later lists include various "positive" rights—the right of citizens to have some good or service provided for them—as well as the more traditional "negative" rights guaranteeing freedom *from* various burdens or forms of oppression.[19]

---

[18] Actually, they're already here, but not available for public use. Tesla has accumulated over 1.2 billion miles on autopilot, although it uses human drivers as a backup (see https://electrek.co/2018/07/17/tesla-autopilot-miles-shadow-mode-report/). Tesla has now highway on-ramp to highway off-ramp capability with fully autonomous driving. Google/Waymo has fully autonomous vehicles that it "teaches" using machine learning, big data sets, and human instructors. The fully autonomous system was not used on public roads until it passed a certain safety rate. Then they moved it to 15- to 25-mile-per-hour-zone driving. Autonomous driving has already reached fatality rates lower than the deaths-per-million-mile rate of an average human driver.

[19] I would argue that government can only *guarantee* negative rights, by abstaining in all cases from behavior that would violate them. Positive rights involve the transfer of resources, at least sometimes without the resource provider's consent, and thus subject to the political process and capable of being denied.

**Figure 13.11 Unalienable Rights: Life, Liberty, and the Pursuit of Happiness**

*Source:* Detail from John Trumbull's "Declaration of Independence." Public domain.

To assess changes in the state of freedom over time, I rely here on studies that compare both the political and the economic freedoms on offer in various countries. I believe that economic freedom, the freedom to earn a living, keep most of the fruits of one's labor, spend or save them as one wishes, and invest for the long run and reap the rewards of that investment, is as important as political freedom. Each set of freedoms, political and economic, is a *sine qua non* without which the other has little meaning. We'll discuss economic freedom in detail later in this chapter.

## Political Freedom Today

We look at the distribution of political freedom both across countries and across time. There are several agencies and organizations that rate countries according to their degree of political freedom. All the ratings are flawed, and a debate rages among political scientists about which are least flawed. For current data, we use the Economist Intelligence Unit (EIU) because it uses finer gradations than Freedom House, the other reliable non-academic source.[20] For change through time, we use sources chosen by Our World in Data to cover long histories.

Figure 13.12 shows the Economist Intelligence Unit's index of political freedoms by country as of 2017. Note that the United States is not in the top-ranked category. Although the underlying data are not shown, the United States is only 0.02 of a unit (out of 10) below the cutoff for the top category. The EIU's index is biased toward parliamentary systems because one of their criteria for "full democracy" is whether the legislature holds the supreme power in a country. As a result, the tripartite American system of government puts the United States at an unfair disadvantage. Moreover, until quite recently the EIU placed the United States in the top category.[21]

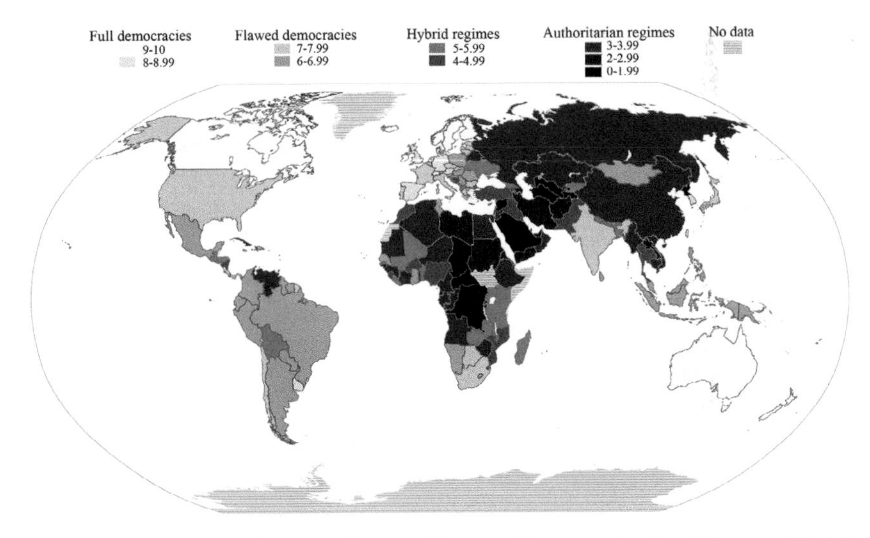


**Figure 13.12 Countries rated according to Economist Intelligence Unit's Democracy Index, 2017.**

*Source:* https://commons.wikimedia.org/wiki/File:EIU_Democracy_Index_2017.svg. Licensed under CC BY-SA 4.0.

[20] The source most commonly used by academics, called polity-4, comes to some startling conclusions, for example that Israel and Zimbabwe are in the same category with regard to political freedom. This alone renders the methodology suspect. Unfortunately polity-4 data are the only ones available over very long historical periods, so they are used in the time series graphs.

[21] An anonymous blog poster stated, only half in jest, that James Madison, in *Federalist #10,* described the division of federal power in the United States between three branches of government, as "a feature, not a bug." I cannot improve on that. The three-way split reduces the power of the legislature below what is regarded as ideal by the EIU.

Because of Russia's vast land area, the map looked better before Vladimir Putin's recent consolidation of power; for a period after the fall of the Soviet Union, Russia was a hybrid regime, not an authoritarian one.[22]

But the map looked much worse at most times in the past, even the recent past. Most of Latin America and most of Eastern Europe only graduated to the flawed-democracy club in the last 30 years. Maoist China was about as bad as it got. We'll look at the long-term history in a moment.

For now, we can say that, as the EIU reports, "Almost one-half (49.3%) of the world's population lives in a democracy of some sort, although only 4.5% reside in a 'full democracy,' down from 8.9% in 2015" due to the EIU'S questionable decision to demote the United States "from a 'full democracy' to a 'flawed democracy' in 2016."[23] That's a remarkable achievement considering that democracy didn't look like it had much of a chance as recently as the 1940s, and was not much more than a pipedream 250 years ago.

The democratization of global political institutions has been as dramatic a change as the Great Enrichment, and has taken place over roughly the same time period. Expressing a thought we've all had, Max Roser writes, "Economic success goes together with political liberation. The countries that have democratized first are mostly those countries that first achieved economic growth. The present rates of economic growth in the poorer countries of the world therefore give hope for further democratization around the world."[24]

The EIU report cautions, "around one-third of the world's population lives under authoritarian rule, with a large share being in China." China has gotten somewhat rich but not free. Still, as Johan Norberg writes,

> The Chinese people today can move almost however they like, buy a home, choose an education, pick a job, start a business, belong to a church (as long as they are Buddhists, Taoist, Muslims, Catholics, or Protestants), dress as they like, marry whom they like, be openly gay without ending up in a labor camp, travel abroad freely, and even criticize aspects of the Party's policy (although not its right to rule unopposed). *Even 'not free' is not what it used to be.*[25]

---

[22] Polity-4 had Russia as a full democracy from 2000 to 2006, and still gives it a high rating, above Brazil. The EIU downgraded it in 2006.

[23] Economist Intelligence Unit (2018).

[24] Roser (2018).

[25] Norberg (2017).

## A Timeline of Political Freedom

The fact that half the world lives in a democracy is a sterling achievement, but only a very recent one. Until 1846, the restrictive polity-4 database asserts that there was only one democracy, the United States. It was procedurally democratic, but suffrage was at first restricted to (mostly) white male property owners.[26] Even Britain was considered undemocratic by polity-4 until 1809, when the franchise was greatly expanded. Thus the second enduring modern democracy was Switzerland, not Britain.[27]

My own standards are more lenient; the American Revolution was essentially an assertion by colonists that they deserved the Rights of Englishmen as declared in 1689, when Parliament achieved effective dominance over the monarchy, so Britain, not the United States, was the first modern democracy, even though most people could not vote. (We had our own problems with that.)

As shown in Figure 13.13, democracy really caught on between 1875 and 1925, mostly in Europe. The huge increase in "closed anocracy"

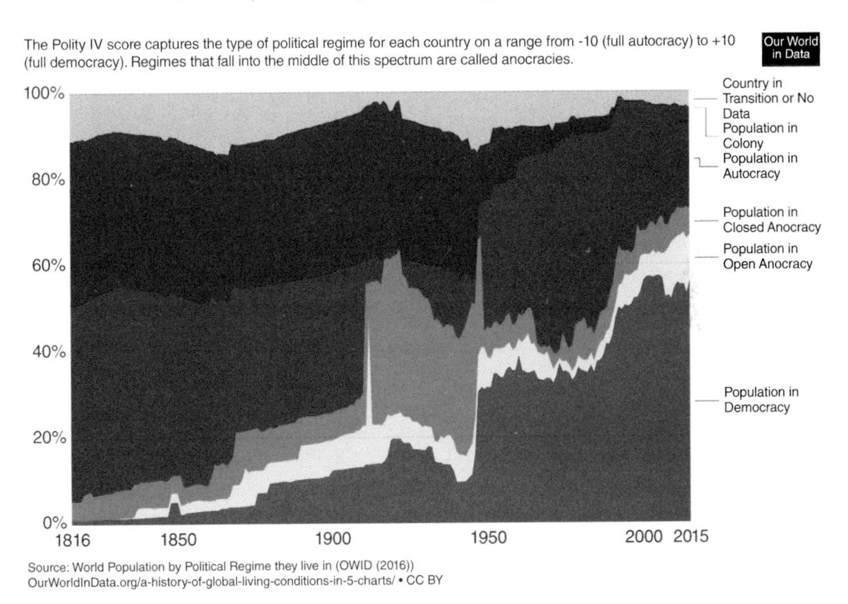


**Figure 13.13 Percentage of world population living in democracy, anocracy, and autocracy or colonialism.**

*Source:* Our World in Data, https://ourworldindata.org/democracy. Licensed under CC BY.

[26] The polity-4 data used to construct the figures in Our World in Data don't even recognize the United States as a democracy until 1808; I disagree. The election of 1800 was a wild free-for-all, the loser stepped down gracefully, and the Democratic-Republican administration of Thomas Jefferson replaced the Federalist administration of John Adams in an orderly manner.

[27] I am not counting Athenian and other ancient democracies or medieval and Renaissance republics with very limited suffrage.

(*anocracy* is an academic term for a hybrid of democracy and autocracy[28]) around 1912 and the corresponding huge decrease around 1950 is due to China, which had a government that was liberal in principle, but ineffectual, after the fall of the empire and before the Communist revolution. The big increase in democracy around 1948 is due to the decolonization of India. The rise in the last 30 years reflects the fall of the Soviet Union and its dominance of Eastern Europe.

Still, Francis Fukuyama's dream of an "end of history" in which all peoples agree on a representative form of government, guarantees of human rights, and peaceable solutions to dispute has not come close to being achieved. Yet, as pointed out earlier, even being unfree is better than it once was, except in a few outposts of misery. The human race can congratulate itself on the question of political freedom up to a point, but we have much work yet to do.

## Economic Freedom

The Heritage Foundation classifies countries according to an extremely complex scheme of economic variables, including these supercategories, each of which has multiple components:

| | | |
|---|---|---|
| Property rights | Judicial effectiveness | Government integrity |
| Tax burden | Government spending | Fiscal health |
| Business freedom | Labor freedom | Monetary freedom |
| Trade freedom | Investment freedom | Financial freedom |

These attributes are what make an economy tick by allowing entrepreneurs and investors to interact, take risks, and reap the fruits of their efforts if successful. Hernando de Soto, the modern Peruvian economist (not the Spanish explorer of old), emphasizes property rights and has dedicated much of his career to providing the poor with clear title to property so they can use it as collateral to borrow and start businesses. Institutionalists emphasize the rule of law and the absence of corruption. Most economists agree on the important of business and financial freedom.

[28] I do not like the word. Its etymology goes back to *anarchy* (lack of rule) and suggests that forms intermediate between autocracies and democracies are anarchic. They usually are not. Anocracies can have strong and effective governance even if the leaders are not (or are not fully) democratically chosen. Singapore, admired by many, is called a *closed anocracy* by polity-4.

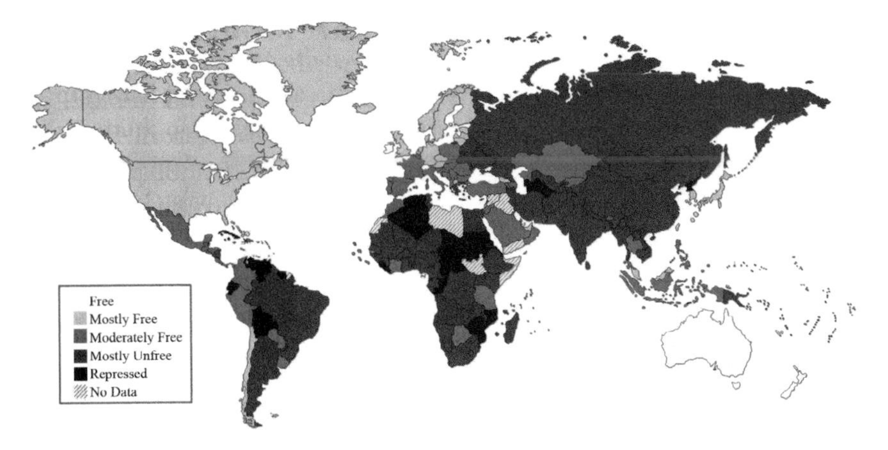


**Figure 13.14 Fraser Institute Economic Freedom Rankings by Country, 2018.**

*Source:* © Fraser Institute, www.fraserinstitute.org/economic-freedom/map?geozone=world&year=2016&page=map. Created by David Stanwick.

Figure 13.14 shows the Heritage Foundation's overall assessment of economic freedom, country by country, as of 2018. Unsurprisingly, the map looks like our earlier map showing political freedom by country: the two are closely linked with some exceptions such as Brazil and India, which have a great deal of political freedom but limited economic freedom. The opposite, economic freedom without political freedom, is almost unknown.

## Changes in Economic Freedom Over Time

We'd like to show the evolution of the map in Figure 13.14 over time, preferably a long period of time when the amount of change is dramatic. However, the data do not exist. Indices of economic freedom started to be compiled around 1970, when the United States, Western Europe, and Japan offered roughly the same amount of economic freedom as they do now. The big changes have been in the developing world, especially China and India, but also in Russia, much of Eastern Europe, and many other countries that have at least partly adopted the values listed earlier in our discussion of the Heritage Foundation index. As with political freedom, at least half the world has a long way to go, and we should not be satisfied with the progress that has been made.

# 14

# *The Alleviation of Poverty*

Until humans came along, poverty was the natural state of *every member of every species that ever existed.* Think about that. If you don't know where your next meal is coming from, and most creatures generally don't, you're poor.

Fifty or even 20 years ago, for half the world's population to be middle class or better was unimaginable. Poverty was the natural state of almost everyone outside of the developed world. Yet in September 2018 it was announced to great fanfare that half the world's population is now middle class or better.[1] Despite the tremendous gain, the alleviation of global poverty is still the most important problem we face.

Half the (human) world is still poor and there is no reason that they should be. Although some might disagree, I'd assert that the half of the world's population that is still poor is no less intelligent, or industrious, or audacious than the half that is not. They are cursed with bad institutions, the burdens of history, unkind geography, and just plain lousy luck.

I've entitled this chapter "The Alleviation of Poverty" in imitation of Milton Friedman, almost always a good person to imitate. His 1962 book, *Capitalism and Freedom*, contained a chapter with exactly that title.[2] His chapter was mostly a pitch for a negative income tax, but the chapter

[1] Kharas and Hamel (2018).

[2] Friedman (1962).

*title* homed in on what I think most of us would most like to achieve: the betterment of those who have little.

## Bottom Up and Top Down

There are two ways to alleviate poverty. One is by making workers more productive by augmenting their skills, their access to capital, and their access to markets, thereby boosting their incomes. We covered that well-documented aspect of economic development in Chapter 6, "Before the Great Enrichment," and Chapter 7, "The Great Enrichment," and we'll return to it later in this chapter. The top-down process will continue. Let Vietnam and Bangladesh and India and Ghana do what they've been doing in the last quarter-century and in a few decades (or sooner) there will be a lot less poverty.

## Bottom Up: The MacGyver Way

The other way to alleviate poverty, the way that is more interesting to read about, is what the authors Peter Diamandis and Steven Kotler call the MacGyver way: to make valuable things out of spare parts and Scotch tape. Okay, I'm being a little silly (so was MacGyver). But developing better and cheaper ways of producing and delivering goods and services, with the improvements often incremental rather than dramatic, is at least as promising a path to betterment as hoping that governments pursue policies that produce a rising tide that lifts all boats.

Bottom-up ways of alleviating poverty are what your grandfather did, or your great-grandmother in Ireland or Taiwan or South Africa. Although the innovations we're going to discuss are on a larger scale than just one person tinkering or working hard, what they have in common is that they bubble up from the minds of motivated and clever individuals or small groups, rather than being imposed from above.

### The Apostles of Abundance

In their book, *Abundance: The Future Is Better Than You Think,* Diamandis and Kotler show how incremental innovation improves the quality of life of the poor, gradually making them not poor. Diamandis is the space-flight entrepreneur who conceived of the X PRIZE (now the

Ansari X PRIZE, awarded for developing and flying a reusable manned spacecraft). Kotler is an author, entrepreneur, adventure traveler, and dog rescuer. Although the authors sometimes come across as dreamers, they're to be commended for presenting both sides of their story, often identifying obstacles to abundance and explaining why some of their hoped-for advances may not materialize.

In addition to Diamandis and Kotler, there are many sources of stories about micro-innovation, from the ridiculous ("chicken farmer Harold Bate discovered an eco-friendly way to power his car using . . . chicken . . . manure," reported the funky *Mother Earth News* in 1971) to the sublime. We'll start with a couple of favorites collected from the news media, then dig more deeply into Diamandis and Kotler's book.

## Headmaster, Here's My Seven Dollars

I heard my very favorite story about the bottom-up alleviation of poverty on NPR a few years ago. Husband-and-wife entrepreneurs Jay Kimmelman and Dr. Shannon May are offering what they describe as topnotch-quality private education to some of the poorest people in the world. Starting in Kenya, the couple has formed a company, Bridge International Academies, to establish low-cost, for-profit schools in a variety of developing countries from Uganda to India. *For-profit.* (Although India has many highly educated people, the quality of public education in some regions is abysmal.)

The global average cost of a Bridge education is, astonishingly, $7 a month.

Parents pay their children's tuition using an app on their mobile phones. The poorest children get scholarships. For comparison, private schooling in the United States costs about $7 every 20 minutes, $30,000 a year.

I interviewed Ben Rudd, international director of communications at Bridge, and asked him how they performed this seemingly cost miracle. He replied,

> When Shannon and Jay were setting up Bridge, they set up the first school in the same way they wanted to set up thousands of schools. They centralized a lot of support services and empowered teachers

> with resources that have been created centrally in their country. We give every teacher a handheld device so they can access top-quality policy guides for every single lesson, for every single subject, for every single grade. Teachers have found these resources to be incredibly powerful.

What, I asked Rudd, does the centralized support group do for the front-line teachers? He responded,

> We have international experts in pedagogy working on questions like: How am I going to divide up my time as a teacher? When am I going to get the kids to learning in small groups or one on one? And when is the best time to walk around the classroom and make sure the kids are working by themselves?
>
> The central support service tries to determine these best practices and then share them with the front-line teachers through the handheld devices. But we don't just leave teachers alone with these devices. There's also a large human workforce supporting the teachers—making sure they get very regular coaching and feedback in the classroom on the job.

It all sounds a little mechanized, but consider the alternative in a country where many teachers are political appointees who can't read. And much of public education in developed countries is mechanized too.

How can they possibly make a profit while charging such low prices? They don't yet, says Rudd, but they will when their expansion plans come to fruition. "The key," Rudd explains,

> is to reach a really large scale. It obviously wouldn't work financially to have a small number of children paying a tiny fee, but it does work when you have a *lot* of children paying a tiny fee. We currently have roughly 250,000 children who are either part of the Bridge system or supported by the Bridge system (through government programs that use Bridge as a resource) and when we reach 500,000 children, Bridge will break even.

Figure 14.1 shows one such student in East Africa. Although she is not a Bridge student (I couldn't get a picture of one), it's hard not to be drawn in by the curiosity and excitement in her face. She's learning, and she's going places.

**Figure 14.1 Enthusiastic Tanzanian school girl.**

*Source:* Anca Dumitrache/Shutterstock.

## Banker, Here's My Five Dollars . . .

Perhaps better known (while we're in East Africa) is M-PESA, a venture started by Vodafone to bring branchless banking to "unbanked" Africans. M stands for "mobile" and *pesa* is Swahili for money. On a smartphone, the user can receive deposits when he or she sells something or performs a service, store the money, obtain cash from an ATM or banking agent (such as a grocery store) and, of course, pay for goods and services. It is just like having a bank account—*exactly* like it, except for the building. There are tens of millions of users, in many African countries, and the enterprise makes a profit.

The benefit to consumers is spectacular. The banking services traditionally provided to the poor, if any existed, were awful. They involved exorbitant fees, unreliable service, and the hazards of transporting cash. Diamandis and Kotler write, "In Uganda (circa 2005), there were 100 automated teller machines for 27 million people. Opening an account in Cameroon costs $700—more than most people make in a year—and a woman in Swaziland can manage that feat only with the consent of a father, brother, or husband."[3]

All that is mostly history, or soon will be, now that M-PESA and similar services are spreading worldwide. Figure 14.2 shows the spread of cell phones, the prerequisite for both phone-based school tuition and M-PESA, in selected African countries. It's quite an achievement in a very short time.

## Harington, Crapper & Gates: Law Firm or Plumbing Contractor?

As Diamandis and Kotler remind us (although the story is fairly well known), in Queen Elizabeth's day only one person had a flush toilet—her. Her godson, John Harington (1560–1612) built it. He was mostly known for his poetry and collections of epigrams, if he was known at all. To his credit, he did either collect or invent the thought-provoking

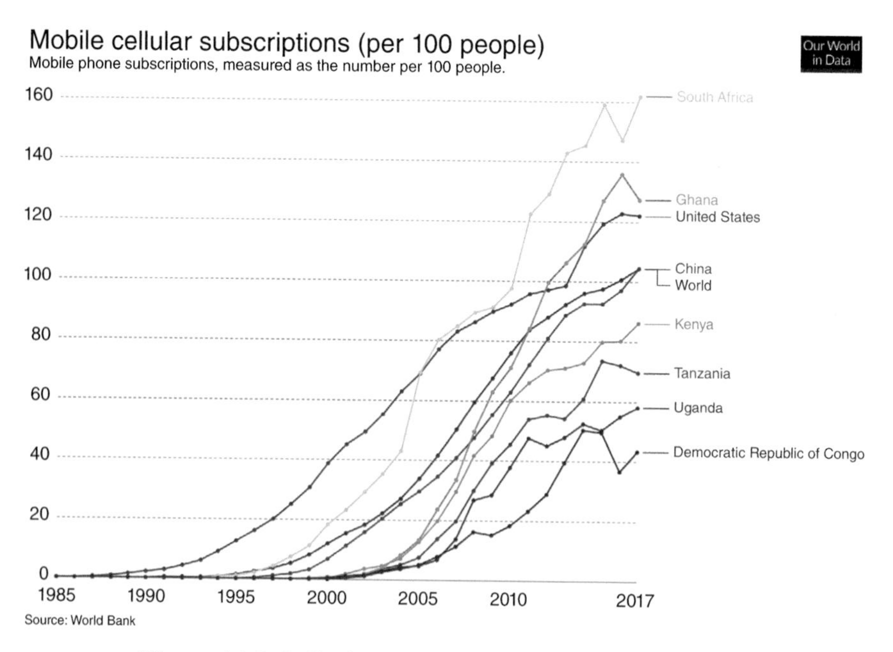


**Figure 14.2 Cell phone ownership surging in Africa.**
*Source:* https://ourworldindata.org/technology-adoption. Licensed under CC BY.

[3] Diamandis and Kotler (2012), p. 145. Swaziland is now called Eswatini.

doggerel, "Treason doth never prosper/What's the reason?/For if it prosper,/None dare call it treason."

Oddly, he never built another toilet. Ten generations of people would have felt more flush (sorry!) if they had not had to wait for Thomas Crapper, around 1860, to commercialize the concept.[4]

Now, flush toilets are cheap enough that four or five billion people have access to one.[5] Other basics of life have become cheap too: a clock, camera, or phonograph used to be a luxury, but now each, or its equivalent, is an app on a phone, provided at zero marginal cost to the user. And there are about five billion mobile phones in the world, one for each person who can use a toilet. As we've seen in other chapters, nutritious food, clean water, energy, clothing, knowledge, entertainment—the list is endless—have all come down in price and risen in availability due to innovation. *This is an integral part of the way the poor become the nonpoor.*

But four or five billion people having access to a flush toilet isn't good enough. There are something close to three billion people who don't. So, having invented the personal computer, Bill Gates is now trying to reinvent the toilet. Harington, Crapper & Gates sounds like a Dickensian law firm, but it is actually a plumbing combine spanning five centuries.

Admittedly, anything involving Bill Gates sounds more top-down than bottom-up. But toilet design isn't macroeconomic policy, and at any rate the Bill and Melinda Gates Foundation is trying to grow a new kind of toilet organically, so to speak. It's distributing grants widely to universities, research institutes, and companies with ideas for toilets that meet Gates's criteria for solving the sanitation problems of the world's poorest people. These include:

- Off the grid—no water, sewer, or electrical connections needed
- Costs five cents (US) a day or less to operate
- Produces fertilizer and other valuable resources
- Scalable (about two billion of them are needed)
- Makes a profit

[4] Not only did Crapper not invent the flush toilet, he wasn't even the first to hold a patent on a modern (post-Harington) version of one. That was Alexander Cumming, a Scottish watchmaker, who in 1775 patented the S-trap, the crucial ingredient that keeps sewer gases from escaping back into the bathroom. Crapper's contribution was to make the toilet commercially feasible and attractive, moving the humble appliance from invention to innovation.

[5] "Toilets flushed in the world today" (web counter) at www.worldometers.info/view/toilets/, citing a United Nations WHO/UNICEF Joint Monitoring Programme for Water Supply and Sanitation page that is no longer available.

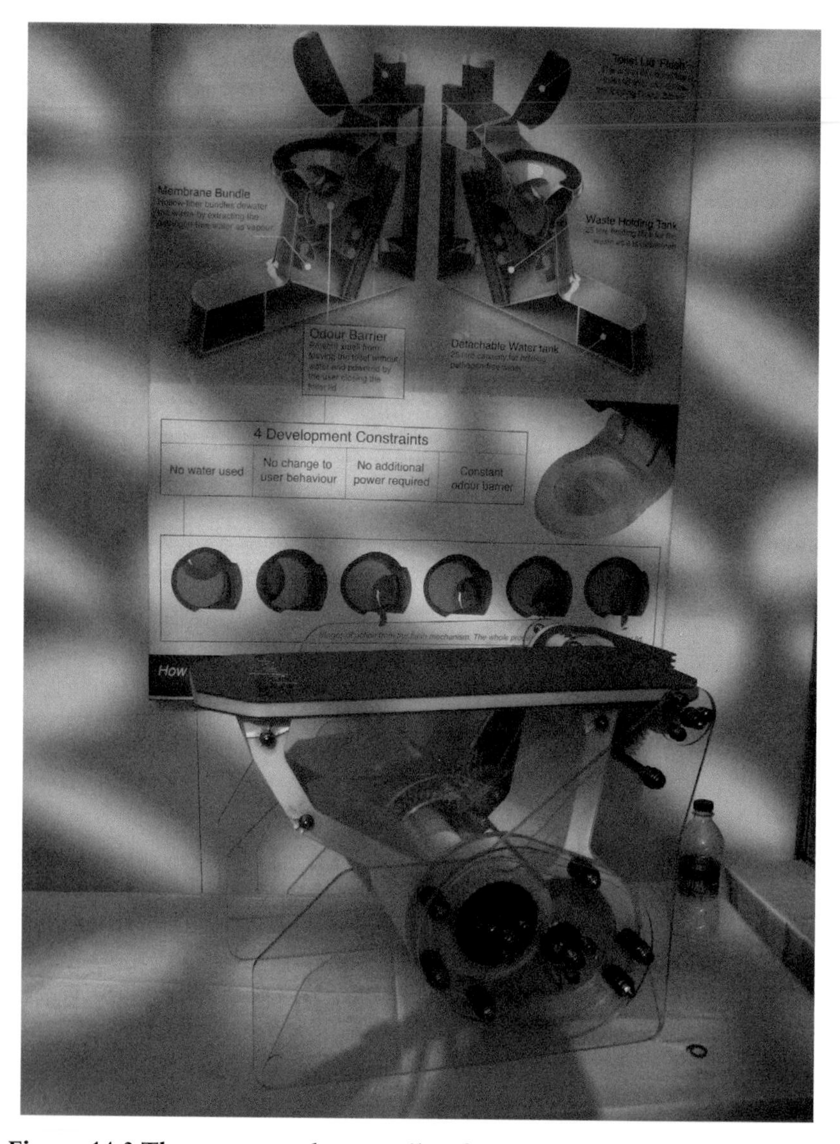


**Figure 14.3 The nanomembrane toilet, from Cranfield (UK) University.**

*Source:* SuSanA Secretariat, 2014, https://commons.wikimedia.org/wiki/File:A_prototype_exhibit_of_the_flushing_mechanism_that_works_without_water_(13359256335).jpg. Licensed under CC 2.0.

"Make no small plans," exhorted the architect Daniel Burnham more than a century ago. Bill Gates seems to have listened.

So far, a number of competing designs have resulted from $200 million in grants, and were exhibited for the first time at the Reinvented Toilet Expo in Beijing in November 2018. Figure 14.3

illustrates a "flushing mechanism that works without water," and includes a membrane bundle, described as "hollow-fiber bundles [that] de-water the waste by extracting the pathogen-free water as vapor."[6]

Diamandis and Kotler are wildly enthusiastic about the effort, saying, "the upside of this toilet is almost incalculable." The cost of curing diseases spread by poor sanitation will fall. The decrease in disease will increase labor productivity greatly, as the decline in malaria in Africa (in large part also due to Gates's efforts) already has. The need for expensive sewage connections and sewage treatment plants will decline. In addition, Diamandis and Kotler remind us, "toilets account for 31% of all water use in America"; although the technotoilet is not targeted to the United States, this percentage could hypothetically become zero. Gates believes the program will save 500,000 young lives per year.

## The Multiplier Effect of Innovation

This multiplicative effect applies to any general-purpose technology. It is one of the reasons that the Great Enrichment is not an inexplicable miracle but the natural outcome of inventors and entrepreneurs pursuing their own interest and that of their potential customers. Thomas Crapper did not set out to improve the world or raise GDP, but to profit personally from realizing that there was a sizable market for well-functioning bathrooms and toilets. The demand for plumbers and plumbing supplies went way up, and so did the demand for whatever these individuals buy.

Bridge International Academies and M-PESA are, likewise, profit-seeking enterprises. Innovative entrepreneurs hire workers who then go on to buy more goods and services, and satisfy customers who then have to do less dull or dangerous work than before, freeing them up for more productive pursuits or for leisure. *That is how the economy grows, and how the alleviation of poverty takes place.*

# The Granny Cloud

In one of their most charming stories, Diamandis and Kotler recount the tale of Sugata Mitra, an Indian professor who teaches at Newcastle University in England (see Figure 14.4). Mitra found that children left to their own devices would, if provided with a computer, figure out how to use it, teach each other computer skills, and learn English and

[6] Description comes from SuSanA Secretariat, 2014, https://commons.wikimedia.org/wiki/File:A_prototype_exhibit_of_the_flushing_mechanism_that_works_without_water_(13359256335).jpg. Licensed under CC 2.0.

**Figure 14.4 Sugata Mitra.**

*Source:* www.flickr.com/people/campuspartybrasil/, 2013, https://upload.wikimedia.org/wikipedia/commons/2/20/Sugra_mitra.jpg. Licensed under CC BY-SA 3.0.

academic material (such as biotechnology) better than those instructed by a teacher. Mitra's project, with the delightfully bottom-up name Hole in the Wall, lacked only one thing: a supply of available mentors and coaches for when the children needed a little something extra.

"That," says the project's website,

> was when the idea of the Granny Cloud was born! The basic, delightfully simple, idea was to have folks [in England] who were native English speakers Skype in with children in these remote and disadvantaged locations [in India] and enable them to pick up English in the way we typically pick up any language—through hearing it spoken around us and using it in conversation.[7]

Before you fixate on the image of a slightly older Mary Poppins, read the site's description of the grannies:

> The "Grannies" are women and men . . . from 24 to 78 years of age. . . . The "grandmother approach" was how the concept was explained to emphasize the warm, encouraging nature of the interaction. The media picked up on the word "grandmother" and . . . the term "Granny

[7] "Granny Cloud" (undated).

> Cloud" developed and stuck! The Grannies are ALL volunteers, who come from many different countries in the world and English is not the native language for all. [They include] educators, engineers, scientists, architects, government officials, and homemakers.[8]

I want to become a granny myself! The idea practically begs for radical expansion. Imagine *abuelitas* (grannies) in Mexico City teaching indigenous children, many of whom do not speak Spanish, in remote areas of Mexico or other Latin American countries via Skype. Comfortable Eastern Chinese helping to bring impoverished Western Chinese children into the country's burgeoning economy . . . you get the idea.

## The Falling Real Cost of Almost Everything That Comes in a Box

The MacGyver way, conceived of broadly, refers not just to clever ways of delivering private schooling to Kenyans or toilets to Indian villagers, but the dramatic improvement in the quality and pricing of industrial goods (and some personal services) that has taken place quietly over the last generation. Some of this improvement is due to technology—a 2019 Toyota is probably safer and more reliable than a 1970 Rolls-Royce—but much of it is due to what the labor economists Dan Breznitz and Peter Cowhey call "incremental product and process innovation."

> [This] consists of improvements in the ways that goods or services are designed, produced, distributed, and serviced. As the economist Joseph Schumpeter observed, [product and process] innovation has the greatest impact on economic growth. . . . Some industries, including cars and personal computers, are less defined by rapid product innovation than by continuous process improvements, which alter cost and performance capabilities. . . .
>
> Taking an idea from concept to marketed product requires an array of incremental product (such as continual improvements in automobile transmissions) and production process innovations. Such innovations are often made on the factory floor by workers with intimate knowledge of their product and process. For example, at Inspur, a Chinese high-end computing company, a regular line worker devised a way—now patented—to control static buildup on the

[8] Ibid.

> production line using plain tap water. This innovation is essential for the production of sensitive electronics.[9]

What is this observation doing in a chapter on poverty? The answer is in Figure 14.5, which shows the declining cost of some basic appliances and gadgets—as measured by the number of hours of labor it took to buy them if one was earning the average US wage for production and nonsupervisory workers—between 1979 and 2015. Although "real" wages for such workers (that is, adjusted for inflation as measured by the Consumer Price Index) did not rise over that period,[10] *what these wages can buy* has grown tremendously in quality and quantity.

Another way of saying this is that the measure of inflation used to convert nominal to real wages has greatly overstated the increase in the true cost of "living" (that is, of receiving desired goods and services). Although real estate, a college education, and health care have indeed become more expensive, televisions and microwaves have become absurdly cheap, and the real cost of an expensive necessity such as a refrigerator has fallen by more than half. *This is one of the ways the poor get richer.*

## Falling Costs Over a Longer Time Frame

Poverty, then, is the lack of goods, services, energy, comfort, and convenience—that is, what money can buy. (Money, which one surely needs, is just an intermediate step, a unit of account.) Extending the analysis back farther in time, some of these benefits are being provided at a tiny fraction of the cost that our ancestors paid, if they could obtain these goods and services at all.

We've already explored some of these improvements. The cost of lighting has fallen by 99.98% (5,000 to 1), as we showed in Chapter 10, "Energy" (Figure 10.9, right-hand scale, 1800–2000). The cost of computation has fallen by much more than that, and much more quickly: a factor of $10^{18}$ in 120 years.[11] The cost of most medical treatments has fallen by a factor of infinity: they didn't exist at all until recently, at any price, although some are still expensive.

---

[9] Breznitz and Cowhey (2019).

[10] Of course, the people earning the average wage in 1979 and 2015 are not the same people. Most people start out at or near the bottom and move up in socioeconomic status over time. So any comparison of the two price levels does not reflect the experience of a typical individual.

[11] www.flickr.com/photos/jurvetson/31409423572.

**Figure 14.5 Change in real cost (in hours of labor at average hourly US wage for production and nonsupervisory workers) of various items, 1979–2015.**

*Source:* www.cato.org/projects/humanprogress/cost-of-living. Reproduced with permission.

Most necessities are not going to fall in price by $10^{18}$ or even 5,000 to 1. But their real costs (in this context, the amount of labor it takes to obtain the needed or desired good or service) will continue to decline, often in unforeseen ways. (Who, other than Maxwell Smart, foresaw the smartphone, no pun intended?) And the lifespan of material objects is generally increasing. Although the money price of goods and services will fluctuate, that's due to the unpredictable value *of the money*, not the cost in labor terms of the goods and services themselves. They will keep getting cheaper.

## Making Necessities Accessible

Accessibility is as important as price. In many places, water is free for the taking (and hopefully clean enough to use; often it isn't) but you have to get the water to where you need it. If you have to walk for four hours to fill a 40-pound jug of water and carry it back on your head, it is anything but free. The opportunity cost of your time is huge in that you can't get anything else done that day, and it gradually ruins your body, especially if you are a small, older woman. Figure 14.6 quantifies the burden of water carrying by the poor in Kenya. All of this wasted effort, with its accompanying pain and suffering, can be eliminated.[12]

We've already seen, in Chapter 10, "Energy," that electricity for lighting has become almost free by historical standards. Yet the further

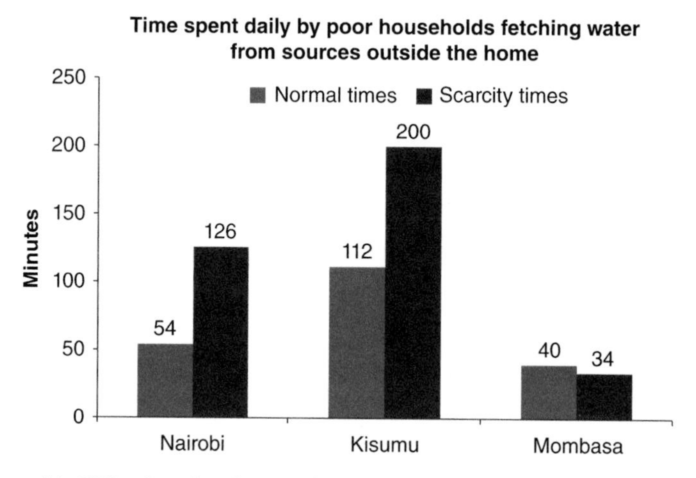


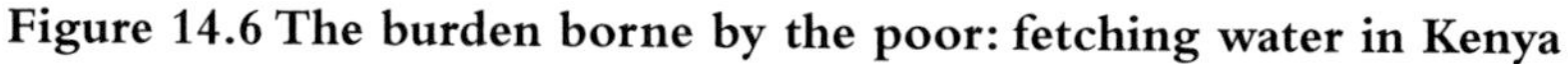

**Figure 14.6 The burden borne by the poor: fetching water in Kenya**

*Source:* https://twaweza.org/uploads/files/Its%20our%20water%20too_English.pdf. Reproduced with permission.

[12] No one wants to carry huge buckets of water. Yet, in an anecdote that would be amusing if it weren't so sad, Dan Bouchelle, a missionary, reported that women in an African village were bypassing newly installed water pumps near their homes to gather at the traditional water sites. They did so because it was the only time they could socialize with one another and get away from their husbands. See www.mrnet.org/news/doing-short-term-missions-well-part-2.

80% energy cost decline that is taking place just now, from being able to light a room with a 20-watt LED bulb instead of a 100-watt incandescent bulb, is significant if you are living on two dollars a day and are trying to study at night.

Even more important is having access to electricity at all, so you can decide whether to turn on the light or not. According to Yinka Adegoke in *Quartz*, "More Africans than ever before have electricity but there are still 600 million on the continent without it. The household electrification rate in sub-Saharan Africa is the lowest in the world, around 42% in 2016."[13]

That, Adegoke adds, is "despite the many programs providing access for 70 million people since 2014." Bringing electricity to 70 million people in two years is quite an accomplishment—the US Rural Electrification Administration did not work nearly that fast—and, at that rate, it would take only eight-and-a-half more years to fully electrify the current population of sub-Saharan Africa. That is almost certainly an overoptimistic estimate, since such projects usually pick the low-hanging fruit first, and the population will also be growing over that time span. The important finding, however, is not that there is much work to do (we knew that), *but that it is getting done.*

## For African Farmers, "No More Hungry Season"

Diamandis and Kotler report that the charitable organization KickStart International develops "tools to end poverty" in Africa. The charity's primary tool is portable, low-tech irrigation equipment. Although irrigation equipment can be expensive and require engineers to install, KickStart does it differently—the MacGyver way.

Among their tools is the MoneyMaker Max, "a unique, high-quality, human-powered treadle irrigation pump that retails across Africa for [about] $170, including hosepipes. The more basic MoneyMaker Hip Pump is a hip operated irrigation pump that retails across Africa for [about] $70."

Why MoneyMaker? Because poor people need to make more money. Why, of all the causes a charity in Africa could support, irrigation? Because

> currently, only 4% of farmland in sub-Saharan Africa is irrigated, which means the vast majority of farmers depend on the unreliable

[13] Adegoke (2018).

> rains to grow their crops. With everyone planting and harvesting at the same time, farming families all end up attempting to sell their crops in over-saturated markets for low prices where supply is high, but demand is low. Up to 65% of the food grown in the rainy season therefore spoils before it is eaten or sold. Unfortunately, just months later, when the rains dry up, these same farming families are left with nothing to eat and without an income, keeping them stuck in a vicious cycle of hunger and poverty.
>
> The tools farmers need to escape from their dependence on rainfall, however, not only exist, they're for sale. MoneyMaker irrigation pumps can break this cycle of subsistence farming and lift millions in Africa out of poverty.[14]

Does it work? The organization reports 331,556 pumps sold, 1.2 million people lifted out of poverty (one person every seven minutes), 250,000 businesses created, 12 million people fed each year, and $210 million yearly in new farm profits and wages. KickStart says that "farming families increase overall household income by 400%, on average." Had a 400% raise lately? If you were as poor as an African farmer, you might appreciate a raise like that even more.

If there is any reason that water keeps coming up, it is that water is a fundamental part of human life. The majority of the world's problems would go away if everyone had clean water, food, clothing, and safety and a good night's sleep.

## The Great Contribution of the Charitable Sector

The KickStart story demonstrates that one cannot discuss alleviating poverty without mentioning the work of the many great charities in the world. An often overlooked upside to the world getting richer is that many entrepreneurs will use their wealth to help fund and operate charities that help some of the poorest people in the world lift themselves out of poverty. The Bill and Melinda Gates Foundation, discussed earlier, is a shining example, but the principle is old. "Without Carnegie, Rockefeller, Rosenwald, Eastman, and others," writes the entrepreneur and author Gary Hoover, "we'd have fewer and worse museums, hospitals, universities, libraries, and schools."[15]

---

[14] Source of all KickStart-related quotes: kickstart.org.

[15] Gary Hoover. Personal correspondence, March 2008.

John D. Rockefeller was one of the great pioneers of philanthropy. The man he put in charge of setting up his philanthropic efforts was Frederick T. Gates (no relation to Bill). Under Gates, Rockefeller's millions created well-funded foundations run by experts. Before that, funds were doled out individually. Rockefeller's legacy includes the University of Chicago, Spelman College (a historically black women's college; Spelman was Mrs. Rockefeller's maiden name) the General Education Board, the Rockefeller Sanitary Commission (which eradicated hookworm and greatly improved productivity in the Deep South) and, of course, the Rockefeller Foundation, which, among many other projects, funded Norman Borlaug's Green Revolution described in Chapter 8, "Food."

If there's a point of intersection between the bottom-up strategies we've been discussing and the top-down policy issues with which we'll conclude this chapter, it is likely to be found in the field of charity or philanthropy. Most charities are intended to better the world, but misplaced charity can destroy lives (during the Ethiopian famine in the 1980s, food aid drove local farmers out of business because they could not compete with free food). The alleviation of poverty can only be accomplished if one understands the three basic principles of economics that govern our lives: the universality of trade-offs, the fact that people respond to incentives, and the cruelest law of all, that of unintended consequences.

## Top Down

Why are some nations so rich and others so poor? So asked the economic historian David Landes in the subtitle to his elegant volume, *The Wealth and Poverty of Nations*.[16] The urbanist Jane Jacobs ventured, if not an answer, a way to look at the question: "To seek 'causes' of poverty . . . is to enter an intellectual dead end because poverty has no causes. Only prosperity has causes."[17]

Jacobs meant that poverty is the natural condition of man and animal, and what needs explanation is not poverty but *wealth*, which is an aberration that should astonish us. This idea goes back to Adam Smith and has been repeated from one author to another until it appears to have no progenitor. It's part of our collective wisdom. Yet, as Steven

[16] Landes (1998).

[17] Jacobs (1969).

Pinker reminds us, "even today, when few people believe that accidents or diseases have perpetrators, discussions of poverty consist mostly of arguments about whom to blame for it."[18]

I've set up an apparent contradiction: poverty is the natural state of man, yet there is no reason that half the world should still be poor. What am I talking about? We now *know how* not to be poor, how to discover, innovate, save, and invest. This is knowledge that mankind did not have for most of its existence. Now that we have it, we can use it to continue the process of democratizing wealth until no one, pathological cases aside, is in serious want. What is stopping us?

At the macro level, as Tyler Cowen said in his recent book, *Stubborn Attachments,* we simply need to pursue policies that maximize global economic growth.[19] These include:

- free markets;
- free trade;
- borders that are as open as practicable;
- sensible regulation (mostly minimal regulation, but with a sensitivity to legitimate concerns about the environment, safety, and honesty);
- and, perhaps most important, the crafting of good laws and institutions that, among other things, convey secure rights in private property.[20]

In addition, people need a multitude of ways to *build their human capital*, the only kind of capital that reliably results in bigger paychecks and reduced misery.

This set of prescriptions needs to be applied all over the world. We live in a time when there has been some backsliding relative to these ideals, but the desire to innovate and build wealth has not gone away. We need to continually remind ourselves and our fellow citizens about both the ideals and the proven best ways of realizing them.

---

[18] Pinker (2017), p. 25.

[19] Cowen (2018).

[20] The Peruvian economist Hernando De Soto is well known for emphasizing the importance of identifiable and enforceable property rights to alleviating poverty. For example, owners of small houses or plots of land can only borrow against their value to start businesses if they can prove to lenders that they are the property owners. De Soto supports blockchain as a possible technology for achieving this goal, but mostly he wants to protect property from confiscation by authorities and non-state actors (gangsters). De Soto (2000).

If such policies are followed, the poor will get richer as quickly as they have in the last quarter century, wherein the incidence of extreme poverty has fallen to 9% of the world population. Although inequality is increasing within countries for reasons we'll touch on in Chapter 15, "Robots Don't Work for Free," it is decreasing *across* countries (the Great Convergence from Chapter 1). The latter is the more important trend. We don't want it to stop.

This last section will briefly discuss these top-down issues, which have been covered in 10,000 dissertations, position papers, and popular books. Since wealth is the only cure for poverty, we'll review the challenges involved in establishing the preconditions for wealth creation listed earlier.

## Institutions That Encourage Wealth-building

In Chapter 19, "Obstacles," I tell the sometimes ridiculous, sometimes macabre, stories of really bad institutional practices that prevented or reversed economic progress. The practices ranged from ordinary stupidity to mass murder. Here, we accentuate the positive: what institutions, laws, and forms of government make it *easier* to accumulate wealth and allow it to be distributed to the many who create it through their labor and inventiveness, not just to the few in power?

The work of the economic historian Douglass North gives some valuable guidance. He was known as an institutionalist. That's the economist's term for someone who thinks that the quality of institutions (not just government, but forms of corporate organization and voluntary association) matters a great deal to economic progress. He won a Nobel Prize in 1993 for his attention to this topic.

### Why the United States Was an Early Success

North recalls the founding conditions that prefigured the spectacular success of the United States:

> [T]he English colonies were formed in the century when the struggle between Parliament and the Crown was coming to a head. Religious and political diversity in the mother country was paralleled in the colonies. The general development in the direction of local political control and the growth of assemblies was unambiguous. Similarly, the colonist carried over . . . fee simple ownership rights [in land] and secure property rights in other factor and product markets.

**Figure 14.7 Douglass North.**

*Source:* Hoover Institution. Reproduced with permission.

North continues,

> British efforts to impose a very modest tax on colonial subjects, as well as curb westward migration, led . . . to the Revolution, the Declaration of Independence, the Articles of Confederation, the Northwest Ordinance, and the Constitution [in quick succession]. . . . [Yet] post-revolutionary history is only intelligible in terms of the continuity of informal and formal institutional *constraints* carried over from before the Revolution and incrementally modified.[21]

Thus, the key elements, at least according to North, were:

- secure rights in both real and chattel property;
- local control of politics;
- carefully crafted legal documents respected by all as authoritative;

[21] North (1991).

- finally, and, most importantly, made possible by local political control, constraints on arbitrary authority and confiscation. Checks and balances. Constitutional and statutory limitations on the power of government.

Some observers also credit religion. Most of the early settlers of North America were Protestants. Max Weber's very popular 1905 sociological treatise, *The Protestant Ethic and the Spirit of Capitalism*,[22] argues that Protestantism—especially in its Calvinist manifestation—encourages hard work, frugality, and deferred gratification. This may also be a factor in the advance of Germany, the Netherlands, and Scandinavia relative to Catholic Europe in the nineteenth century and the first half of the twentieth.[23]

Although this feels like conventional, high school history–class wisdom, North accumulates a great deal of evidence that these factors are in fact critical for economic development. Good laws, especially, are underestimated (by nonlawyers) as harbingers of success. If you can enforce contracts, rely on the authorities to back up honest dealing and punish chicanery, and be secure from confiscation or ruinous taxation by the local potentate, your incentive to invest for the far future (or even the near future) is enhanced and multiplied.

## Beyond the United States

But the United States, or the Anglo-American system more broadly, is not the only possible model of economic enhancement. Europe, with its monarchies and wars, also got rich, albeit later; and Japan and, much later, China and to some extent India have achieved economic success using very different models. What did they do right?

Not much, at first.[24] The imagined Europe of wealth and power is not matched by reality until the last half-century or so, unless you were an aristocrat, merchant, or learned professional. Magnificent cathedrals,

---

[22] Weber (1904).

[23] An example of the polar opposite would be Buddhism, in which the object is to be reborn (as another person or animal) as few times as possible, to reduce the suffering the soul will experience over multiple lifetimes. To *avoid being reborn* is the goal. Many people believe this philosophy is inconsistent with capitalism (although Buddhists in Vietnam and China are turning out to be very competent capitalists!). One should remember that Buddhism emerged under conditions of extreme deprivation, in the Indian subcontinent in the sixth century BC.

[24] An exception is the Netherlands, which followed (or even prefigured) the Anglo-American model.

museums, government buildings, and the mansions and townhouses of the wealthy do not mean the common people were rich in the centuries in which those artifacts were built; mostly, as we saw in Chapters 6 and 7, laborers and farmers were shockingly poor, even hungry. (The best buildings were preserved and the meager dwellings of the poor were not, so we see only the nice ones. Selection bias.) Figure 14.8 shows the state of the European economies in the Depression year of 1938, only 80 years ago.

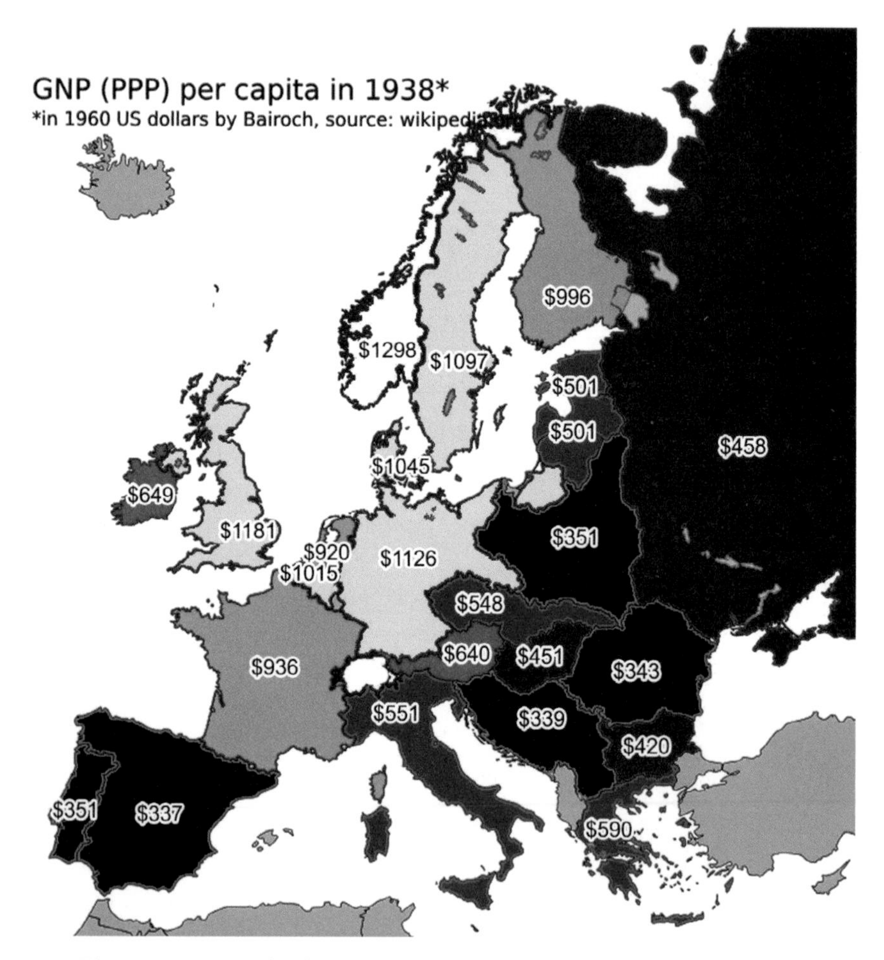


**Figure 14.8 Purchasing Power Parity (PPP) GDP per capita of European countries in 1938 (expressed in 1960 dollars).**

*Note:* German *per capita* GDP may overstate standard of living because arms production was high in that year.

*Source:* Adapted from Ale Irez. Posted at www.quora.com/Why-is-Europe-so-developed, using data from Paul Bairoch.

To convert to current dollars (we use 2017 dollars), multiply by 8.4. These are annual, not monthly, incomes per person. Admittedly 1938 was a bad year, but the numbers are amazingly low, even in the richest countries where incomes averaged $9,000 a year in today's money. That's only 25% higher than India's PPP-adjusted per capita GDP today, and just over half of China's. Europe was not rich in 1938.

Yet Europe was not, by any means, an economic failure by the standards of the times. As we saw in Chapter 7, Germany's ascent in the mid to late 1800s was faster than that of any country before it. After World War II, France, then Italy, then Spain were described as experiencing economic miracles. Scandinavia has been prosperous for a long time, achieving by 1938 what *Life* magazine called "the world's highest standard of living."[25]

The riddle of Europe is not that it eventually got rich but that it took so long. By 1800, Europe had a gargantuan head start over the rest of the world, having accumulated a great deal of social and physical capital, unevenly distributed but of immense value. This included:

- A built environment that reflected an exceedingly rich history (see Figure 14.9).
- A university system that, after some delay, rewarded medical, legal, and humanistic scholarship as well as religious study.
- A more literate population than the rest of the world.
- Extensive trade.
- The rudiments of democracy, which became more real after the revolutions of 1848 and then again after the wars of the first half of the twentieth century.
- The Republic of Letters, as we'll see in Chapter 17.
- Kings and queens who were not all bad.

With all those advantages, it's a wonder that Europe, not North America, was the laggard in bringing wealth to the common people. This observation suggests that Douglass North was right: institutions and laws matter, a lot, and in those the North Americans excelled.

[25] "King Gustaf of Sweden" (1938).

**Figure 14.9 The European head start: Genoa in 1481, by Cristoforo de Grassi (*View of Genoa,* 1597).**

*Source:* https://upload.wikimedia.org/wikipedia/commons/f/fa/Genova_1481_%28copy_1597%29.jpg. Public domain.

## The Rest of the World

I don't have the space or the knowledge to explain the rise of Japan and the Asian Tigers (South Korea, Taiwan, Hong Kong, and Singapore), then China, and finally India and a number of other of today's growth engines. Japanese industry was first known as an expert copyist, then as an unsurpassable machine of growth, now as a bit of a laggard.

China is little short of a miracle, although the most recent news out of that country is discouraging due to an increase in authoritarianism. It suffices to say that Deng Xiaoping's mysterious comment, "[T]he market economy happens under socialism, too . . . planning and market forces are both ways of controlling economic activity," frames the riddle that those who would understand the growth of China's economy need to solve.[26]

Apparently, there is more than one way to go from poor to rich. Before China in the last third of a century, we did not know that. We knew of one way: market capitalism. Now we know there are others, although some scholars believe that China uses many more of the tools of market

[26] Cited by Gittings (2005).

capitalism than the Chinese government, nominally communist, would like you to believe.[27]

## Why Do Rich Countries Have Poor People?

Ironically, poverty is easier to remedy in poor countries than in rich countries. The poor in poor countries are accustomed to working very hard for very little. If you increase the wage of an African laborer from $3 a day to $6 a day for the same work, say, because you've opened up his product to a new, richer market, he's experienced a tremendous gain in utility.

But the poor in rich countries are mostly those who are least able, least adaptable, least motivated, least mobile. They are the poor who will always be with us, says the New Testament,[28] because human abilities vary. (Some may also be poor because of oppressive institutions and practices; that can be fixed.) They may never have worked for pay.

Statistically, it is very difficult to be a member of the global poor in a rich country. Only the homeless come close to qualifying. According to Matt Ridley, of those classified as poor in America, 99% have electricity, water, a toilet, and a refrigerator; 95% have a television; and 88% have a mobile phone.[29] The value of the social benefits available to the poor in rich countries is a multiple of the average income of working people in the poorest countries.

However, poverty is experienced in a relative, not absolute, framework. Thus, even though a welfare recipient in Kentucky may be many times richer in dollars than an African laborer making $6 a day, he will always feel poor, and may adapt the manners and habits of the poor, keeping him poor indefinitely.

If there is one policy change I'd make that really will reduce poverty in rich countries, it's to eliminate what I call "the cruelest tax," the tax on going from nonworking poor (and receiving substantial government benefits) to getting a low-paying job and becoming part of the working almost-poor. You lose valuable benefits so quickly that the marginal tax

---

[27] I believe that Chinese Premier Xi Jinping is a committed communist, but the Chinese economy operates as a "socialist market economy" in which the supply-demand-price system provides much of the resource allocation function, a situation that many would call capitalism (and, in addition, there are many purely privately owned businesses).

[28] Matthew 26:11.

[29] Ridley (2010), p. 16.

rate on this admirable step toward self-sufficiency is over 75%.[30] When you get a job you may lose medical insurance, housing assistance, and cash benefits all at once; food assistance is a little more forgiving. Michael Falk has written eloquently about the need to make social safety nets more like trampolines so that escaping poverty becomes an achievable goal.[31]

In the United States, I'd add one more policy change (and I'm going to step on some toes here): eliminate all racial preferences, quotas, and set-asides. They are deeply divisive, arouse mostly unwarranted suspicion about the abilities of those in preferred classes, and usually hurt those whom they're intended to help. Our Constitution, as repaired in 1865 to 1870, is color-blind; our institutions should be too.

## Conclusion

The alleviation of global poverty is intimately interwoven with the advancement and democratization of technology. It's been hard to keep the two topics separate, and it's been even harder to limit this discussion to ideas that can fit in a chapter rather than an encyclopedia. In the next chapter we'll discuss the interplay of technological advancement, creative destruction, and the labor market prospects for people of all skill levels.

---

[30] Giannarelli and Steuerle (1995).

[31] Falk (2016).

# PART IV
# EXPLORATIONS

# 15

# *Robots Don't Work for Free: A Meditation on Technology and Jobs*

Are robots going to take away all our jobs? Will we have a Robot Apocalypse? No, we won't. Like all the technological changes that came before them, robots and related technologies are just ways of enhancing productivity.

Robots are a convenient and colorful metaphor for any new technology that makes work easier and workers more productive. First, let's say what I mean by a robot. In a very real sense, all labor-saving devices are robots, and all devices are labor saving or we wouldn't use them. To a prehistoric farmer, an ox hitched to a plow is a robot. To a man with a horse, a car is a robot. To the women who were called calculators and depicted in the movie *Hidden Figures*, an electronic computer is a robot.

So robots are going to take away almost all of our jobs, as they have in the past, and *give us new jobs* that we couldn't previously imagine existing. That's the short version.

The long version is that what we now perceive as "real robots"—advanced electromechanical devices that appear to "think" for themselves—can, in fact, do some of the jobs we are accustomed to doing for ourselves. This substitution will increase.

How will this transformation affect our lives in the future?

What has been waggishly called the *robot apocalypse* is really just an extension of the ongoing, millennia-old transition from human muscle power to animal power to machine power, making work easier and multiplying productivity a thousand-fold.

## Automation, Robotics, Artificial Intelligence

The last time we panicked about losing our jobs to machines, the culprit was Automation and the date was sometime in the 1960s. None other than Martin Luther King Jr. said in 1961, "New economic patterning through automation is dissolving the jobs of workers in some of the nation's basic industries. This is to me a catastrophe."[1]

But automation began the first time somebody thought of putting a feedback loop into a mechanical device. Industrial & Manufacturing Automation Experts Alberta says 1771 was the date, and Richard Arkwright was the culprit: he invented the first fully automated spinning mill. "This," says the Alberta group, "this is one of the first examples of a fully automated industrial process."[2]

Robotics is the use of automation to do something you've never seen a machine do before. And artificial intelligence is the use of machines to calculate something (usually using statistical inference techniques developed by Thomas Bayes and William Gosset 250 and 125 years ago) so quickly a person would have trouble doing it.

Okay, I'm oversimplifying. But my point is that machines doing work that used to be done by humans is (1) not new, (2) not magic, and (3) not surprising, since people are always looking for better, faster, and cheaper ways to get things done. All of these transitions raise humanitarian and policy issues, which we'll get to.

## How Low Can You Go?

But something about the current wave of automation feels like *this time is different*. What feels different is that today's robots are doing—or look like they're doing—some cognitive work, which we associate with

[1] American Federation, n.d.

[2] Henrie (2018). Henrie adds a precursor: "1589 automated knitting machine: Queen Elizabeth denies William Lee a patent for automated knitting contraption citing fear it would deprive women of ability to make a living."

"humans only." For example, they're providing medical diagnoses and legal advice. *But they are still only doing what they're programmed to do.*

Robots are also getting cheaper, as mechanical looms and horseless carriages and electronic computers did in ages past. But there is a point of cheapness beyond which humans can't really compete with robots on cost, say Erik Brynjolfsson and Andrew McAfee, the authors of *The Second Machine Age*. The reason is there's only so low a wage can go; the environmental scientist Jesse Ausubel calls that the "wolf point," the point at which the wolf is at the door.[3] Below the wolf point, a would-be worker will find something else to do: keep looking for a better-paying job, beg, borrow, steal, live off their savings, move in with parents, seek charity, or obtain public assistance.

The problem with this analysis is that all replacements of workers by some sort of machines involved *exactly* the same dynamic. That is why we don't see elevator operators, shorthand writers, or pencil-and-paper-wielding human calculators. They were outcompeted on cost and efficiency by the machines that were explicitly designed to replace them. We have been seeing this force at work since the possibly legendary Ned Ludd (see Figure 15.1) smashed mechanical looms in the late 1700s or early 1800s and we always will.

What is different about today's generation of robots is that they are doing what we perceive to be fairly sophisticated work that we associate with intelligent humans, and it will take a *long* time before most job functions, especially cognitive, nonroutine, and emotional ones, can be fulfilled by a robot that outcompetes qualified humans on a combination of quality and price that gives the robot the job. It is unlikely ever to happen so widely as to become a robot apocalypse.

In fact, overall employment may not decline at all. It's more likely that robots will take that *part* of a job that the person doesn't want to do, such as writing down courtroom proceedings by hand. The court reporter now uses a machine. Maybe we'd all be better off if robots, instead of stressed and low-paid workers, loaded boxes onto trucks at Amazon.

## Brand-new Robot, for Sale Cheap?

Why do I think humans can compete with robots? To start with, robots don't work for free. Try buying one. It costs a large fortune to design a usable robot (in the sense that we now use the word) and build a

[3] Ausubel (2004).

**Figure 15.1 Artist's conception of Ned Ludd, enemy of the robots of his day.**

*Source:* Wikimedia Commons, https://commons.wikimedia.org/wiki/File:Luddite.jpg. Artist unknown; captioned "drawn from life by an officer." Published in May 1812 by Messrs. Walker and Knight, Sweetings Alley, Royal Exchange, Cornhill, London.

prototype, and a smaller fortune to make each copy. As Brynjolfsson and McAfee said, they'll get cheaper, perhaps surprisingly so; in 2008 a single transistor cost a millionth of a penny,[4] and I'm sure it's less now.

And it's true that robots don't call in sick, bring their children or pets to work, have psychological problems, get in fights, or call Human Resources (or have HR called on them, or deserve to) with alarming regularity.

But they do break down. And require expensive repairs by expensive repairmen, or by expensive repair-robots. And how do you repair a repair-robot? At some point in the supply chain, there's a human being, or, more likely, lots of them, with expensive engineering degrees.

---

[4] www.kurzweilai.net/average-transistor-price.

In addition, robots rarely make useful "incremental and process" innovations (the little improvements made by line workers who are intimately familiar with the way things are designed and built). Human workers do it all the time, and such innovations are a vitally important contributor to economic growth.[5]

Finally, the cost measure that economists consider relevant in almost all circumstances is *opportunity cost*, what you could do with the resources deployed for a given use if you put those resources to their next-best use. For robots, that cost includes the expertise (and large paychecks) of the engineers, materials scientists, programmers, and countless others who make robots appear to "think" and to do the work they're assigned. There is nothing cheap about robots.

## The Endless Supply Chain

When I get in my car, I am putting Henry Ford, Thomas Edison, and a zillion other inventors, mechanics, engineers, factory workers, and car salespeople to work. I'm *hiring little bits* of each of these people, a point also made by Matt Ridley and mentioned in Chapter 1, "Right Here, Right Now." Most of them have already been paid all they'll ever be paid, because they're dead. Part of the reason that some of them made so much money is that they left behind a legacy of knowledge that now enables me to drive a car that was partly designed, manufactured, or sold by them.

The supply chain goes much deeper, all the way to the farmer who grew the food that these people consumed, and to the inventors and manufacturers of the farmer's implements—and, for that matter, through a myriad of almost invisible connections, to almost everyone in the world who is not living self-sufficiently in the wilderness. But, now that I've mentioned this, it's obvious, so I won't go into any further detail. I mostly just wanted to highlight the economist's habit of looking all the way through to the root causes of things rather than just focusing on what can most easily be seen.

## I'm a Pencil, and Nobody Knows How to Make Me

This idea is closely related to, but not identical with, the brilliant point made by Leonard Read 70 years ago in *I, Pencil*, and later popularized by

[5] We made this point in the previous chapter, but it's worth repeating; see Breznitz and Cowhey (2019) in Adler and Siegel (2019).

Milton Friedman in his 1980 TV program on PBS, *Free to Choose*. I'm a pencil (says the pencil)—I'm a simple device that anyone can afford and use, but *there is no one in the world who knows how to make me*.

What were Read and Friedman talking about? They were showcasing the genius of the global supply-demand price system, which calls forth a supply of graphite ("lead") from Sri Lanka, wood from Canada, eraser rubber from Malaysia, and a little bronze ferrule (the thing that holds the eraser) from God-knows-where.[6] Read and Friedman were trying to teach the principle of interdependence through one lens, the coordinating and motivating power of the price system; and I am doing it through another—the interplay of skills and talents across geographies and generations.

This interplay is particularly remarkable in that most of the people involved in it will never meet each other, don't speak the same language, don't know they are helping to make a pencil, and don't particularly want to make a pencil. They are just doing their jobs, what the supply-demand-price system "told" them to do.

It's tempting to say the system of interdependence is making the whole world work together as a robot, but maybe that's a bridge too far.

All human-invented devices, then, are robots. They do work we could do but would rather not. So far, most robots (and other machines) perform physical labor, but some robots appear more intelligent than others ("artificial intelligence") and save us from having to perform *mental* labor. They still only do what they are programmed to do; if the task is pattern recognition or language translation or some other form of apparent learning, they may amaze with their abilities—but they are still, as someone said a long time ago about computers, very fast idiots.

## The First Robot Apocalypse: Czechoslovakia, 1920

By the way, the word *robot* comes from the Czech word for *forced worker*. The Czech literary man and satirist Karel Čapek popularized the word in his 1920 play *R.U.R.*, which stood for Rossum's Universal Robots. Like robots in many people's imagination today, Rossum's—I mean Čapek's—robots are malign and rebel against their human masters, almost wiping

[6] The humble ferrule has an interesting history. Until World War II, pencil manufacturers were inventive and stylish with their ferrules—one of the ways they differentiated their product was with elaborate brass designs. With the war shortages, manufacturers switched to a basic plastic and cardboard band, and after the war they never returned to the elaborate, only differentiating by ferrule color. See Weaver (2017).

them out. The play is almost universally considered worthless except for the invention of the word *robot* itself.

## Understanding the US Labor Market

Since this chapter is a meditation, not on technology itself, but on *technology and jobs* and how they interrelate, let's take a fairly deep dive into the US labor market and explore some ideas, starting with the concern about technologically driven unemployment and moving to what appears to be wage stagnation, except for highly compensated professionals and owners of capital.

### Unemployment Always Converges to 4%

Will robots put us all out of work, as I asked at the outset?

They will certainly put *somebody* out of work. Low-skill jobs have been at risk from automation for so long that, now, middle-skill jobs are also at risk, the low-hanging fruit having already been picked. Middle-skill jobs include occupations such as bookkeeping, bus and truck driving, and machine operating. What my father did for a living, shorthand reporting, is at the upper end of middle skill—it's very hard to learn—but the cost of acquiring the skill is not always equal to its value. Electronic recording devices do a better job.

As we all know, technology also creates new jobs. The problem is that the people losing the old jobs are not usually the same people as those getting the new ones. It's hard to retool a human being. Of course, we can think of exceptions: the blacksmith who became a car mechanic, the coffee barista who became a professor. But such happy transitions are the exceptions, not the rule. Humans hate change and most are not very good at it. This circumstance raises the policy issues that will be discussed later.

*But unemployment always seems to revert to 4% in every business boom.* See Figures 15.2 and 15.3. This fact is crucial to understanding the interplay between technology and jobs. Deirdre McCloskey writes:

> If the nightmare of technological unemployment were true, it would already have happened, repeatedly and massively. In 1800, four out of five Americans worked on farms. Now one in 50 do, but the advent of mechanical harvesting and hybrid corn did not disemploy the other 78 percent.

In 1910, one out of 20 of the American workforce was on the railways. In the late 1940s, 350,000 manual telephone operators worked for AT&T alone. In the 1950s, elevator operators by the hundreds of thousands lost their jobs to passengers pushing buttons. Typists have vanished from offices. But if blacksmiths unemployed by cars or TV repairmen unemployed by printed circuits never got another job, unemployment would not be 5 percent, or 10 percent in a bad year. It would be 50 percent and climbing.[7]

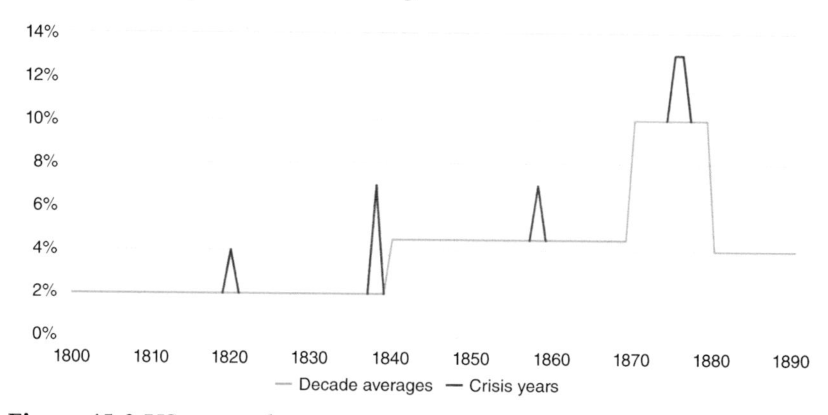


**Figure 15.2 US unemployment rate, retrospective estimates, 1800–1890.**

*Note:* Retrospective estimates are those made later by researching past data, rather than being estimated at the time.

*Source:* Underlying data are from Lebergott (1964), pp. 175–188.

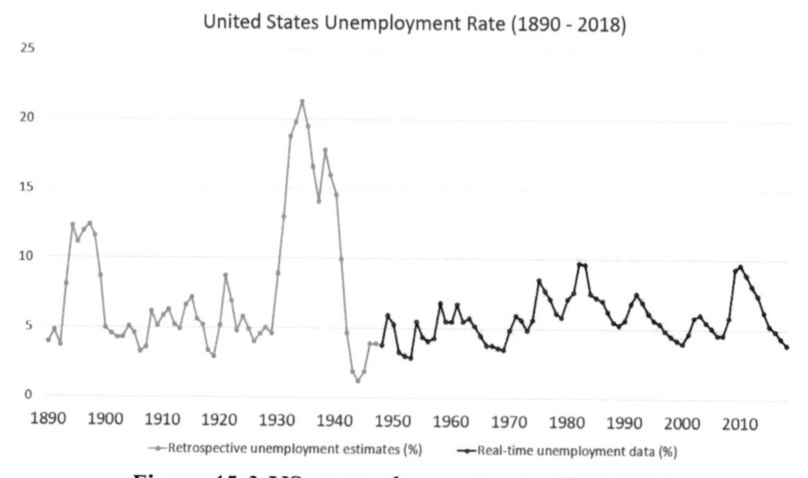


**Figure 15.3 US unemployment rate, 1890–2015.**

*Source:* Chart based on data sourced and organized by Wikipedia user Peace01234. 1930–2009 data are from Bureau of Labor Statistics, Employment status of the civilian noninstitutional population, 1940 to date. 1890–1920 data are from Romer (1986). 1920–1930 data are from Coen (1973).

---

[7] McCloskey (2017).

## "From This You Make a Living?"

Notice, if you haven't already, that this relatively steady-state unemployment number has taken place beneath the backdrop of a tremendous increase in population. In 1800 there were 5.3 million people in the United States; in 1890, 63 million; today, about 330 million. The obvious conclusion—so obvious that it seems a little silly to state it—is that net job creation has been wildly positive, even as technology has eliminated the need for millions of existing jobs.

This will continue, even if we can't exactly predict what the new jobs will be. (Imagine a software engineer or a biotech worker explaining her job to her great-grandmother.) The old will continue to ask the young, "From this you make a living?"

## "But the Labor Participation Rate Changes!"

You bet it does. Your forebears, if you go back far enough and are not descended from royalty, worked from sunup to sundown at backbreaking work that barely yielded enough food for those doing it, and for their children. Actually, the children were working about as hard as the adults. The labor participation rate, or number of people employed as a percentage of the total population, was, in round numbers, 100%. In the more humane primitive societies, the halt and lame, the very old, and the very youngest children were supported. But the basic rule was that if you didn't work, you didn't eat.

From that inauspicious start, the only possible direction for the labor participation rate to go was down. This rate matters because unemployment is calculated as a percentage of those considered to be in the workforce, not of the whole population. Thus, it's conceivable that the unemployment rate could go down or stay steady even while people are losing jobs, because people who cannot find work or do not want to make the needed adjustments (say, by relocating) are leaving the work force. This has happened on some occasions.

So, what became of the labor participation rate as the United States and much of the rest of the world industrialized? In the United States, the rate headed toward zero for children and the very old, a marker of progress. The "market economy" participation rate was clocked at 2.5% for married white women and 38.4% for unmarried white women in 1890, the oldest date for which data are easily available, before rising

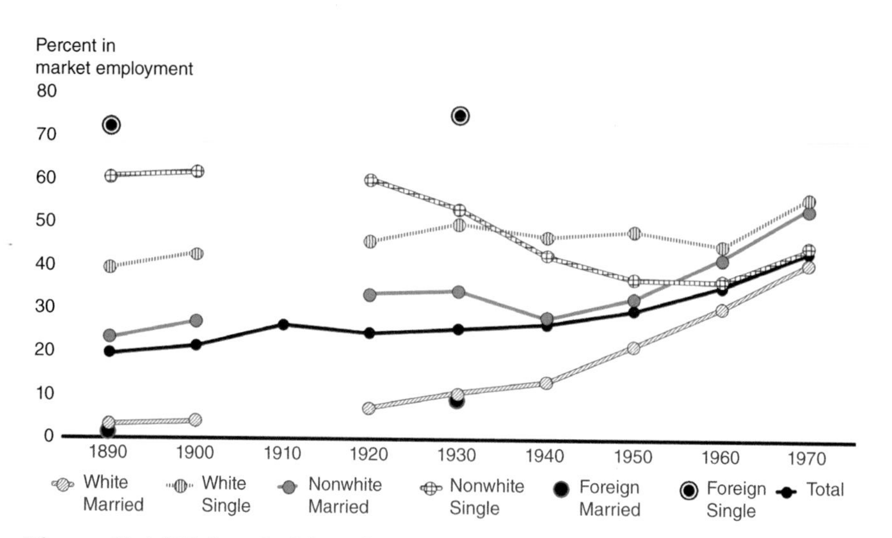


**Figure 15.4 US female labor force participation rate by race and marital status, 1890–1970.**

*Note:* The 1910 labor force figures are too high relative to those for other years because the census reported certain women employed in agriculture as in the labor force rather than at home.

*Source:* Adapted from Goldin, Claudia. 1977. "Female Labor Force Participation: The Origin of Black and White Differences, 1870 and 1880." *Journal of Economic History* 37 (1): 87–108.

dramatically for white women.[8] (Nonwhite women always had much higher rates.) Figure 15.4 shows these trends.

The data in the figure do not imply that women were taking it easy, writing romantic poetry like Emily Dickinson and sipping on mint juleps. Housework and childrearing were much more difficult in 1890 than they are now (remember that families were, on average, much larger then.) In addition, the data in Figure 15.4 are just for nonfarm households. Farming is so physically taxing that the fieldwork more or less had to be done by men. Women did the rest of the work, which was voluminous. My wife's grandmother somehow prepared 357 meals a week—three a day for each of the 17 members of her family, including herself and her husband.

Still, someone has to earn a money income, and that role has traditionally fallen to men. Figure 15.5 shows the adult male labor force participation rate in the United States over a more recent time span.

---

[8] The data in the figure go only through 1970, but by then the largest part of the increase in the female labor participation rate had already occurred. It rose further from about 40% to about 60% by the end of the twentieth century. The data are just for nonrural households. That is very impressive in terms of workload reduction, since the move from farms to cities and towns meant that the extremely difficult nonmarket work by women in farm families was disappearing.

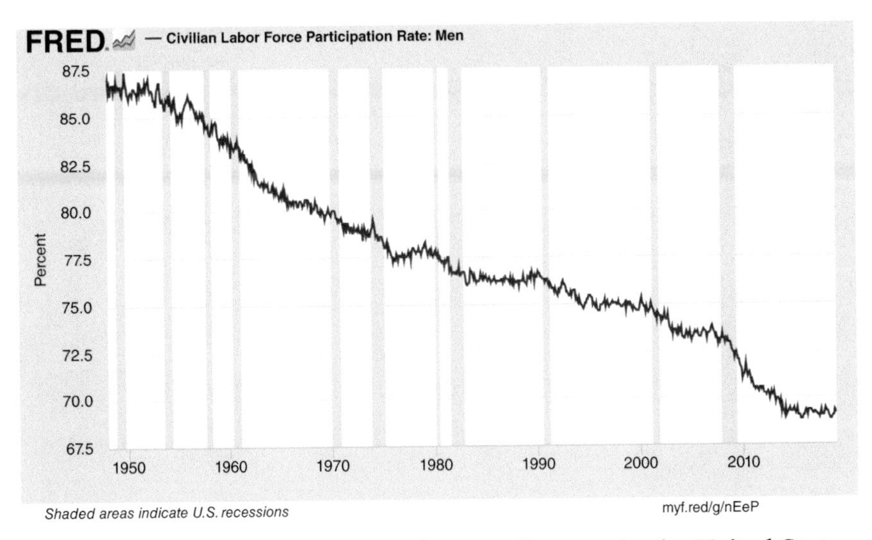


**Figure 15.5 Labor force participation rate for men in the United States, 1948–March 2019.**

*Source:* FRED, Federal Reserve Bank of St. Louis.

On the surface, the trend toward not working (for pay) shown in Figure 15.5 looks like a catastrophe—and, in some individual cases, it is. But let's put the pieces together.

What happened, as shown in Figure 15.6, is:

- Total labor force participation remained fairly steady, around 60%, with a bulge around the year 2000. (An early 1990s *Time* magazine cover bore the caption "Work, work, work.")
- Women entered the labor force in massive numbers.
- Men left it, although not as rapidly as women entered it.
- Over the whole period, there was an increase, not a decline, in the percentage of the working-age population that was working for pay.
- After the year 2000, everybody started working less, but not at a dramatic rate.

The shift of US women into market work—and its counterpart around the work—is as profound a social change as can be imagined. It was liberating for women, but Economics 101 dictates that the change couldn't possibly be equally good for everyone. It's not surprising that, with a huge influx of women into the market economy, at the same time as increased integration of minorities into the workforce and massive

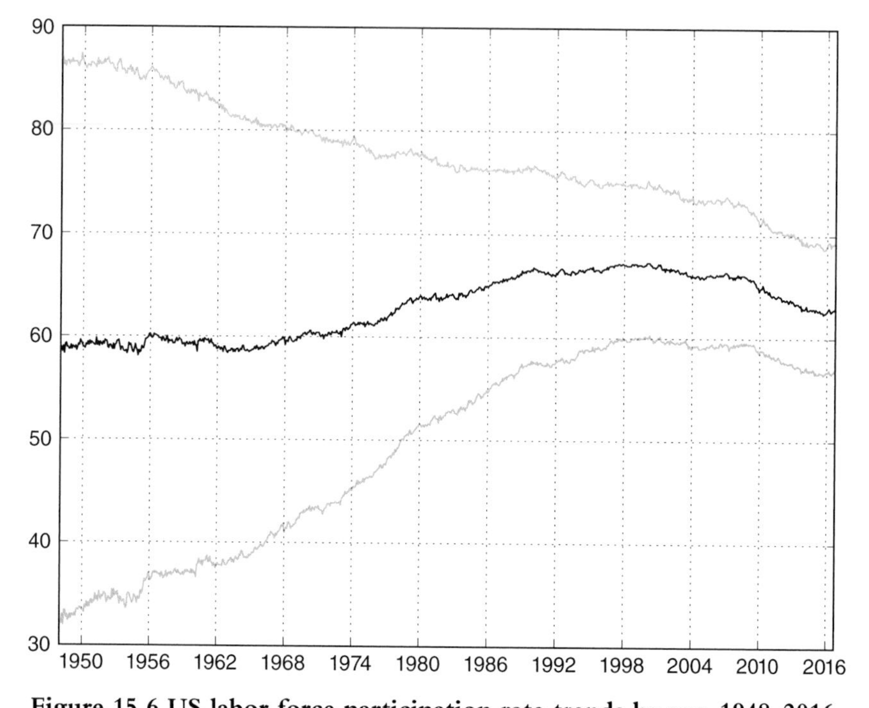


**Figure 15.6 US labor force participation rate trends by sex, 1948–2016.**

*Source:* Wikimedia Commons, https://upload.wikimedia.org/wikipedia/commons/4/42/US_Labor_Participation_Rate_by_gender.svg. Upper line—men, lower line—women, middle line—total working-age population.

immigration in the wake of the Immigration and Naturalization Act of 1965, that wages for low- and middle-skilled men would grow more slowly than they otherwise would have. They might even decline. It's nobody's fault; it's just something that happened.

Incomes of high-skilled workers have, of course, risen greatly for people of all sexes, colors, and places of national origin.

## The Inequality Riddle

***By David L. Stanwick and Laurence B. Siegel***

A commonly held belief in the United States is that the middle class is in decline due to wage stagnation, with the increase in GDP being captured by owners of capital. This analysis reflects some reality (real wages have stagnated or even declined for industrial and certain service jobs) and some error (most of the increase in GDP is being captured by high-skill workers, although the share going to capital has also risen).

It's worth noting that a mass *upper* middle class has emerged in the last two generations, consisting of an unprecedentedly large population of professionals, managers, senior civil servants, academics, highly paid technical workers, and a few artists and scientists. Richard Florida calls them the Creative Class, David Brooks calls them Bobos (bourgeois bohemians), and John Kenneth Galbraith, long ago, called them the New Class. (Pundits spend a lot of time making up cute names for the upper middle class.) Most of the members of this class were recruited out of the traditional middle-class population, shrinking its numbers.

We don't mean to minimize the pain caused by the loss of manufacturing jobs, but the existence of a (somewhat) well-paid corps of industrial workers occurred only for the briefest moment in human history, roughly the 30 years after World War II and, even then, only in a few selected countries, in particular the United States, which was the only major industrial country that was not devastated by the war.

Now that the US labor market has absorbed a massive influx of new workers, wages are likely to rise simply because the Baby Boomers are retiring, and young workers are getting older and acquiring more skills. Increased income disparity in developed countries is a serious issue not to be discounted, but should be viewed as a natural and expected consequence of *reduced* income disparity across countries. If an American accountant is competing with a Chinese accountant, the incomes of the two will converge; if a British sheet metal worker is competing with an Indonesian sheet metal worker, the incomes of the two will converge. This means that income disparity will increase in all four countries but decrease globally! Do the math—you'll see that my statement has to be correct. It's all part of the Great Convergence (see Chapter 1).

Of course, understanding both local and global income dynamics doesn't put another dollar in a person's pocket or solve local issues, where increased disparity can lead to increased crime, poisonous politics, and slower-than-necessary economic growth. These issues need to be addressed. But it might help one to sleep better at night knowing that average incomes across the world are both on the rise and becoming more equal.

## Chasing the Golden Age

Why do so many people seek to return to a Golden Age of American industry, when Marty McFly was an up and coming young man and rock music was actually good? When, in the 1950s and 1960s, employment was stable and wages were high? Let's stop and ask if the Golden Age really existed—and, if it did, why it did and what caused it to end.[9]

I'd note first that just about all peoples in all times have imagined that there was a past Golden Age, better than the present, when "giants in the earth" performed heroic deeds, and to which they have always sought to return.[10] This collective misremembering seems to be hard-wired into the human psyche. But sometimes things do get worse and the memory of a better past is real. Let's investigate this thought in the context of the history of the US labor market.

The destruction all the major industrial economies except the United States experienced in World War II bequeathed to the United States the high ground of large-scale industrial manufacturing in the postwar years. From trains to tractors to rockets, the United States reigned supreme. American physical and human capital created an economic moat that showered wealth on workers, consumers, and shareholders.

Some workers did really well. Union members who worked for large corporations such as the auto and steel companies tended to make good money, bolstered by contracts that kept both wages and employment high, with the consumer surplus still large enough that high labor costs could be borne comfortably. The companies' ability to make a profit, despite such costs, was made possible by the dominant position of US industry at the time. But all such moats are weakened by time and competition, gradually at first and then very rapidly when unexpected shocks occur.

For the United States, the two shocks were (1) the Arab oil embargoes of 1973 and 1979, and (2) the Clean Air Act, both of which caused much of US automakers' capital equipment to become "stranded" assets. (Stranded assets are those that a company has paid for but, for whatever reason, can't use.) Because Japanese and other non-US auto manufacturers were not saddled with these costs, and because Japanese management techniques had progressed while US companies' management stagnated during the period when they were dominant, the US auto industry suddenly faced serious competition from fuel-efficient imported cars. The

[9] This theme is developed further in Sexauer and Siegel (2017).

[10] "Giants in the earth": Genesis 6:4. See Chapter 21.

foreign competition extended to steel, heavy equipment, and consumer appliances.

The newly competitive environment affected everyone in the supply chain. Profits fell. Productivity fell (clean air was a positive externality, not expanded output that the companies could charge for). The US companies could not sustain both high wages and large workforces. Layoffs were massive, and (using two-tier labor contracts) newly hired workers did not make the high wages that their predecessors had earned. The change from high wages and abundant jobs to falling wages and disappearing jobs was exquisitely painful. For the stranded workers, the Golden Age was gone forever.

A new age of high-tech, high-productivity jobs in the US manufacturing sector eventually arose, with far fewer jobs but competitive pay and benefits. And new economic sectors other than manufacturing would create a different class of high-paying jobs. But the new industrial age employs a different generation of people, and the Golden Age workers would never recover.

## The Other Side of the Tracks

The high wages of some US industrial workers in the 1950s and 1960s, however, were not typical. Even while the Golden Age lasted, high-paying union jobs were generally reserved for white men, handed down from one generation to the next ("Dad will get you a job at the factory"). African Americans eventually broke into this market—just before it fell apart.

Meanwhile, on the other side of the tracks, many industrial workers fared horribly. The songwriter and musician Sixto Rodriguez, featured in the Oscar-winning movie *Searching for Sugarman*, was a nonunion laborer in Detroit; he and his family had to move 22 times while his children were growing up—not because he faced discrimination but because he couldn't afford to pay the rent. (Rodriguez is a native-born American citizen.) Although his case may be extreme, we should not look back on the 1948–1973 period, when the United States enjoyed a quasi-monopoly in world markets, as a halcyon time for all US workers.

And, although I'm fully aware of the data showing that median real wages have not increased (or have barely increased) since the end of that period, it doesn't ring true: housing-type wealth has grown hugely, with much working-class housing of the time now considered so substandard that the government can't even give it away to the poor. The house I grew up in, in Cleveland, in a lower-middle-class neighborhood once filled with factory workers, was recently sold for back taxes of $200.

Moreover, as we saw in Figure 14.5 of Chapter 14, "The Alleviation of Poverty," the cost of many goods has plummeted, so that a person with the same dollar income, adjusted for inflation, can buy much more useful appliances and other possessions (and more of them). Marty McFly got a big laugh out of saying his family had two televisions; nobody in the 1955 of *Back to the Future* believed him. His character was not supposed to be poor. Now, even many of the poor have two televisions, plus air conditioning, a smartphone, and, of course, a car.

## The High Cost of Living Alone

If there's a single factor that best explains wages seeming not to go as far as they used to, despite having kept up more or less with inflation, it's the dramatic reduction in household size. The number of households has grown much faster than the population, with single adults now making up 45% of the US adult population. Since there are some economies of scale in managing the cost of living, the same *per capita* income does not go as far when the income is spread out over many different households, with a large fraction of people living alone.

A simple arithmetic example illustrates. Let's say US GDP is $20 trillion and there are 330 million people in the country, living 4 to a household; each household thus gets an average of $242,424 (including taxes, corporate profits, etc.). Now say they live one to a household, each individual in their own house or apartment; each household then gets $60,606. Because of the economies of scale, the standard of living is substantially higher in the first arrangement.

By joining forces, either through marriage, cohabitation, or household sharing (having roommates), then, people can choose to do something about sluggish wage growth, especially for young or less skilled workers. Notice that very few married couples are poor. In 2017, only 3.5% of white, non-Hispanic married couples (including those with children) were living below the poverty level.[11]

## The Adaptability Dilemma

We see, then, that *in the aggregate* people adapt to change stunningly well: the economy having transitioned from agricultural to industrial to service-based to information- and knowledge-based, the unemployment

[11] Duffin (2019).

rate has converged to 4% basically every time. And this took place in the context of a rapidly growing population!

That's the good news. The bad news is that the experience of individuals cannot be gleaned from the aggregate. Individual people don't adapt to change well at all (some do, but it's the exception, not the rule) and may fail entirely at making the transition. They may join the long-term unemployed, and we've even seen falling life expectancy—for the first time in our history other than in wartime or during an epidemic—among mature white men, the demographic group most adversely affected by the latest changes.

## Lots and Lots of Industrial Productivity—Good or Bad?

Figure 15.7 shows what happened in the US manufacturing industry over a (fairly) recent 64-year period. We manufacture more goods than ever (did you know that?), but without using workers. Well, we use a few, but the decline in the number of industrial workers is dramatic and represents a *huge* decrease in the percentage of the population so employed, because the total US population grew from 144 million to 311 million over that period.

Is this good or bad? It depends on whether you're one of the workers (bad, if you're not adaptable), or a consumer, earning your living in

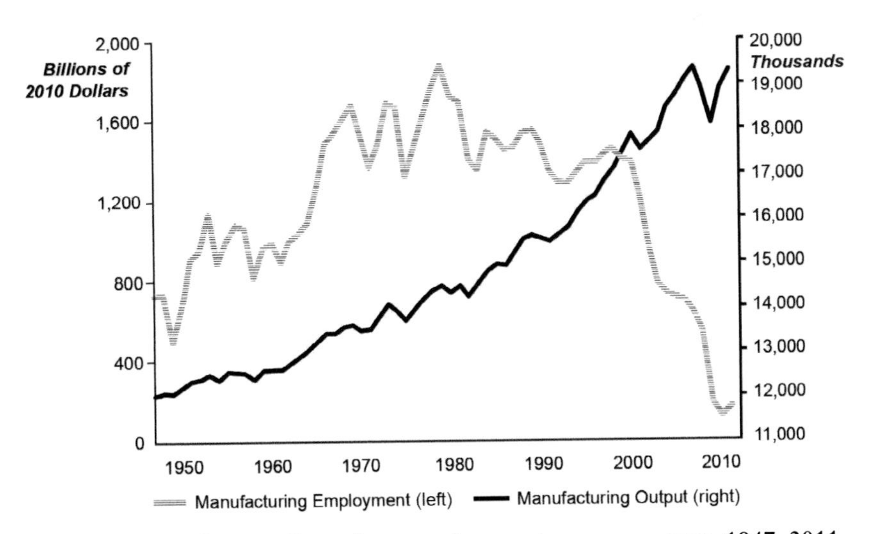


**Figure 15.7 US manufacturing employment versus output, 1947–2011.**

*Source:* Adapted from Perry (2012).

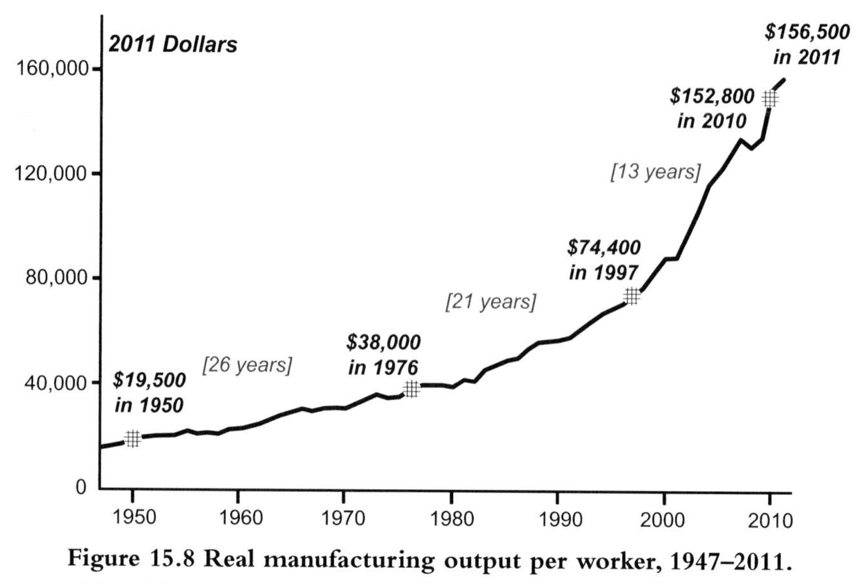


**Figure 15.8 Real manufacturing output per worker, 1947–2011.**
*Source:* Adapted from Perry (2012).

a different line of work and enjoying the plunging real price, that is, the price in terms of number of hours of labor, of all kinds of necessities and luxuries (this is good). Figure 15.8 shows the same data, but expressed as output per worker, also called *productivity*.

Let's hope you're adaptable. More about that later.

## What Can and What Cannot Be Seen (Again)

But with all those jobs lost, and with unemployment hovering around 4% in an expanded population, where did all the new jobs come from? Most people have relatively little intuition regarding this question, because of the Bastiat what-can-and-what-cannot-be-seen problem.

When jobs disappear, they are often lost in mass layoffs or business bankruptcies, which throw a lot of people into the job market at the same time and cause a great deal of misery and anxiety. Such job losses also catch the attention of the media.

But a lot, maybe most, of the new jobs created appear one or two or three at a time. A couple of middle managers go to the boss and say, "The staff is working like crazy and they're not getting everything done. I need money to hire three more people." This doesn't make the newspaper. It also doesn't spark protests, strikes, or angry speeches in Congress, because the newly hired people aren't unhappy. Just the opposite.

## Sunrise, Sunset

So what are these new jobs? Figure 15.9 shows the US Department of Labor's best guesses for the 10 years ending in 2026.

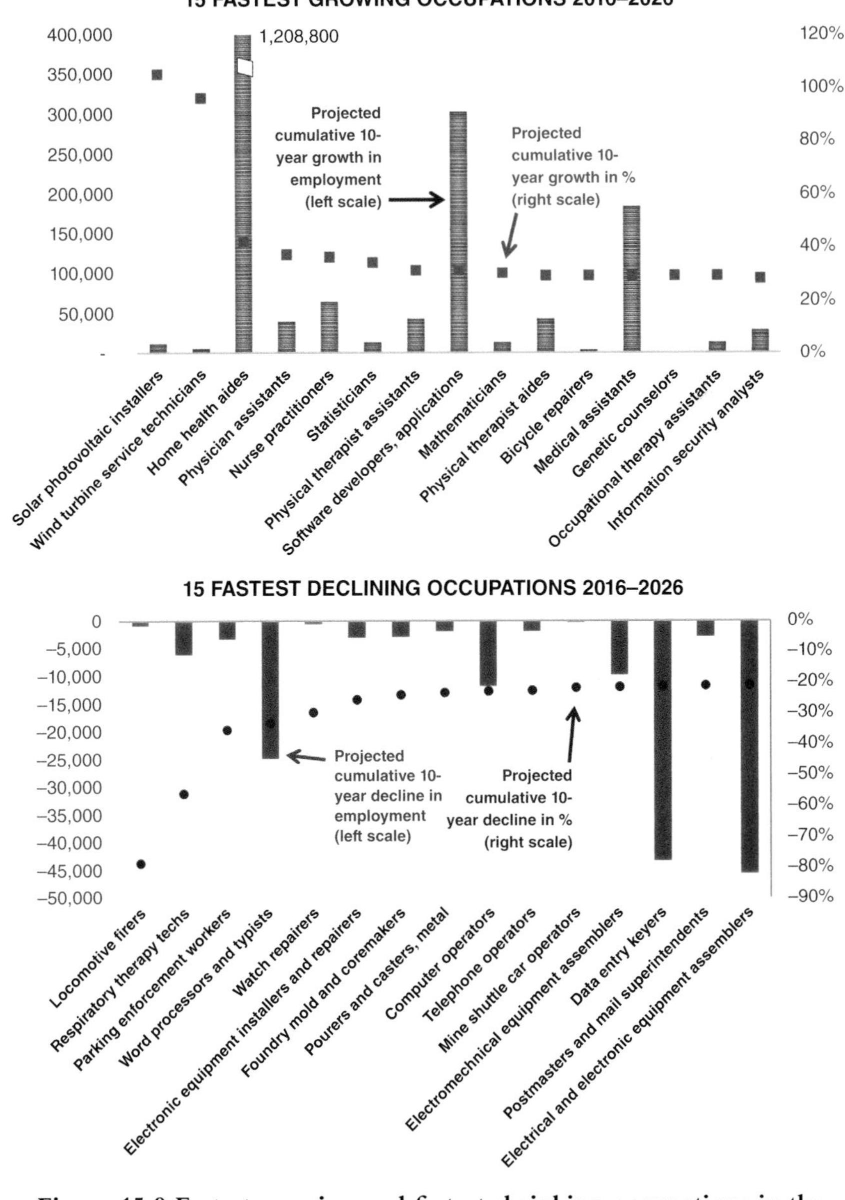


**Figure 15.9 Fastest growing and fastest shrinking occupations in the United States: projections from 2016 to 2026.**

*Source:* US Department of Labor. Occupational Employment Statistics program, US Bureau of Labor Statistics, author's calculations.

Note that the biggest numerical increase is not in technology workers (they're second at 302,500) but in home health aides (a whopping 1,208,800). A large increase in care workers is consistent with an aging society. (I really can't explain the presence of bicycle repairers.) Note that eight of the 15 fastest growing occupations are in health-related fields; what will they do after the Baby Boomers have died and the country is young again?

Now about those sunset industries: about half the jobs sound terrible. Parking enforcement workers, begone! (Lovely Rita, meter maid, was a figment of Paul McCartney's fertile imagination.) There are some exceptions—electronic equipment installers probably have reasonably pleasant jobs—but most locomotive firers, to say nothing of mine shuttle car operators, are probably glad to be doing something else or are retired. The best generalization one can make is that routine, noncognitive jobs are on their way out, and nonroutine, cognitive jobs are on their way in.[12]

This certainly sounds like progress, at both an aggregate and individual level, but what if your natural ability, and training and experience, make you best suited to be a postman or watch repairer? Is there a place for you in the economy of the future?

Almost certainly. To assume there is a large fraction of the population with nothing to contribute, and that people cannot adapt to changing opportunities, is arrogant and dangerous. David Autor, an MIT professor who is considered one of the world's top labor economists, wonders out loud why there are still so many jobs, considering how much automation has already contributed to productivity:

> Why hasn't automation already wiped out employment for the vast majority of workers? [The reason is that] tasks that cannot be substituted by automation are generally complemented by it. Most work processes draw upon a multifaceted set of inputs—labor and capital; brains and brawn; creativity and rote repetition; technical mastery and intuitive judgment; perspiration and inspiration; adherence to rules and judicious application of discretion. Typically, these inputs each play essential roles; that is, improvements in one do not obviate the need for the other.[13]

Thus, workers who possess skills that are complemented by automation will fare well: the emergence of a mass upper-middle class

[12] This theme is treated more fully in Sexauer and Siegel (2019).

[13] Autor (2016).

is testament to that phenomenon. (Consider the transformation of engineering—think locomotive engineer—from a job to a profession over the last century.) Workers who possess skills that are largely substituted by automation will fare poorly. There will be winners and losers, although even low-paid workers benefit from the decline in the real prices of goods and services (that is, the amount of effort required to obtain them) that comes with greater overall productivity.

## Preparing for the Work Environment of the Future

We do not know what skills will be needed in the future. (Who knew that social media marketers would be a thing, much less e-sports content managers, or GoPro experience designers?) We can't know that any more than a teacher in 1875 could know that some of his or her students would be designing telephone circuits or fastening wings onto airplane fuselages. So we cannot teach specific skills with an eye to making someone employable in 2030 or 2050.

Although we cannot eliminate, by fiat, the existence of a bell curve or normal distribution of human abilities, we can moderate its effects through well-considered educational strategies. To do so, we need to combine two ideas: Howard Gardner's 1983 theory of multiple intelligences, and David Ricardo's 1817 law of comparative advantage. It's an odd coupling, so let's go over it carefully.

### Multiple Intelligences

Howard Gardner, a Harvard professor of education, famously noted that general intelligence (IQ), although not useless as a measure of ability, fails to capture the differences in specialized capacities between individuals. He proposed a now-familiar set of "multiple intelligences":

- Linguistic intelligence ("word smart")
- Logical-mathematical intelligence ("number/reasoning smart")
- Spatial intelligence ("picture smart")
- Bodily-kinesthetic intelligence ("body smart")
- Musical intelligence ("music smart")
- Interpersonal intelligence ("people smart")

- Intrapersonal intelligence ("self smart")
- Naturalist intelligence ("nature smart")[14]

It is obvious that there is some truth to this classification scheme. Now let's run it through one of the most basic "machines" that economics provides: the theory of comparative advantage. Originally applied to countries (Portugal should produce wine and Britain wool, even if Britain is better at producing both, as long as they can trade with each other), we'll apply it to *people.*

## Comparative Advantage

No one has an absolute advantage over everyone else in each ability or type of intelligence. And everyone is good, or at least better than somebody else, at something; no one is completely disadvantaged (pathological cases aside).

To maximize the usefulness and adaptability of each individual, then, our educational practices should help young people understand and develop their areas of comparative advantage *while exposing them to other areas of life, work, and thought at which they don't have an advantage,* because (1) someday they might, (2) they will have to interact with people who do, and (3) you might have incorrectly identified their area of advantage.

If everyone has a comparative advantage in *something,* it follows from that observation, plus one of the most basic economic principles, Say's Law (supply creates its own demand), that *there is no one in this world who is economically worthless.*[15]

## Why Are Low-skilled Workers So Poorly Paid?

What, then, is stopping low-skilled workers from getting ahead? Here are some factors:

- A mania for occupational licensing (to keep the riffraff out, when that's whom we're purportedly trying to help). Did you know you need a license to wash someone's hair (for pay) in Tennessee?
- Payroll taxes, unemployment taxes, mandated benefits, human resources overhead, and unbelievably complex labor laws that are

[14] Armstrong (undated). The list of intelligences is from Gardner (1983). The annotations ("word smart," etc.) are by Armstrong.

[15] Again, excepting special cases like the totally disabled, who must be supported.

supposed to help the worker, or workers in certain "protected" classes, but actually keep them from being hired.

- In particular, the high cost of the mandated healthcare benefit for workers in businesses with 50 or more employees. To limit their costs, some businesses keep their staff sizes below 50, which keeps workers from being hired. Others pay for the health insurance but that limits the cash compensation they can afford to give their workers. Although health insurance is a good thing, the cost is spectacularly high and could be lowered by allowing interstate insurance company competition and by permitting employees to choose what coverage they want and need (keeping the cash difference if they choose a low-cost plan).

Sadly, some of the low status of unskilled workers is the workers' fault. When you're less skilled, the most important attribute to have, other than the ability to learn and become more highly skilled, is *conscientiousness*. Low-skilled workers who don't show up, use huge amounts of sick or personal time, don't communicated planned absences, come in drunk or stoned, or are perpetually late aren't worth much money. A few high-skilled workers can get away with erratic behavior because their specific skills are indispensable to a given employer, but when you're easily replaced, that dog doesn't hunt. *Teaching diligence as a core value to the young* is one way to combat this self-imposed disadvantage.

## Conclusion: The Arrogance of the Pessimists

The last word in this discussion should belong to David Autor. In a TEDxCambridge talk, he eloquently explained why we should be optimistic, not pessimistic, about the future of jobs in an age of rapidly changing technology:

> Now, you may be thinking [that I have] told . . . a heartwarming tale about the . . . past . . . but probably not the future. Because everybody knows that this time is different. Is this time different? Of course this time is different. Every time is different. On numerous occasions in the last 200 years, scholars and activists have raised the alarm that we are running out of work and making ourselves obsolete: . . . the Luddites in the early 1800s; US Secretary of Labor James Davis in the mid-1920s; Nobel Prize–winning economist Wassily Leontief in 1982; and of course, many scholars, pundits, technologists and media figures today.

> These predictions strike me as arrogant. These self-proclaimed oracles are in effect saying, "If I can't think of what people will do for work in the future, then you, me, and our kids aren't going to think of it either." I don't have the guts to take that bet against human ingenuity. . . . I can't tell you what people are going to do for work a hundred years from now. But the future doesn't hinge on my imagination. If I were a farmer in Iowa in the year 1900, and an economist from the 21st century teleported down to my field and said, "Hey, guess what, farmer Autor, in the next hundred years, agricultural employment is going to fall from 40 percent of all jobs to two percent purely due to rising productivity. What do you think the other 38 percent of workers are going to do?" I would not have said, "Oh, we got this. We'll do app development, radiological medicine, yoga instruction, Bitmoji."
>
> I wouldn't have had a clue. But I hope I would have had the wisdom to say, "Wow, a 95 percent reduction in farm employment with no shortage of food. That's an amazing amount of progress. I hope that humanity finds something remarkable to do with all of that prosperity."[16]

"And," concluded Autor, "by and large, I would say that it has." I agree. The new Golden Age is manifested by the great gains in health, wealth, and cultural variety that productivity growth has showered on us. It should endure for a long time if the basic forces of adaptation and innovation, combined with property rights, are allowed and encouraged to continue.

[16] www.ted.com/talks/david_autor_why_are_there_still_so_many_jobs/transcript?language=en#t-1105332.

# 16

# *The Mismeasurement of Growth: Why You Aren't Driving a Model T*

Over the last 250 years, real incomes grew by a factor of about 100 in the richest countries and 30 on average across the world. Do those numbers overstate or understate human progress over that period? The answer depends on what you think is important.

This chapter focuses on why the measured growth rate is an understatement. The most obvious source of understatement is that, for much of the period, if you had certain needs—a shot of penicillin in "poor" Nathan Rothschild's case—you were out of luck. It hadn't been invented or discovered yet. So having a hundred or even a million times more money wouldn't help—likewise if you needed a computer, a ride on an airplane, an air conditioner, or a conversation with your mother who lives in another state.

Chapter 20, on hedonic adaptation, talks about why the mammoth rise in incomes over the last 250 years might *overstate* the amount of human betterment. Life has always been a mixture of good and bad, happy and sad. That has not changed and it is not about to change, and the ratio of happy to sad has not gone up by a factor of 30 or 100. Human nature, the ability to adapt to some degree to whatever circumstances you find yourself in, causes that ratio to be far more stable than is suggested by money incomes or the quantity and quality of possessions or even health "outcomes" (which are, of course, incomes—each day of good health is a benefit).

## Punkahwallah, Please

Many inventions simply bring to the masses services that have long been available in some form to kings, queens, and captains of industry. The rich have always had some sort of computer, or home theater system, or air conditioner.

Fans of the movie *Hidden Figures* know that the word *computer* used to refer to human calculators. A person who needed calculations performed could hire a roomful of them. In 1606, King James I of England (he who was also James VI of Scotland) "could sit at home and watch a bloody audiovisual drama about witches," the economist and blogger Bradford DeLong reminds us. "He had William Shakespeare and the King's Men on retainer."[1] If you were the lady of the manor in a tropical country, you could summon a couple of fellows, preferably handsome natives (punkahwallahs in Hindi), to operate a simple mechanical fan (a punkah) and keep cool.

Some inventions, of course, really are new, with no acceptable substitutes available to the rich in olden days. Think air travel and long-distance telephony. And penicillin.

## Supply Side and Demand Side

We can separate material progress into a *supply side* and a *demand side*.

The supply side refers to what's available and at what cost.

The demand side refers to what people can afford.

The two sides are linked by the price. What is unaffordable when the good or service is scarce and expensive is affordable when it's abundant and cheap. And a good or service that is unaffordable when people don't have much money is affordable when they do.

Replacing the labor-intensive solutions I just described with cheap, universally available machines is one of the two main forms of

[1] DeLong (2017).

supply-side material progress. Inventing truly novel devices, processes, and experiences is the other. We tend to think of the new gadgets as the really important innovations, but doing old things in slightly better ways, with the small improvements accumulating over time, is just as important. That is how the Model T Ford mutated into the modern car, a process we'll explore in some detail.[2]

Changes in the demand side come from increases in income, but where do the increases in income come from? They occur because people are producing more with less, *being more productive*, creating things that were not there before or that were there but at a higher price. In this sense, all economic progress comes from changes in supply, a point made by "supply side" economists from John Stuart Mill to Arthur Laffer but rarely fully understood.

## Chitty Chitty Bang Bang

Martin Feldstein, who is about as accomplished an economist as you can get—Harvard professor, chairman of President Reagan's Council of Economic Advisers, and head of the National Bureau of Economic Research—has written an influential article detailing the mismeasurement of long-term economic growth. In "Underestimating the Real Growth of GDP, Personal Income, and Productivity,"[3] Feldstein is one of many scholars who argue that the quality and performance of many products and services have increased more than we realize. As a result, real GDP, the usual measure of economic growth, understates the actual progress that has been made.

Take cars, for example. A 1908 "flivver," or Ford Model T, did one thing that modern cars do—it drove. But it could only be started using a hand crank, had no doors, no heat, no gas gauge, and the tires frequently

---

[2] This progressive mutation over time is an example of a series of sustaining (not "disruptive") innovations. The work of Professor Clayton Christensen of Harvard Business School addresses the difference between these two types of innovations. See Christensen et al. (2015). In part because of Christensen's popularity, many innovations are now incorrectly described as disruptive when, according to Christensen's theory, they are actually sustaining (they enhance the value of an existing idea that has already been substantially developed and implemented).

[3] Feldstein (2017). Reprinted in Adler and Siegel (2018).

**Figure 16.1 Model T Ford—1912 Speedster.**

*Source:* imageBROKER/Alamy Stock Photo.

went flat and wore out in 5,000 miles at best.[4] The model pictured in Figure 16.1, a 1912 Speedster, also had no windshield.

But it was a convertible! At least until it started raining. ("Convertible" does not mean open top—it means you get to choose whether the top is open or not, something you couldn't do with this beauty.) The unreliability of cars became a standard comedy trope, from the early movies of Buster Keaton to the unfortunately named *Chitty Chitty Bang Bang* a half-century later.

It was a real slog, however, to drive 5,000 miles before there were any paved roads (see Figure 16.2). Early drivers had to navigate horse trails, or no trails; the 1909 New York to Seattle automobile race was won in 23 days, much slower on a mile-per-hour basis than the Pony Express, which ceased operation in 1861.

The real value of a car wouldn't be fully realized until a generation later, when a network of paved, but mostly two-lane, roads crisscrossed the United States. The network, inaugurated in 1926, was called the US Numbered Highway System, and, along with state highways built

[4] https://modeltfordfix.com/wp-content/uploads/2017/08/Riverside.jpg.

**Figure 16.2 Lincoln Highway between Ames, Iowa, and Nevada, 1918.**[5]

*Source:* https://iowadot.gov/history/transportationiniowa. Public domain.

around the same time, was quite good; much of it survives today.[6] That is also about when a car radio would have become useful—the first radio station began broadcasting in 1920, and the country was blanketed with stations by the end of the decade.

## Take Your Car in for a Hedonic Adjustment

Henry Ford was known for pricing his cars affordably, but in 1908 he was only able to get the price down to $850, a year's pay for a common American laborer. Good cars today also cost about a year's pay for a laborer, so we can make a hedonic comparison. (By 1925 the price of a Ford runabout, or economy car, reached the stunningly low level of $260 at a time when wages had risen greatly from their 1908 level; now that's progress! But let's stick with 1908 for a moment.)

What do you get for a year's pay in 2019 that you didn't in 1908? Since the list of quality improvements and added features is way too long

[5] This photo also appears in a book I highly recommend, Meinig (2004).

[6] But it was not a national system—the "US" highways were built by the states, and a national association of state highway officials coordinated the numbering so that a driver could cross the country on a series of connected highways by following the signs for, say, Route 66. The first truly national road system, the Interstate Highway System, was begun in 1956—which is not that long ago.

to list, let's try a different approach, one favored by economists: Having bought a 2019 car, how much would you *demand to be paid* to give up some specific attributes?

- The reliability, speed, and smoothness of a modern car, compared with one built in 1908.
- Two or four doors, a roof, and a windshield.
- A heater and air conditioner.
- A gas gauge (if you're as absent-minded as I am, it's the most important of the many gadgets that have been added.)
- A GPS navigation system (now also built into your mobile phone), along with roads that make the GPS worth having.

You get the idea. (Your answers are what an economist would call the *reservation price* for the attributes.) After subtracting the reservation price of just a handful of these characteristics and features, you'd be valuing the 1908 car at less than zero.

But the 1908 car was far from worthless. Actually, it was a miracle. It freed its owner from the boredom of country life or the rigidity of train and streetcar schedules in the city. It especially helped to liberate women. "Rural women, in particular, welcomed the possibility of relieving their isolation by driving into town to shop, to sell their farm produce, or to attend farm clubs," writes British historian Margaret Walsh.[7] It was probably worth more than the modern car in terms of the difference it made in the owner's quality of life.

Yet GDP statistics say that a good or service is worth what you paid for it. These statistics value a 1908 Model T at $850 and a representative 2019 car at $30,000. As I pointed out earlier, these are roughly the same amount of money when adjusted for wage inflation. (This is an exception to the fact that, as we've seen in several contexts, most products other than cars have fallen tremendously in terms of the amount of effort required to buy them.) So, is a laborer who gives up a year's wages—amortized, of course, over the life of the car—better off in 1908 or now?

Now, obviously—by a large multiple; one that nobody knows how to estimate. But the GDP statistic, if not "hedonically adjusted," that is, modified to account for the differing amounts of utility provided by the car in 1908 and today, says that no gain has taken place.

[7] Walsh (undated).

So how large a hedonic adjustment should we make? Economists have developed a technique, called hedonic regression, which produces estimates of smaller gains in utility from product improvements, but that technique is useless for change as profound as the evolution of a 1908 car into a modern one. Sorry—I can't even provide an order-of-magnitude guess. If nobody knows how to estimate the number, we might as well just say it's large and move on.[8]

Repeat the same mental exercise for 1970 and 1990—or whenever your memory of car quality and performance begins—versus a car being produced and sold today. You'll be surprised at how much improvement there has been, even recently.

### The Cambrian Explosion of Auto Designs

While we're on the topic of cars, let's take a very quick side trip into evolutionary biology. The automobiles of the 1930s were indescribably gorgeous, and very different from one another. Today's cars all look more or less the same. The reason is that the 1930s were a period of adventurous experimentation to come up with the best designs—reminiscent of the Cambrian Explosion, 540 million years ago, when nature came up with a wild variety of animal body plans. These later settled down into a few, the rest having become extinct because they were inefficient in some way.

In the 1930s, gasoline was cheap. When it became expensive and auto manufacturers faced mileage-per-gallon restrictions, they moved toward fuel-efficient designs that limit the use of ornamentation and unusual body shapes. It's even said that all the auto companies now use the same optimizer (a program for figuring out how to maximize fuel efficiency)! No wonder today's "evolved" cars all look similar. (See Figures 16.3 and 16.4.)

(*continued*)

[8] Note that I said the 1908 car was probably worth more than the modern car in terms of a lifestyle upgrade. One might naively conclude from that comment that the hedonic adjustment should be in the other direction, old car better than new car. That is not correct—it is not the right comparison for determining the amount of economic growth that has taken place. A piece of bread is worth more to a hungry man than a pound of steak is to a well-fed one; they are different people in different circumstances. But no one would argue from that observation that the bread is worth more than the steak. It's the same with the cars. The modern car is still a much better car, and the early car purchaser seeking liberation from her farmhouse would have strongly preferred it if it had been available.

*(continued)*

**Figure 16.3 Bugatti Type 57 Atlantique (three were sold).**

*Source:* https://commons.wikimedia.org/wiki/File:RL_1938_Bugatti_57SC_Atlantic_34_2.jpg. Licensed under CC BY-SA 3.0.

**Figure 16.4 1938 Delahaye 165 Cabriolet.**

*Source:* michaelsimkov@123rf. Reproduced with permission.

## Adjusting for Wholly New Goods (Caution, Geeky)

What we've done on a macro scale (macro because a 1980 and 2019 car are so different, and the years so far apart) needs to be done for every product improvement or degradation, as well as for every service (much harder), in order to obtain a useful series of *per capita* incomes or productivity statistics. Wholly new goods present a special problem. Let's take a look at it.

The automobile is, again, a good example; when introduced, it fit the description of a wholly new product pretty well. It is not a souped-up horse and carriage, nor is it a type of train.

The auto industry shows up in GDP calculations from the start (or would have if the concept of GDP had existed at the time) because GDP at least theoretically adds up the value of all the transactions in the economy. So far, so good.

But, to arrive at real GDP, which is the most commonly used indicator of economic progress, nominal GDP numbers are adjusted by an inflation measure, for example the Consumer Price Index (CPI). (To ascertain changes in the standard of living, real GDP is further adjusted for changes in population to arrive at real GDP *per capita,* a number we studied closely in Chapter 7, "The Great Enrichment"; but let's ignore this last step for now.)

The rub occurs when moving from nominal to real GDP in the face of a new product, in this case cars. Cars didn't make it into the CPI "market basket" until 1935, when most households already owned one, so most of the early and dramatic decline in auto prices—remember, $850 to $260 for a better car—didn't enter into the inflation measure. This represents a considered decision by the Bureau of Labor Statistics, which constructs the CPI. If they had put cars into the market basket much earlier, which would have captured the price decline, the basket would have been unrepresentative because most people did *not* have cars then. There is no ideal solution to that riddle.

So nominal GDP went up and up as car manufacturing became an increasingly important part of the economy between, say, 1900 and 1935. But the inflation rate used to convert nominal to real GDP was overstated because the falling price of cars wasn't counted in inflation. So the growth rate of real GDP was *understated.*

And that number is what we focus on when comparing standards of living over long periods of time.[9] This understatement occurs every time a novel product is introduced at a high price that later falls, as with televisions, computers, and nearly every other technological innovation.

## Computers, the Internet, and Microprocessors in Everything

Everybody has heard of Moore's Law—computing power doubles every 18 months—but the consequences of anything growing that rapidly are hard to think about. (To be precise, it's not overall computing power that doubles every 18 months, but the number of transistors per square inch on an integrated circuit or "chip"—details, details.)

What this growth rate means, when extrapolated over decades, is that your smartphone has millions of times more computing power than NASA's whole battery of computers had in 1969—the whole agency, not just one of its spacecraft. Actually, the *battery in your laptop* probably has more computing power than the Apollo 11 spacecraft did, although the spacecraft's computer was, thankfully, more reliable and crash-proof. Did you know that your laptop battery has a little computer in it? It does.

To cover the economic consequences of the information technology revolution would take a whole book. Erik Brynjolfsson and Andrew McAfee's *The Second Machine Age* is a good one.[10] Here, we'll just hit a few high points.

## Is Information Technology Worth What You Pay for It, or More?

It doesn't take a genius to note that the tech industry is many times more productive than it was a generation ago. But the amount of money spent on technology is, in real terms, about the same as it was in the 1980s—it's just that you get a lot more for it now. Is there a measurement problem here?

---

[9] Because there are a lot of moving parts to this comment, here's a simple numerical example. Suppose nominal GDP grew at 7%. If inflation-as-measured was 3% but true inflation, lower than measured inflation, because it includes the falling price of cars, was only 2%, then the true growth rate of real GDP was 5%, not 4%.

[10] Brynjolfsson and McAfee (2014).

There certainly is. When calculating GDP, the contribution of information technology to the economy is based on cost, not value—the GDP method implicitly assumes that value is equal to production cost for any good or service. So a sector that has become more "efficient" (this seems like a lame way to describe the explosion in technology, but it will do for now) in delivering value for the price paid has its influence understated.

Although information technology has displaced other sectors, such as book and record sales, labor in factories, and tellers in banks, those declines are in the GDP calculation and the rise in the value of technology is not. So GDP growth is understated in this way too.

James Suriowiecki reports on finance and technology for *The New Yorker.* Writing up an interview with *The Second Machine Age* co-author Erik Brynjolfsson, he notes:

> "We're underestimating the value of the part of the economy that's free," [Brynjolfsson] said. "As digital goods make up a bigger share of economic activity, that means we're likely getting a distorted picture of the economy as a whole."
>
> Digital innovation can even shrink GDP: Skype has reduced the amount of money that people spend on international calls and . . . , thanks to Google and Apple Maps, [the GPS device company] Garmin's sales have taken a severe hit, but consumers, who now have access to good directions at no cost, are certainly better off.

What are the consequences for measurement of economic progress? Suriowiecki continues,

> New technologies have always driven out old ones, but it used to be that they would enter the market economy, and thus boost GDP—as when the internal-combustion engine replaced the horse. Digitization is distinctive because much of the value it creates for consumers never becomes part of the economy that GDP measures. That makes the gap between what's actually happening in the economy and what the statistics are measuring wider than ever before.[11]

But there's a downside: the job mix changes, to the detriment of old-economy workers: "Wikipedia is great for readers. It's awful for the

[11] Alll quotes in this section are from Surowiecki (2013).

people who make encyclopedias," Suriowiecki writes. There may also be a decline in the absolute number of jobs. In the long run, I don't think so—we've been through two other industrial revolutions that, as we saw in Chapter 15, "Robots Don't Work for Free," created far more jobs than they destroyed—but that's little consolation if you have one of the old jobs and are poorly equipped for one of the new ones. And people also have to eat in the short term.

## Other Products

This tale could be repeated, less dramatically, for just about any industry or product. Three-dimensional printing is the source of a number of gee-whiz stories about increase productivity; my favorite is the way that NASA obtained a ratchet wrench on the International Space Station. They didn't fly one up—they *beamed* it up, by sending a computer file to the station's 3D printer, which then deposited 140 layers of plastic in the shape of a wrench and its handful of moving parts. Despite a wrench having a very Victorian feel, looking like it should have been made in a blacksmith's forge, when NASA needed one, they *printed* it. In *space*.[12]

Information technology, and its cousin, robotics, have had a profound impact on productivity in a wide variety of settings. Consider Figure 16.5. US industrial output hits a new high every year or comes close to it, while manufacturing employment is lower than it was in 1947 (and industrial employment is a drastically lower percentage of the population because our population is much bigger).

What happened, starting about 1980, is that machines of varying degrees of sophistication began to replace workers much more rapidly than they had before. Around 2000, the trickle of workers out of manufacturing became a flood.

Is this progress? Sure. The less work it takes to make or build something, the better off we are in aggregate. And tens of millions of jobs have been created in other fields since 1980. But the human problems created by the transition should not be minimized. (We discussed this at greater length in Chapter 15.)

[12] NASA Explores (2014).

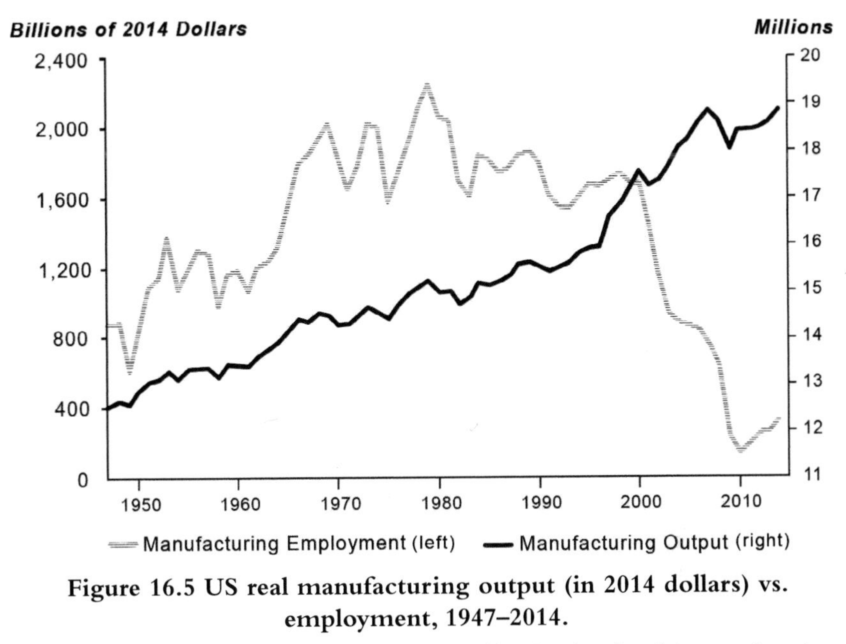


**Figure 16.5 US real manufacturing output (in 2014 dollars) vs. employment, 1947–2014.**

*Source:* Adapted from Perry (2015), www.aei.org/publication/october-2-is-manufacturing-day-so-lets-recognizeamericas-world-class-manufacturing-sector-and-factory-workers/.

## Services: It's Hard to Find Good Help These Days

A common perception is that many services have gotten worse. That's almost certainly true of personal services, which used to be performed by a moderately skilled class of people. Now that many middle-skilled people are instead trained to be nurses, technicians, tradespeople, and so forth, it's harder to find good "help." (Even so, they use more and better machines than they used to; a vacuum cleaner beats a broom.) But let's look at the most widely used service, the one taking up the biggest slice of the economic pie: medicine.

### Don't Like Your Genetics?

Like information technology, the intertwined fields of medicine and biology need a whole book to describe the speed and extent of their advances in the last century and especially in the last quarter-century. Consider this: *CRISPR is a tool, to be specific an enzyme, that can be used to edit the genome of human beings or anything else that has a genome.* Don't like your genetics? Are they making you sick? In 50 or 100 or 1,000 years, you'll be able to change them. With a computer doing the driving.

A hundred years ago, nobody knew we had a genome. A computer was a person with a pencil, an adding machine, and a knack for math.

This change might come more quickly. As of 2017, "the CRISPR genome-editing method, [which] was only developed in 2012 . . . , is proving so powerful and effective that around 20 trials in humans have already begun or will soon. Almost all of these involve removing cells from an individual's body, editing their DNA and then putting them back into the body," says Michael LePage, writing in *New Scientist*.[13] "Conventional" gene therapy, which began around 1990, is already in widespread use, but has much more limited application than what is theoretically possible with CRISPR.

It sounds like science fiction, but so does any technology with which one is unfamiliar. "Any sufficiently advanced technology is indistinguishable from magic," wrote the celebrated science fiction author Arthur C. Clarke in 1968.

## Doctor, I Can't See . . .

Using cataract surgery as his example, Martin Feldstein, the economist whom we met at the outset of this chapter, complains that GDP and related statistics don't capture these gains:

> None of these measures of productivity attempt to value the improved patient outcomes. As one concrete example, when Triplett and Bosworth (2004, p. 335) wrote about the remarkable improvement in treating cataracts—from more than a week as an immobilized hospital inpatient to a quick outpatient procedure—they questioned whether. . . "the increased value to the patient of improvement in surgery . . . belongs in national accounts if no additional charges are made."[14]

Those mercenary medical types! Oops, Barry Bosworth is a respected economist and senior fellow at the Brookings Institution, and was director of President Jimmy Carter's Council on Wage and Price

[13] Le Page (2017).

[14] Feldstein (2017), quoting Triplett and Bosworth (2004).

Stability. Jack Triplett, also at Brookings, is an expert on economic measurement and the CPI in particular. Beware of conventional thinking.

## Doctor, I'm Too Fat . . .

How would you like to add three years to your life expectancy? Here, take some statins.

Feldstein again:

> [C]onsider statins, the remarkable class of drugs that lowers cholesterol and reduces deaths from heart attacks and strokes. By 2003, statins were the best-selling pharmaceutical product in history . . . [b]ut the official statistics never estimated any value for the improvement in health that came about as a result of the introduction of statins.
>
> To understand the magnitude of the effect of omitting the value of that single healthcare innovation, [consider that] . . . researchers . . . found that taking a statin caused a 35 percent reduction in cholesterol and a 42 percent reduction in the probability of dying of a heart attack. . . . [T]he combination of reduced mortality and lower hospital costs associated with heart attacks and strokes in the year 2008 alone was some $400 billion, which was almost 3 percent of GDP in that year.

Feldstein didn't say this, but I'll venture that the part of the $400 billion representing decreased spending on heart attack and stroke treatments *was* counted in GDP—as a negative. Since the increase in well-being was not counted as a positive, that's just one more example of the understatement of progress by economic data.

## Medical Innovation Summary

Although there are many diseases, especially cancers and dementias, that have not yet yielded to medical research, I would rather be born now than in 1954 when I was in fact born—I might be able to avoid a great deal of future medical trouble. And I would *much* rather have been born in 1954 than in 1908, when Calvin Coolidge Jr. began his very brief life, or in 1777, when Nathan Mayer Rothschild started his sumptuous, but not all that long, journey.

## Conclusion

Martin Feldstein concludes,

> [D]espite the various improvements to statistical methods that have been made through the years, the official data understate the increase in real output and productivity. The measurement problem has become increasingly difficult with the rising share of services that has grown from about 50 percent of private sector GDP in 1950 to about 70 percent of private GDP now. The official measures provide *at best a lower bound on the true real growth rate* with no indication of the size of the underestimation. [my italics]

But Simon Kuznets, the Nobel Prize–winning Russian-American economist, invented the GNP (forerunner of GDP) measure of economic performance, so he should get the last word. He said about his brainchild, "The welfare of a nation can scarcely be inferred from a measure of national income."[15]

Kuznets did not know that his comment would inspire a whole subspecialty in economics, that of hedonic pricing and adjustment. He was right, of course, but for comparisons of standards of living across long periods of time and among countries, it is the best measure we have. It's good to know that GDP growth is highly likely to have understated the long-run rate of improvement in well-being.

[15] Organisation for Economic Co-operation and Development (2006).

# 17

# The Discreet Charm of the Bourgeoisie: *Deirdre McCloskey, Capitalism, and Christian Ethics*

Luis Buñuel, *auteur* of the film with the above familiar title, was joking: he despised the bourgeoisie. So do many other artists and intellectuals. It's mostly jealousy, evidenced by the easy lifestyles into which most artists and intellectuals glide when they get a little money. But critics of the bourgeoisie have a wisp of a point: both the *petite bourgeoisie* (little middle class) of shopkeepers and functionaries and the *haute bourgeoisie* (high middle class) of businesspeople and professionals can be hidebound, stuffy, materialistic, silly, and sometimes more than a little bigoted.

But what Buñuel and his allies don't realize, and what the economist and philosopher Deirdre McCloskey powerfully reminds us in the series of books that is the basis for this chapter, is that the bourgeoisie is the *foundation on which a civilized society is built.*

Really? Yes. Great achievements are important, and I'll get to them in a moment. But, for a society to thrive, a large number of people have to live decent, productive lives doing ordinary things. The shoe repairman, the software developer, the grade-school teacher, the bank vice president—the bourgeoisie *haute* and *petite*—are as important to a functioning civilization as the great man or woman of science. As Mister (Fred) Rogers reminded us, you do not have to do anything spectacular to have a worthy and meaningful existence.

## Deirdre McCloskey, Oracle of Bourgeois Virtues

McCloskey, a former teaching assistant of Milton Friedman and until 2015 a distinguished professor at the University of Illinois at Chicago, is the author of three books on the contributions of the bourgeoisie, one on the rhetoric of economics, and many others across diverse fields. Perhaps uniquely among the professors of the world—we'll never know—she simultaneously held appointments in six departments: economics, history, English, communication, philosophy, and classics. (In the last two, she was an adjunct.) I met her when she was teaching at the University of Chicago, in the mid-1970s.

Like many highly accomplished academics, McCloskey is both impossibly charming and an odd duck. She defends capitalism and middle-class values, so she has attracted the admiration of many conservatives. She protests, however: "I'm a literary, quantitative, postmodern, free-market, progressive-Episcopalian, Midwestern woman from Boston who was once a man. Not 'conservative'! I'm a Christian libertarian."[1] That's a lot of adjectives, suitable for describing a complex person.

Among noneconomists she is best known for having transitioned at age 53 from male (Donald McCloskey) to female, and writing a book about it, *Crossing: A Memoir*, that became a New York Times bestseller. Her professional announcement of the change, in an economics journal, was better titled: "Some News That at Least Will Not Bore You."[2]

Some critics find McCloskey's *Bourgeois* trilogy—*The Bourgeois Virtues*, *Bourgeois Dignity*, and *Bourgeois Equality*, all weighty volumes—rambling and too clever by half. I disagree. The books are a stunning accomplishment in philosophy, history, and ethics, not "just" economics.

---

[1] www.deirdremccloskey.com.

[2] *Eastern Economic Journal* (Fall 1995).

## *Épater la bourgeoisie:* Buildings That Make You Cringe

But before taking a deep dive into McCloskey's philosophy and why I think she's on to something, let's have a little fun. What happens when the bourgeoisie does *not* rule?

We get the absurdity illustrated in Figure 17.1.

In Figure 17.1, an aerial photograph of the Left Bank, an almost literal "f—— you" rises from the honey-gold legacy of ten centuries of Parisian

**Figure 17.1 View of Paris's Left Bank. The background, with its churches, stately mansions, and commercial buildings, was built by the bourgeoisie; the skyscraper (the Tour Montparnasse) was built by an architect who hated them.**

*Source:* Steven Strehl, 2004, https://commons.wikimedia.org/wiki/File:Tour_montparnasse_view_arc.jpg. Licensed under CC BY-SA 3.0.

aesthetics. The Tour Montparnasse, built in 1969, just after the high point of Modernism, is the target of repeated attempts by Parisians to have it torn down. Yet it survives and probably will until 2069, a monument to hatred of the bourgeoisie (and love of technocratic efficiency).

In *Current Affairs*, a quirky journal of left-leaning thought, Brianna Rennix and Nathan J. Robinson explain why most modern architecture is so terrible: "A generation of architects with both socialistic and fascistic political leanings saw ornament as a sign of bourgeois decadence and cultural indulgence." Dishing snark on a particularly ugly building designed by the architect Peter Eisenman, the authors "note the total lack of plant life. Plant life might accidentally make you feel happy and comfortable, and happiness is a bourgeois illusion."[3]

At their best, then, the bourgeois not only want to be happy and comfortable; they want to make other people comfortable, through such frivolities as decorative art and traditional architecture. Their opponents think (I suppose) that, in a world full of sorrow and woe, it's dishonest to build something pretty that makes you feel good.

Better to share in the architect's misery-by-proxy. (The architect himself, or herself, lives in comfort but imagines that no one else does.)

Reality is, of course, painted in shades of gray, not black and white, and there are aspects of technocratic efficiency that make your life better and aspects of the bourgeois way that do not. But you'll have to admit that bourgeois architecture just looks better.

## Invasion of the Body Snatchers

Six hundred miles east, in Graz, Austria, is an even weirder attempt to *épater le bourgeoisie* (shock the comfortable, what many artists aim to do). Forgive my poor grammar; the English and French parts of my last sentence don't combine well. The indescribable object in the middle of Figure 17.2 is the Kunsthaus Graz, the city's art museum, built in 2003. Well, almost indescribable—one observer did a pretty good job of it, saying it resembled the beating heart of the monster Cthulhu in H. P. Lovecraft's 1928 horror story.[4]

---

[3] Rennix and Robinson (2017).

[4] Currently pronounced "k'THOO-loo" by fans, although Lovecraft wrote that it should be pronounced "khlûl′-hloo," which I can't pronounce at all. Lovecraft (1928). The description of the building is from Eric Grundhauser's contribution to *Atlas Obscura*, www.atlasobscura.com/places/kunsthaus-graz.

**Figure 17.2 Kunsthaus Graz (Graz Art Museum), Graz, Austria.**
*Source:* Photo by Marion Schneider and Christoph Aistleitner, 2006, https://commons.wikimedia.org/w/index.php?curid=1406917. Licensed under CC BY-SA 2.5.

What were the creators of the *Coeur de Cthulhu* trying to do or say? A clue is in the title of co-architect Colin Fournier's essay, "Urban Transgression and Metamorphosis and a Friendly Alien: Kunsthaus Graz."[5] Transgression means sin or offense; forgive us our transgressions, say some versions of the Lord's Prayer. So what Fournier and Sir Peter Cook, the other architect, were trying to do to us is not printable. They achieved it. Let's move on.

## So, Back to the Value of Bourgeois Values

With our modern sensibilities, which seem strangely classical and anti-democratic, many of us suppose that artists, writers, composers, doctors, scientists, architects, and philosophers are the frame. It's true that we'd be in deep trouble without them. But they do not work for free, nor, in a free society, can they all work for the state or the church. They have to

[5] In Feireiss (2003).

have customers, someone who needs the work done, people who want paintings made, symphonies composed, buildings built.

Usually these customers or patrons are drawn from the merchant classes, ranging from the Medicis and the Rockefellers, at the high end, to you and me in our roles as readers, concertgoers, patients, and students. The grand engineer Isambard Kingdom Brunel (1806–1859) would have remained a tinkerer in his workshop if there were no one who wanted to, and could afford to, ride across England in a train or pay the toll on a suspension bridge across the River Avon at Clifton.[6]

Following this logic, McCloskey also proposes that bourgeois values are *morally* good: diligence, persistence, moderation, modesty. (How boring! We want heroes! But she's right. *Vide* Rogers and his Neighborhood.) This is different from saying that the bourgeois are useful in building and populating a civilized society.

And, in her most important contribution, she identifies the newfound "dignity and liberty" of those engaged in business in the 1600s in the Netherlands, and then in the 1700s in Britain and America, as the catalyst that enabled the Great Enrichment to begin. "In other words, it was *words*" that sparked the transformation—words that freed up imaginations and entrepreneurial spirits.

## McCloskey's *Bourgeois* Trilogy

In what she calls an exordium (somewhere between a preface and a précis; the author loves words, especially Latin ones), McCloskey distinguishes among the three books:

> 1. first, in *The Bourgeois Virtues: Ethics for an Age of Commerce* (2006), [I say] that, the commercial bourgeoisie—the middle class of traders, inventors, and managers—is on the whole, contrary to the conviction of the "clerisy" of artists and intellectuals after 1848, pretty good.
>
> 2. second, in *Bourgeois Dignity: Why Economics Can't Explain the Modern World* (2010), [I argue] that the modern world was made not by the usual material causes, such as coal or thrift or capital exports or imperialism or good property rights or even good science, but by . . . technical and . . . institutional ideas among a newly revalued bourgeoisie.

[6] Actually his father's workshop. Marc Isambard Brunel (married to Sophia Kingdom, thus young Isambard's delightful name) was also a great engineer.

3. and third, in . . ., *Bourgeois Equality: How Ideas, Not Capital or Institutions, Enriched the World,* [I suggest] that [the advancement] arose . . . from a novel liberty and dignity enjoyed by all commoners.

4. The upshot since 1800 has been a gigantic improvement for the poor . . . and a promise now being fulfilled of the same result worldwide—a Great Enrichment for even the poorest among us.[7]

## Virtues Masculine and Feminine, Christian and Pagan

*The Bourgeois Virtues* expands, across many chapters, on the "pretty good" qualities of "the entrepreneur and the merchant, the inventor of carbon-fiber materials and the contractor remodeling your bathroom, the improver of automobiles in Toyota City and the supplier of spices in New Delhi." McCloskey is no Ayn Rand, implausibly making saints out of businessmen and businesswomen. *Pretty good*, not flawless or always admirable.

McCloskey also defends the bourgeoisie against the absurd charges that have been leveled against it by the "clerisy," not just organized religion but today's priesthood of secular intellectuals, especially those who follow in the footsteps of Marx.[8] Capitalism (which she says is a misnomer) is not exploitation but organized cooperation, people working together in order to more effectively produce, compete, and improve.

But the main thrust of *The Bourgeois Virtues* is surprising for a celebrated professional economist and historian: it's mostly about ethics, Christian ethics in particular. McCloskey classifies the seven classic virtues as masculine (prudence, temperance, justice, and courage), especially admired in the pagan or pre-Christian era; and feminine or Christian (faith, hope, and charity). A complete life involves all the virtues.

McCloskey notes that the word *charity*, in Latin *caritas*, also translates to the Greek *agape*, roughly spiritual love, one of the three forms of love the Greeks recognized. They are *eros*, or sexual love; *philia*, friendship; and *agape*. So *charity* does not just mean "generosity." Elsewhere, McCloskey

---

[7] All quotes in this section come from McCloskey (2016).

[8] A provocative and passionate article on the history of hostility to business and businesspeople is in Goldberg (2018). Aristotle, Plato, Cicero, and Jesus all figure in this history. Even more than Marx, Martin Luther comes off particularly poorly, something of a contradiction given his role in promoting the primacy of the individual relative to central authority (in his case, the Church); although Luther was a religious fanatic, not a humanist or skeptic, many historians have drawn a direct line from the Protestant Reformation to the Enlightenment and modernity. I guess he didn't consider Jews to be fully human.

indicates that love is universal, not "feminine." So there are three masculine or pagan virtues, three feminine or Christian ones, and love, which has no gender. A perfect balance!

In an economics book?

Yes. There is more to life than economics can comprehend, a concept that pervades McCloskey-the-economist's work. She introduces a fellow named Max U to make this point. Max is a perfectly rational economic agent with the last name U, punning on the phrase "Maximize Utility" (shortened to *max U* in formulae). Nerd humor.

McCloskey's point, of course, is that Max U is not mirrored in the real world and that "the economist's theory is not complete." To hear her tell it, the economics profession believes that if your mother loves you it's because she is maximizing her own utility, getting something out of it. I don't think most economists believe this, but maybe some do. At any rate I think we'll all agree that there is more to life than that. Thus, McCloskey's emphasis on Christian ethics and various other aspects of moral philosophy.

## Max U, Where Are You?

But her campaign against Max U is a little overwrought. She picks on several thinkers whom I respect, especially Steven Pinker ("a P man," P for prudence-only, or Profane as opposed to Sacred). I *know* that Pinker does not believe people make decisions based on cold utility calculations, or that they should. His expansive view of humanity pervades everything he writes.[9]

Likewise with the late Gary Becker, whom McCloskey admires but nevertheless describes as "seem[ing] to think in this [Max U] fashion." Becker, a deserving Nobel Prize winner if ever there was one, applied economic reasoning to family life and practically every other aspect of existence. Because Becker found it helpful to employ a tool, in this case economic analysis, of which he was an expert user does not mean he thought it was the only tool worth using. Becker had as complete a view of the human spirit as any economist, ever.

Just as economists must achieve a balance between formal analysis and a broader understanding of the human condition, we all must find a balance between economic activity and other matters. The last word

[9] To the extent that it is possible to know a man by reading thousands of pages of his writing. I do not know him personally.

on this question should probably belong to John Adams, who in 1780 wrote to his wife,

> I must study Politicks and War [the P-virtues] that my sons may have liberty to study Mathematics and Philosophy. My sons ought to study Mathematicks and Philosophy, Geography, natural History, Naval Architecture, navigation, Commerce and Agriculture, in order to give *their* Children a right to study Painting, Poetry, Musick, Architecture, Statuary, Tapestry and Porcelaine [the S-virtues].[10]

(My parentheticals; spelling as in the original; S for sacred, in McCloskey's lingo.)

## Explanations for the Great Enrichment

In addition to setting forth an "ethics for an age of commerce," as the subtitle of her first book proposes, McCloskey had another purpose in writing the *Bourgeois* trilogy: to document the fact that the Great Enrichment happened, to say how important and surprising it was, and to ask *why*. *Why* at that particular time and place, northwestern Europe in the second half of the eighteenth century, when opportunities for self-sustaining economic growth had occurred in antiquity in Egypt, the Fertile Crescent, Greece, and Rome, at various times in China and India, and in many places in Europe during the Middle Ages and early modern times?

It's an important question, given that, according to McCloskey, the beginning of the Great Enrichment was *the most important event in human history.* Before, almost universal poverty; after, an immense improvement in living standards, continuing to this day and extending to almost every corner of the world. The second and third volumes of the *Bourgeois* trilogy explore these questions.

Historians, economic and otherwise, have had a field day thinking up explanations for this most surprising hinge of history. Some popular explanations include:[11]

- *Capitalism*—that is, private ownership of the means of production, with relatively free markets in capital, goods, and labor: the old, conventional

[10] Adams (1780).

[11] The Northwestern University economic historian Joel Mokyr, a friend of McCloskey's, has written an excellent summary of this dispute and the various points of view within it; see Mokyr (1993).

explanation. The first Arnold Toynbee (uncle of the better-known historian) wrote in 1884 that "the essence of the Industrial Revolution is the substitution of competition for the medieval regulations which had previously controlled the production and distribution of wealth."[12]

- *Capital accumulation*—not capitalism *per se*, but a large *amount* of capital (owned by whomever), built up over time, to which workers could apply their skills. Thomas Piketty supports this view.
- *Property rights* and equitable, enforceable laws, which are necessary conditions for capitalism and well-functioning markets. Daron Acemoglu and James Robinson [2012] are the latest in a long line of proponents of this story.
- *The Enlightenment*—much admired by Steven Pinker.
- *The scientific revolution.* This explanation is associated with Walt W. Rostow, who, in 1960, described the economic "takeoff" into "self-sustained growth" that, in his telling, started in England around 1820.[13] Rostow's analysis was widely followed by policy makers for a generation after the publication of his book, *The Stages of Economic Growth: A Non-Communist Manifesto.*
- *Improvements in institutions.* The late economic historian and Nobel Prize winner Douglass North promoted this idea, broadening the property-rights theory to include innovations such as patents and incorporation laws. "When it is costly to transact, institutions matter," wrote North. Thus, better institutions lowered transaction costs and made rapid economic development possible.
- *Political fragmentation in Europe.* The great economic historian Joel Mokyr, much admired by McCloskey, writes, "Interstate rivalry . . . constrained the ability of political and religious authorities to control intellectual innovation. If they clamped down on heretical and subversive (that is, original and creative) thought, their smartest citizens would just go elsewhere (as many of them, indeed, did)."[14]
- And my favorite, *The Republic of Letters.* Not just "letters," as in the college of arts and letters, but *letters* that you put in an envelope; this

[12] Toynbee (1884).

[13] Rostow (1960). Rostow, an economist and political scientist, served as Special Assistant for National Security Affairs to President Lyndon B. Johnson.

[14] Mokyr (2017). The article also appears as a chapter in VoxEU.org (2017).

Republic of intellectual scribblers was the Enlightenment's proto-Internet. Joel Mokyr argues that this small group of influencers was a critical factor in the Great Enrichment.

It's a great image, Isaac Newton and Benjamin Franklin and Samuel Johnson, quills in hand, changing the world. This story is not that far removed from McCloskey's invocation of "words" as the deciding factor in the Great Enrichment, but the words had to spread among the common people to have the observed impact, not just be read by a few intellectuals after opening well-traveled envelopes. And they did spread, gradually, as the words of the letter writers were published in newspapers, magazines, and pamphlets accessible to anyone who could read.

McCloskey sums up, then, what the causes were *not*:

> . . . coal, thrift, science, transport, high male wages, low female-and-child wages, surplus value, human capital, geography, institutions, infrastructure, nationalism, the quickening of commerce, the late medieval run-up, the First Divergence, the Black Death, the original accumulation of capital, eugenic materialism, high science, or property rights.[15]

"Such causes," she says, "had been fairly routine in a dozen of the leading organized societies from ancient Egypt and China down to Tokugawa Japan and the Ottoman Empire." Accepting these explanations is like invoking oxygen as the cause of a fire: they're necessary but not sufficient conditions.

And my earlier recitation of potential explanations is more like a pile of puzzle pieces than a solved puzzle. So we need a new explanation, one unique to the Great Enrichment's time and place, and capable of adding explanatory power to an already rich set of factors.

Deirdre McCloskey's reason for the Enrichment, as you already know, is the *newfound dignity and liberty of the bourgeoisie*.

This new respect proceeded, of course, from a change in the intellectual and social atmosphere—originating, perhaps, in the Republic of Letters. It's fair to say, however, that thoughtful and widely read champions of freedom had been writing pamphlets and publishing in newspapers since the days of Erasmus and Spinoza, 200 and 100 years earlier,

---

[15] McCloskey (2016).

respectively. What was new, McCloskey writes, is an egalitarianism that infected the common people:

> The more fundamental definition of equality, praised in Scotland after it awoke from its dogmatic slumbers, is the egalitarian opinion people have of each other, whether street porters or economic philosophers. Adam Smith, a pioneering egalitarian in this sense, described the Scottish idea as "allowing every man to pursue his own interest his own way, upon the liberal plan of equality, liberty and justice."[16]
>
> How did this mutual respect promote economic growth?
>
> The double ideas of liberty and dignity, summarized as Scottish "equality," or political "liberalism" in a mid-nineteenth-century definition, mattered . . . more than wars or trade or empire or financial markets or high wages or high science. The Bourgeois Revaluation led to a Bourgeois Deal. "Let me creatively destroy the old and bad ways of doing things, the scythes, ox carts, and factories without robots, and I will make you-all rich."
>
> The Deal crowded out earlier ideologies, such as ancient royalty or medieval aristocracy or early-modern mercantilism or modern populism. The bettering society . . . was not ruled by the great king or the barons or the bureaucrats or the mob, all of whom took their profits out of zero sum. It was ruled by the betterers, taking their profits out of a positive sum produced by water-powered saw mills and hand-puddled wrought iron. A creatively bettering bourgeoisie then invented the steam ship and the widespread secondary school, the telephone and the Internet, and enriched us all.

## Conclusion

Novelists are the canary in the coal mine, and Theodore Dreiser, introducing (in two different novels) Jennie Gerhardt and "Sister" Carrie Meeber as country girls gone to the city to seek their fortune in early-twentieth-century America, was one of the more perceptive ones. Writing in 1917, the usually acid-tongued H. L. Mencken noted that escape

[16] All quotes in this section come from McCloskey (2016).

from poverty is good for the soul.[17] McCloskey recalls: "Mencken, no softie, noted . . . à propos Jennie Gerhardt's and Sister Carrie's good fortune that, 'with the rise from want to security, from fear to ease, comes an awakening of the finer perceptions, a widening of the sympathies, a gradual unfolding of the delicate flower called personality, an increased capacity for loving and living.'"

And return to poverty is bad for the soul. In his affecting morality tales, Dreiser, no softie, sent his fictional creations, Carrie the actress and Jennie the good-hearted mistress, back to poverty and unhappiness as their later lives unfold. A reminder of the tragic vision is a needed offset to the relentless optimism of an economics book that documents the Great Enrichment. McCloskey does not provide the reminder, but of course she knows it well.

To conclude: the social critic Irving Kristol wrote in 1979, "bourgeois society . . . is a society organized for the convenience and comfort of common men and common women."[18] It is not perfect, but it is the best that millennia of human striving have come up with. Not exploitation at all. Just very useful work, done pretty well.

[17] Mencken (1917)

[18] Kristol (1979).

# 18

# *Simon and Ehrlich: Cornucopianism versus the Limits to Growth*

The debate between optimists and pessimists on the future of the human race was, for a time, dominated by two large personalities: Julian Simon (1932–1998), a marketing professor in a business school, and Paul R. Ehrlich (1932–), an entomologist by training. A more unlikely pair of adversaries could hardly be imagined. But, boldly departing from their original fields of expertise, they became a generation's spokesmen for "cornucopianism," the idea that resources expand to meet needs, and catastrophism, respectively.

## The Cornucopian

Julian Lincoln Simon could be hard to love. He delighted in taking unpopular positions and defended them combatively. He was not an economist, environmental scientist, or demographer. He did not think much of the theories or evidence behind global warming. Astonishingly, he did not want the population explosion to end. His most compelling thought was that more people meant more brains working on solutions to problems—more Newtons, more Edisons, more Einsteins. Describing *people* as "the ultimate resource," he advocated for continued population growth. Worrying about what people were going to eat or where they'd live was for somebody else. At least in the recent past, population growth had usually meant better lives.

Simon was right about many things but wrong about an important one: the ability of the world's population to grow indefinitely without negative consequences. No physical quantity can grow at a positive rate forever. Simon was never clear on how many is too many. In his mind, the marginal contribution of each person to the world's well-being had to be, almost as if by magic, larger than that person's marginal cost to the environment and to other people.

However wrong Simon was about some things, he was right about two: (1) we've gotten richer as our population has grown. Nobody disputes that. And, (2), useful natural resources have gotten cheaper in real terms, that is, in inflation-adjusted dollars or the amount of labor needed to acquire them—and will continue to.

## The Famous Bet

At the time Simon made the latter claim, almost everyone disbelieved it, because we were experiencing an oil shortage and the prices of other resources were rising too. Paul Ehrlich, tiring of being mocked by the abrasive Simon, eventually agreed to a $10,000 bet that a portfolio of metals, to be chosen by Ehrlich, would either be cheaper in 1990 than it was in 1980 when the bet was made (in which case Simon would win) or more expensive (Ehrlich would win).

Simon won.[1] Figure 18.1 is a chart of the composite price of a basket of "all commodities," including but not limited to those in the Simon–Ehrlich bet, from 1913 to 2019 with the period of the bet highlighted.

Metals prices rose again after Ehrlich paid Simon the $10,000. For a while, the prices surpassed their 1980 starting point but then fell again. From October 1980 to June 2015, the real (inflation-adjusted) combined price of the resources in the Simon–Ehrlich bet fell by 26.6%, a much slower rate than over the 10-year span of the bet itself. And there were huge price spikes upward in the prices of many commodities, including metals, in 2007 and 2011.

But, over the very long run, the chart shows that the real prices of raw materials have fallen dramatically on average, as theory says they

[1] A detailed account of the Simon–Ehrlich bet is in a 2010 *New York Times* article by that newspaper's redoubtable science editor, John Tierney. Tierney (2010). See also Simon (1998).

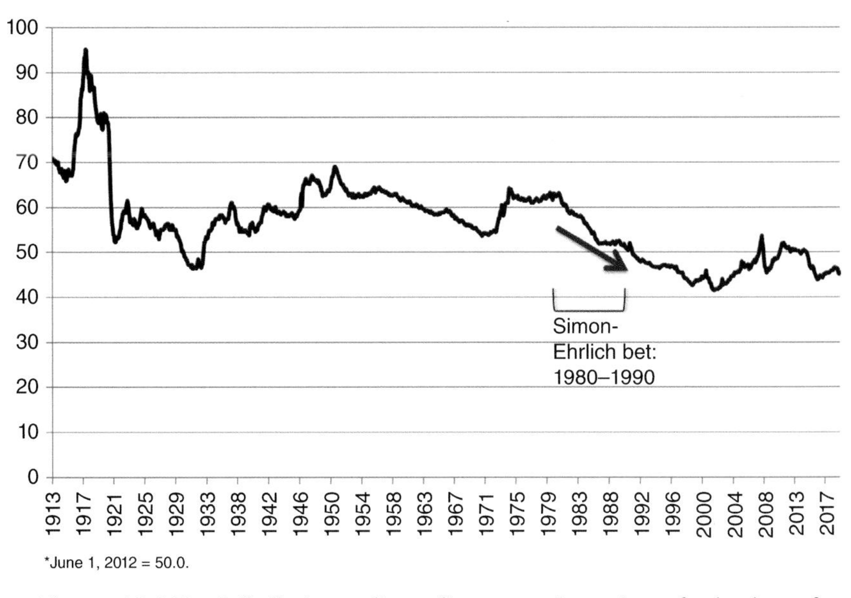


**Figure 18.1 Real (inflation-adjusted) composite price of a basket of "all commodities," 1913–2019.**

*Source:* Constructed by the author using data from FRED, Federal Reserve Bank of St. Louis.

should. When something becomes scarce and expensive, producers are motivated both to look for more supply and to find substitutes. Meanwhile, consumers are incentivized to use less. Finding substitutes is the most important effect: we make "tin" cans out of aluminum, "silver" photographic plates out of silicon, and so forth. These forces put continual downward pressure on real prices. We should expect this phenomenon to continue.

## The Doomsayer

Paul Ehrlich, in contrast, is by all accounts funny and pleasant. Julian Simon described him as "a treasure-trove of snappy quotes." Even Ehrlich's insults are memorable, having called Simon the leader of a "space-age cargo cult."[2] Ehrlich believed, with all his heart, that the world was—and maybe still is—on a path to self-destruction caused by population growth and the despoilment of the natural environment.

[2] Simon (1998) (both quotes).

When you believe that, and think you have the evidence to prove it, you'd better speak up. It's more important than if your steadfast belief is that everything is pretty much OK. Ehrlich's passion recalls Martin Luther's unforgettable words (if he ever spoke them): "Here I stand—I can do no other."

Ehrlich wrote bestselling books, most notably *The Population Bomb* (1968), and endorsed others (*Famine 1975!* written by William and Paul Paddock in 1967), that would cause any well-meaning soul to panic if the forecasts were any good. Ehrlich's case was weakened by the wild nature of his claims—he said in 1970 that "in ten years all important animal life in the sea will be extinct" and that he "would take even money that England will not exist in the year 2000."[3]

His warnings, which fit in well with the apocalyptic *zeitgeist*, got a lot of attention, as any persuasive person would if they seriously made the case that the world was coming to an end.

Ehrlich's forecasts were profoundly and almost laughably wrong. The Green Revolution was in full flower, launching world *per capita* calorie consumption on an almost unbroken upward path since the time *The Population Bomb* was written. In addition, birth rates had already started to fall. But, like that other hedgehog, Julian Simon, he was right about one big thing.

(Calling Simon and Ehrlich hedgehogs is not an insult. Drawing on a saying of the Greek poet Archilochus, the modern philosopher Isaiah Berlin contrasted "foxes," who know many little things, with "hedgehogs," who know one big one. The world needs both.[4])

Ehrlich was right about this: if population simply kept growing exponentially at then-prevailing rates, the peak rate being 2.1% per year (which occurred right around the time *The Population Bomb* was published), it would have almost certainly outrun the food supply sooner or later. At that rate of growth, the world would have had 28 billion people by 2068 and 1.8 trillion by 2268. It won't.

Then why were Ehrlich's forecasts so wrong? Three reasons:

1. Population growth results from the combined effect of fertility and longevity, and death rates aren't going to fall to zero.

---

[3] Oceans: An Ehrlich speech on Earth Day, April 22, 1970. England: Ehrlich's speech at the British Institute of Biology, September 1971. Both as reported by Bailey (2010).

[4] Berlin (1953). The Archilochus quote ("the fox knows many things, but the hedgehog knows one big thing") is often misattributed to the much better-known Aesop (he of the *Fables*).

2. The food supply increased faster than he expected, mostly due to Green Revolution–type innovations, which he mostly disregarded.
3. Most important, as an entomologist and not an economist, he didn't seem to understand the most basic of economic mechanisms: people responding to incentives. As the price of something (in this case children) goes up, or as the benefit goes down, they "buy" less of it. And, as we saw in Chapter 4, "Having Fewer Children," families reacted to these changing incentives very quickly.

Stewart Brand, the wide-ranging thinker and activist whom we'll meet in Chapter 24, "We Are as Gods," said of Ehrlich that his prophecy, while wrong, was valuable because it was *self-negating*. The opposite of self-fulfilling. Because the catastrophists made the general public aware of a looming disaster, Brand reasons, it became possible to do something about it. Disaster was averted. I think that is too kind, but it's possible that the response to Ehrlich's catastrophism contributed at least a little to today's rosier prospects.

Mostly it just scared the bejeezus out of me, and of most of the people I knew.

**Figure 18.2 Ehrlich (left) pays Simon (right) part of the latter's winnings.**

*Source:* © Ken Fallon. Reproduced with permission.

# 19

# *Obstacles*

Why does human life sometimes fail to advance? Stupidity, inertia, a lack of imagination, and a shortage of money are always good candidates. So is plain old human nature: we may be the "paragon of animals," but we're still startlingly imperfect. At any rate, a book on betterment should address its opposite, the failure of mankind to make progress on occasions when it should have been easy to do so.

## Off With Their Heads!

One perennial obstacle is *bad government.* It can masquerade as bad laws or bad people enforcing them.

Peter Diamandis and Steven Kotler (see Chapter 14, "The Alleviation of Poverty") tell the story of an ancient Roman goldsmith who somehow discovered a way of extracting aluminum from bauxite, a technology that was completely unknown at the time. (Modern metallurgists couldn't do this efficiently until 1886, using advanced concepts from chemistry—so this guy must have been quite a goldsmith!) He brought an aluminum dinner plate to the emperor, who was concerned that the national stores of gold and silver would lose value if this useful new metal became widely available.

But, instead of rewarding the goldsmith, the emperor had him beheaded. Now that's *really* bad government! The emperor's barbarism not only cost the Roman Empire a brilliant artisan who should have gotten a medal and a pension—it cost the human race 1,900 years' enjoyment of the benefits of aluminum.

## Trade if You Dare

The temptation to protect entrenched interests with the death penalty did not end with the Roman Empire. In France starting in 1726, the penalty for importing "printed calicos," cotton fabrics or garments on which patterns were printed, was death. The purpose was to protect French wool and silk manufacturers and workers from foreign competition. Printed calicos made their way into France anyway, largely from India, and the heinous law was repealed in 1759. No wonder there was a revolution in France shortly afterward![1] Not that the Jacobins and Napoleon, who took power after the revolution, were much better.

Japan went one step further. Under the Tokugawa shogunate, starting in the 1630s, any Japanese returning from a foreign country was subject to execution. Is it any wonder that Japan remained poor for two more centuries?

## The Travails of Galileo

A church is a different kind of government. Galileo Galilei fought the Catholic Church over his discoveries of the moons of Jupiter, the rings of Saturn, and his accumulation of new evidence (the phases of Venus) for Copernicus's heliocentric conception of the solar system. Why did the Church so vehemently oppose this new knowledge?

It was not because anything about his discoveries was disrespectful to religion or could not be absorbed into Church teaching. No, the priests opposed him because they claimed a monopoly on *all knowledge*, derived from their reading of the Bible and of prior ecclesiastic study of Scripture through the ages. If one man with a telescope could overturn all that accumulated effort, they had reason to feel threatened: *they had no natural right to rule.*

Johan Norberg has suggested that the proximate cause of the Great Enrichment was the newfound ability of discoverers and inventors, beginning in the Renaissance, to keep their hard-won learning away from the grubby paws of knowledge-destroyers such as those who imprisoned Galileo. This idea is related to the "political fragmentation" theory of early progress, promoted by the economic historian Joel Mokyr, and discussed in Chapter 17, *The Discreet Charm of the Bourgeoisie.*

---

[1] Postrel (2018).

But I'd note that the modern capitalist system, which typically involves concentrated rather than dispersed political power, nevertheless encourages individual invention through patents and copyrights. The system keeps discoveries away from naysayers. You can invent, promote, and sell anything you want, as long as you send the patent office a description of it, which they put in a file cabinet. Across countries in modern times, there's a substantial correlation between intellectual property rights and economic success.[2]

As a closing thought, Galileo could have saved himself a lot of trouble if he had halfheartedly recanted and gone back to teaching and experimenting—his opponents were not the brutes who burned Giordano Bruno at the stake for similar activity a half-century earlier. They offered the respected Galileo many chances to preserve his freedom. But Galileo was a hothead who made enemies easily, and he stood on principle. If he had recanted, his work might have languished in obscurity until some other scientist—maybe Newton—revived or rediscovered it decades later, but Galileo decided to be a hero, and cosmology progressed pretty quickly after that. Hurray for hotheads.

## Bad Government, or Good?

Our examples of bad government may be laughable or tragic—but economically sensible, they're not. However, it's important to keep in mind that almost all governments survive because they are expressing the will of some of the people, who think they are getting *good* government that acts in their interest. Some Romans wanted to keep aluminum off the market; some Japanese wanted to keep people who had interacted with foreigners out of the country. Today, in the same spirit, we see "populist" movements on the rise, eschewing free trade in favor of protection for established interests. It is economic suicide.

### Bastiat's Law: What Can and Cannot Be Seen

The reason that these kinds of policies persist over the millennia goes back to the nineteenth-century French economist Frédéric Bastiat's maxim about what you can see and what you cannot see. He wrote,

> [A]n act, a habit, an institution, a law, gives birth not only to an effect, but to a series of effects. . . . [The] first [effect] . . . is immediate; it

[2] See, for example, Gould and Gruben (1997).

> manifests itself simultaneously with its cause—*it is seen*. The others unfold in succession—*they are not seen*: it is well for us if they are *foreseen*.
>
> [T]his constitutes the whole difference . . . between a good and a bad economist—[a bad] one takes account of the *visible* effect; [a good one] takes account both of the effects which are *seen* and also of those which it is necessary to *foresee*. Now this difference is enormous, for it almost always happens that when the immediate consequence is favorable, the ultimate consequences are fatal, *and the converse*.
>
> Hence it follows that the bad economist pursues a small present good, which will be followed by a great evil to come, while the true economist pursues a great good to come, at the risk of a small present evil.[3]

This fragment, excerpted from one of the greatest of all economic essays, expresses a fundamental principle of human action. What Bastiat attributed to good and bad economists, I ascribe to well-informed and poorly informed voters and their representatives. The thought is not mine; it's the foundation of public choice theory, the branch of economics that seeks to explain political behavior. In the case of today's protectionists, what you can see is jobs saved and profits guaranteed; what you cannot see is the jobs that could have been created and the rising standard of living you could have had from allowing the principle of comparative advantage from trade, first set forth by David Ricardo in 1817, to work its magic.[4]

## Forgetfulness

Mister (Fred) Rogers once said that his mother responded to scary news by telling him, "Look for the helpers." So when it seems the world is going to hell in a handcart, look for the innovators, the ones who are steadily creating a way out of our next disaster.

The greatness of these creators comes with a humanly generous dose of foibles and shortcomings. They're capricious (Howard Hughes), hotheads (Nikola Tesla), misanthropes (Beethoven), or wiseacres (Ben Franklin), and often seem a wee bit crazy. Richard Feynman, the Nobel

[3] Bastiat (1850). Italics in the translated original. These ideas, which Bastiat developed out of François-René de Chateaubriand's writings on the nature of history, are echoed in the better-known (to English speakers) *Economics in One Lesson* by Henry Hazlitt. Hazlitt (1946).
[4] Ricardo (1817).

Prize–winning physicist whom we met in Chapter 12, tells a story in his book *Surely You're Joking, Mr. Feynman!* of a time when he was working on the Manhattan Project at Los Alamos. Leaving through a well-guarded gate, waving to the guard, he quickly made his way back into the compound through a hole in the security fence. He retraced his route several times until a guard got suspicious and called his superior. Feynman was not trying to cause trouble or test the security system—he did this just for the hell of it. To quote an Apple ad, such wonderfully human innovators "change things. They push the human race forward."

Elon Musk, as human as the rest, is making a grand endeavor of his time on earth. He is most widely known as the founder and current CEO of three companies: Tesla (accelerating the creation and adoption of electric cars), SpaceX (the first company on earth to land and reuse rockets—the firm hopes to make humans a multiplanetary species by settling Mars), and the Boring Company, a tunneling company that is working to connect cities with "hyperloops," vehicles for passenger and freight transport that operate in a partial vacuum to reduce air resistance to achieve very high speeds. I wonder what he does when he is not busy!

In a recent TED conversation, after reflecting on the downward trend of the United States's space program, Musk pauses, and says,

> People are mistaken when they think that technology just automatically improves. It does not automatically improve. It only improves if a lot of people work very hard to make it better. And actually it will, I think . . . degrade . . . by itself actually. You look at great civilizations like Ancient Egypt, and they were able to make the pyramids and they forgot how to do that. And the Romans, they built these incredible aqueducts. They forgot how to do it.[5]

One major obstacle to progress is the loss, sometimes for a very long time, of existing knowledge—even knowledge that had been widely put to practical use. One of the best examples is ancient Roman central heating. The Romans had a heating system called a hypocaust, from the Greek words for *underneath* and *burning*.[6] Figures 19.1 and 19.2 show two of them.

The first design is related to the forced air system many of us now use in our homes, and the second is even more closely related to a newer

[5] Sourced at www.inverse.com/article/31049-elon-musk-best-quotes-ted-2017.

[6] Forbes (1966). Forbes says the hypocaust was invented by Sergius Orata in 80 BC, but some other sources say the hypocaust was inherited from the Greeks.

**Figure 19.1 Roman hypocaust, "forced air" version: a furnace heats air that then flows through ducts into living spaces.**

*Source:* Bogatyr Khan (www.deviantart.com/bogatyrkhan). Commissioned by Laurence B. Siegel.

design, subfloor radiant heating. In both, a furnace, safely tucked away from the home dweller and not emitting toxic fumes into the house, either warms the floor itself or else air that then passes through ducts into the living area. It's not rocket science, and we'd like to think we could come up with the design ourselves if necessary, but *that's because we've seen one*. Necessity is the mother of invention but knowledge is the father.

The hypocaust was popular among rich Romans. Rarely mentioned in medieval documents, it may have been used in the Arab world for a while, but it then seems to have disappeared completely, and heating technology reverted to "direct heating" (fireplaces and stoves). It took James Watt, in Scotland in the late 1700s, to bring back central heating in the form of steam heat, and forced air heating is credited to William Strutt in England in 1805. That's a long wait.

The hypocaust is only one of a very large number of forgotten and rediscovered technologies. At least in the conventional telling, the

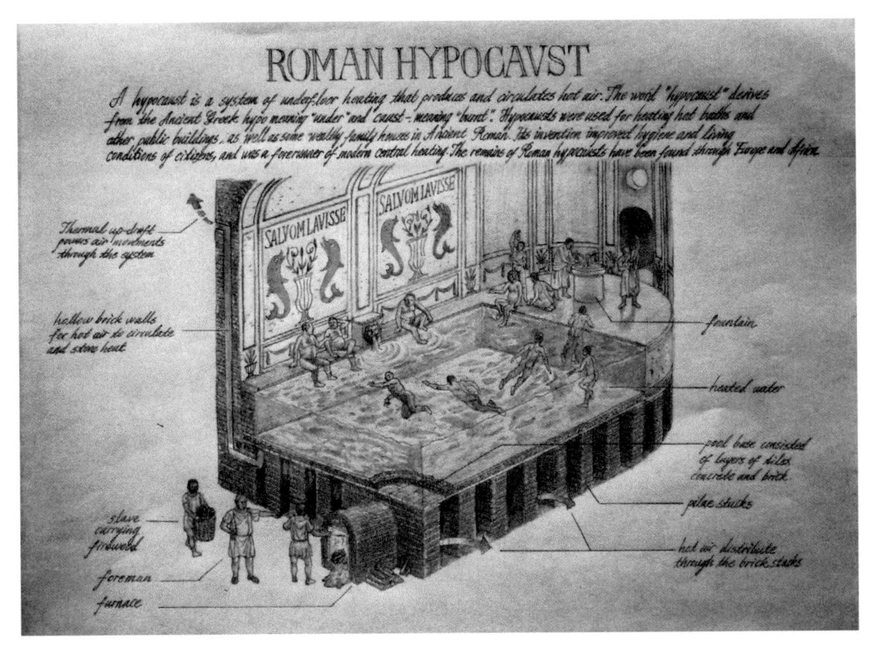


**Figure 19.2 Roman hypocaust, "radiant heat" version: a furnace heats a bath in the room above it. There are also air ducts.**

*Source:* Bogatyr Khan (www.deviantart.com/bogatyrkhan). Commissioned by Laurence B. Siegel.

Middle Ages were basically a Great Forgetting, in which old knowledge of technology, art, literature, and philosophy was lost until their rediscovery in the Renaissance or even later.

There have been other Great Forgettings: the Athenian dark age, various Chinese dark ages including the Mao era,[7] and many more. That does not mean there was no progress during those times; many economic historians admire the Middle Ages for their ground-level innovations, but many valuable ancient technologies really were lost for a long time.

What keeps technology alive is civilization: writing, cities, travel, exchange, scholarship, and practical implementation by artisans. A discussant in an Internet community, musing on the back-to-the-land or survivalist movement, raised the interesting question of what technologies we would lose if the movement succeeded and we all lived in villages of 200 people or fewer. Would we remember how to recreate wireless telephony? Almost certainly not. Double-entry bookkeeping, a

[7] From 1949 to1976, not that long ago, lest we forget.

nearly thousand-year-old-technology? Maybe. Almost all advanced human knowledge would be lost.

About a half-century ago Tom Wolfe thought we were headed toward another Great Forgetting:

> In 1968, in San Francisco, I came across a curious footnote to the psychedelic movement. At the Haight-Ashbury Free Clinic there were doctors who were treating diseases . . . that had disappeared so long ago they had never even picked up Latin names . . . the mange, the grunge, the itch, the twitch, the thrush, the scroff, the rot. And how was it that they had now returned? . . . [T]housands of young men and women had migrated to San Francisco to live communally in what I think history will record as one of the most extraordinary religious experiments of all time.
>
> The hippies . . . sought nothing less than to sweep aside all codes and restraints of the past and start out from zero. At one point Ken Kesey organized a pilgrimage to Stonehenge with the idea of returning to Anglo-Saxon civilization's point zero . . . and heading out all over again to do it better. Among the codes and restraints that [were] swept aside . . . were those that said you shouldn't use other people's toothbrushes or sleep on other people's mattresses . . . or that you and five other people shouldn't . . .
>
> . . . take tokes from the same cigarette. And now, in 1968, they were relearning the laws of hygiene by getting the mange, the grunge, the itch, the twitch, the thrush, the scroff, the rot. [T]he relearning . . . follow[ed] a Promethean and unprecedented *start from zero.*[8]

The Jacobins of the French Revolution also tried to start from zero. It did not work out well. Off with their heads. We believe ourselves to be quite advanced, but *we have not forgotten how to forget.*

## Ridiculous Comments by Famous People

We've all heard of forecasts, real or apocryphal, by highly regarded people who got the future completely wrong. Perhaps the most notorious is the supposed comment by Charles H. Duell, head of the United States Patent and Trademark Office in the late nineteenth century, to the effect that everything that can be invented has been invented. He didn't say it and actually said something almost exactly opposite.

---

[8] Wolfe (2000). Italics are mine.

But respected people really do sometimes say the silliest things, and we tend to think of them as obstructing progress by failing to think about the possibilities that await us. Thomas Watson, president of IBM and a great business leader, said, in 1943, "I think there is a world market for maybe five computers." And Ken Olsen, of Digital Equipment Corporation, said in 1977, "There is no reason anyone would want a computer in their home." My favorite is Harry Warner, one of the Warner brothers of Hollywood fame, ruminating on the possible end of silent movies: "Who the hell wants to hear actors talk?"

But did these comments impede progress, or are they just echoes of your aunt Edna who says her black AT&T desk phone is the only phone she'll ever need? Ken Olsen withheld capital from the personal computer revolution because he thought it was a bad investment, and Aunt Edna withheld a smaller amount of capital from the mobile phone revolution for the same reason. But neither one actually did anything to *stop others* from inventing, developing, and selling the improved products. I wouldn't worry about ridiculous forecasters.

## A Nation of Slugs?

Here's a paradox: We live in an age of migration worldwide, yet Americans who've found limited opportunity in their home towns are unprecedentedly staying put instead of moving in search of better prospects. What's going on?

The current global migration is intense, but it's not unprecedented. My parents and grandparents also lived in an age of migration. Their families left Central and Eastern Europe somewhere between three and one generations ahead of Hitler or I would not be telling this tale. The Holocaust consumed six million Jews but a three and a half million survived, mostly by getting the hell out of Dodge (if a little graveyard humor is acceptable). Millions more Jews, such as my direct ancestors, avoided the Holocaust entirely by coming to the United States, or Britain, or Canada before the first stirrings of Nazism, and mostly for economic opportunity. It was one of the world's great movements of people.

We're in another such period, less dire but just as impressive in terms of the number of people in motion. Despite some slowing of net immigration to the United States, you'd have to be blind to deny that people of all skill levels, from all over the world, are hoping to move here, and some are taking considerable risk to do so, leaving their birthplaces and families and

| | Adults, in millions |
|---|---|
| China | 16 |
| India | 15 |
| Brazil | 11 |
| Ethiopia | 8 |
| Nigeria | 7 |
| Mexico | 7 |
| Bangladesh | 5 |
| Philippines | 3 |
| Japan | 3 |

**Figure 19.3 Countries with the largest number of people who want to move to the United States.**

*Source:* Constructed by the author using data from Gallup World Poll, https://news.gallup.com/opinion/gallup/212687/coming-america.aspx.

starting over with almost nothing. According to a Gallup Poll, 147 million non-US nationals would move here if they could, as shown in Figure 19.3.[9]

(By the way, I think our bountiful land could handle 147 million more people, and we could use their talents. However, people are fruitful and they multiply, so the potential population increase from immigration could be much larger, and our resources are not unlimited.)

Yet, unlike previous generations of the less fortunate among native-born Americans, today's Americans who are not making it are stuck to their home states and counties like glue, preferring to collect meager social benefits—or live off more productive family members—instead of striking out in search of fresh opportunity. To exaggerate a little, "Go west, young man" has become "Go to the basement, young man."[10]

## Staying Home[11]

Did you know that Americans have (almost) stopped moving house? "The interstate migration rate has fallen 51% below its 1948–1971 average,"

[9] Clifton (2017). The numbers seem low. I don't know China or India well, but I have traveled in Mexico enough to say that if only nine million Mexicans wish they could live in the United States, then I'm Bonnie Prince Charlie.

[10] For the youngsters reading this, it was the *New-York Tribune* editor Horace Greeley who, in 1865, said to an advice-seeker, "Go west, young man, and grow up with the country." (It is sometimes disputed whether he was the original source of this quote.)

[11] This section is from my article, "Tyler Cowen and the Fallacy of American Laziness," *Advisor Perspectives*," May 3, 2017, http://larrysiegel.org/tyler-cowen-and-the-fallacy-of-american-laziness/, with minor edits.

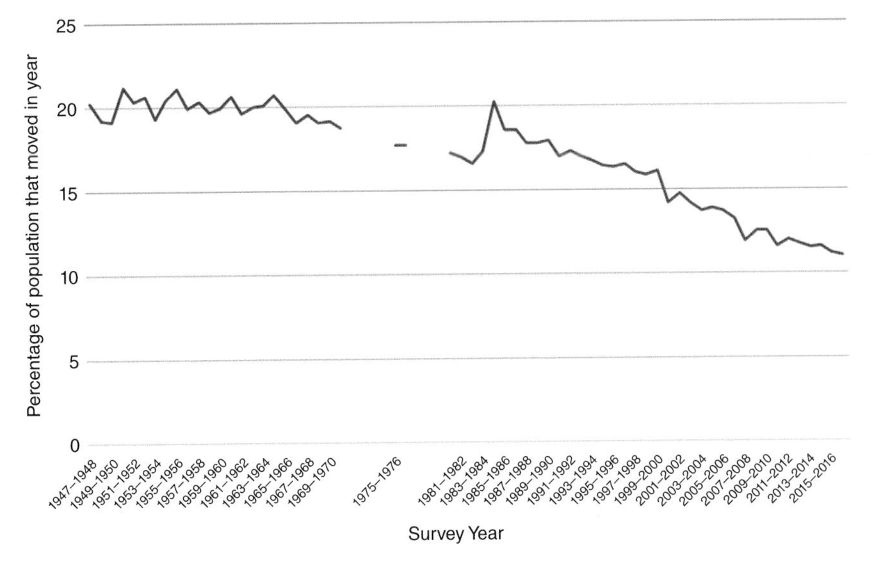


**Figure 19.4 US geographical mobility rate (percentage of population that moved in past year), 1947–2016.**

*Source:* Constructed by the author using US Bureau of the Census data, www.census.gov/data/tables/time-series/demo/geographic-mobility/historic.html, Table A-1. Annual Geographical Mobility Rates, By Type of Movement: 1948–2017.

Tyler Cowen writes. During the peak years, about 20% of households moved each year. Now, it's not much above 10%, as shown in Figure 19.4.

A group of Germans with whom Cowen spoke thought he was kidding or exaggerating when he quoted the 20% number. In Germany, buying and moving into a house, often one's parents' house or a house nearby, is usually a lifetime decision. But Americans, in search of opportunity, a change of scenery, or better weather—or to escape from bad conditions—really did move at the rate described.

The new, more sluggish rate of relocation keeps many people from getting ahead. A quarter-century ago, Michigan residents who had lost their jobs in the auto industry migrated quite aggressively to Texas. Today, people in similar situations tend to stay behind, adjusting to the lower pay of new jobs or of social benefits. It's hard for the economy to grow that way, and it's not healthy to have a resentful and dependent population concentrated in one area. Today, there are many such areas.

People who rarely move can be either rooted or stuck. Cowen suggests they are mostly stuck, although they may be comfortable in communities to which they are economically and culturally matched. Although there are many reasons why people get stuck, Cowen points out that NIMBYism, Not In My Back Yard, is one of them, because high-growth areas can be numbingly expensive, reducing the incentive to move for better pay.[12]

I suspect, however, that the most common reason is that the lifestyle made possible by government benefits is survivable, as it should be, and that the people who are stuck have started to find such a straitened level of living acceptable, as it should not be. I don't really think we've become a nation of slugs, but people respond to incentives, and we need to re-examine the incentive structures for people trapped in unpromising locales.

## Misery

I don't feel the need to present both sides of every issue. Mostly, I want to persuade you of a point of view, and let you do your own research if you want to refute me.

Occasionally, however, I read something that makes me wonder just how much progress human beings can make, due to something called human nature, which isn't all good. A psychiatrist who writes a popular and respected blog under the pseudonym Scott Alexander says:

> One "advantage" of working in psychiatry is getting a window into an otherwise invisible world of really miserable people.
>
> I work in a wealthy, mostly-white college town consistently ranked [as] one of the best places to live in the country. If there's anywhere that you might dare hope wasn't filled to the brim with people living hopeless lives, it would be here. But that hope is not realized.

Alexander presents two examples of the bizarre problems he hears about every day:

> A perfectly average patient will be a 70 year old woman who . . . has no friends within five hundred miles and never leaves her house except to go to doctors' appointments. She has one son, who is in jail, and one daughter, who married a drug addict. She also has one grandchild, her only remaining joy in the world—but her drug-addict son-in-law uses access to him as a bargaining chip to make her give him money from her rapidly-dwindling retirement account so he can buy drugs.[13]

---

[12] It can be hard to distinguish complacency from a rational response to incentives. Casey Mulligan, in *The Redistribution Recession* (Oxford University Press, 2012), argues that much of what appears to be the former is in fact the latter, and argues for reform of policies that have the effect, intended or unintended, of reducing job mobility.

[13] Alexander (2015).

The woman whose retirement account is being depleted by the son-in-law thinks she will spend her last years on the street, homeless. Alexander continues:

> Here's another. Sixty-year-old guy who was abused as a child, still has visible scars. Ran off at age 15, got a job in a factory, married let's say a waitress. There was some kind of explosion in his factory, he got PTSD, now he freaks out every time he steps within a hundred meters of a place where manufacturing is going on. . . . He started beating his wife, put in jail for a year or two for domestic violence, came out, by this point his wife has run off with another man and took everything he owned with her.

In one of the richest towns in the richest large country in the world! I'm sure you know people like that—I certainly do.

Of course, Alexander realizes there's some selection bias at work, because psychiatrists see people who are "disproportionately the unhappiest and most disturbed." But, he says, he heard the same tales of woe when he worked in general medicine, where there's less selection bias. Alexander concludes: "People were still much worse off than I thought"—a sobering reality check on the hopefulness evidenced by the data in this book.

## Conclusion

There are times when humanity does fail to advance: bad government and sclerotic institutions temporarily prevent human life from moving forward; technologies and knowledge are lost and forgotten. Eventually, those governments are replaced when they continuously fail to express the will of the people, and over time the technologies and knowledge are rediscovered. Despite the setbacks, the long-term trend is a steady advance in well-being.

The anecdotes I've presented in this chapter aren't supposed to be an encyclopedic list of obstacles to progress. They're vignettes, intended to show that I can see both sides of a question, and that I sympathize with the struggles of those who aren't participating in the Grand Betterment or feel that they're only participating in certain aspects of it while remaining miserable. Modern life is remarkable, but it is not all wine and roses.

# 20

# *"He Shall Laugh": Why Weren't Our Ancestors Miserable All the Time?*

## On Hedonic Adaptation

If less than a century ago young Calvin Coolidge Jr. died of a blister on his foot, after receiving the best medical care in the world, because there were no antibiotics; if Grimm's fairy tales were not horror stories intended to entertain but realistic depictions of medieval German life; if Stephen Jay Gould's graves of dead children (Chapter 3) were the rule rather than the exception—why weren't our ancestors miserable all the time? How could life be any fun?

In thinking about this question, I am reminded of the Old Testament story of Abraham and Sarah. Aged 100 and 90 (although their years probably did not correspond exactly to ours), and barren up to that point, they produced a son whom they called Yitzhak, or Isaac in English. We do not think much about the original meanings of familiar words, but Yitzhak in ancient Hebrew means "he shall laugh."

The modern conception of life in the Holy Land of 3,500 years ago is one of extreme deprivation and struggle in the desert sun. But on learning of the meaning of Yitzhak our mental image of Isaac immediately shifts to that of a man seated at a primitive dinner table with his friends, sharing a joke and laughing his ass off. Life has always been a mixture of the sweet and bitter. Isaac had fun.

Let's consider Geoffrey Chaucer, the incomparable fourteenth-century English poet. He wrote:[1]

Whan that Aprill, with his shoures soote
The droghte of March hath perced to the roote
And bathed every veyne in swich licour,
Of which vertu engendred is the flour;

When April, with its showers sweet
The drought of March has piercèd to the root
And bathèd every vein in such liquor
As gives the strength to generate the flower

Chaucer oozes delight in simply observing nature:

Whan Zephirus eek with his sweete breeth
Inspired hath in every holt and heath
The tendre croppes, and the yonge sonne
Hath in the Ram his halfe cours yronne,

When Zephyr too, with his sweet breath
Has inspired in every wood and heath
The tender crops, and the young sun
Has, through the Ram, half his course a-run

He can't stop:

And smale foweles maken melodye,
That slepen al the nyght with open eye-
(So priketh hem Nature in hir corages);
Thanne longen folk to goon on pilgrimages

And small fowl make melody
That sleep all the night with open eye
(So Nature pricks them in their spirits)
Than folk long to go on pilgrimages

Finally! Chaucer gets to the point, which is that the verses are a prologue to an epic poem about pilgrimages. But we are fools if we don't

[1] The modernization (not a translation! It's our own language) is mine, and closer to the original than is customary—I'm trying to preserve the connection between Middle English and the modern words, sometimes at the expense of meaning. But Chaucer's original is not that difficult (except that *vertu engendered* has frustrated almost everyone, not because the words are hard but because the meaning of *vertu, or virtue,* has changed). His work is no farther removed in time from Shakespeare's than, say, Jefferson's crystalline prose is from ours.

get his other point, which is that April is wonderful, and life itself not so terrible. Now compare T. S. Eliot, a twentieth-century poet of almost equal skill:

> April is the cruellest month, breeding
> Lilacs out of the dead land, mixing
> Memory and desire, stirring
> Dull roots with spring rain.

Ugh—beautiful words, ugly thoughts. Was life in the late 1300s more pleasant than life in the early 1900s? Of course not! The 1300s are often described as the worst of centuries, with repeated epidemics of plague, massive famines, and, in nearby France, the Hundred Years' War. (See Figure 8.3.) But Chaucer wrung every bit of pleasure he could out of his grungy century, as did his famously randy characters.

Maybe the difference is just one of temperament: Chaucer delighted in every detail of the natural and human world, whereas Eliot was notably gloomy. But it's more plausible to observe that the difference is due to *hedonic adaptation*, the tendency to adjust one's perception of pleasure and pain to what is available and achievable given one's circumstances. (*Hedonic* means "relating to pleasure"; economists also use the word to refer to the study or attempted measurement of pleasure.)

Hedonic adaptation is the characteristic of the human mind that evens out bad experiences and times by accentuating the positive (apologies to Mr. Bing Crosby) and good ones by emphasizing the negative. It was first described by psychologists trying to make sense of the observation that most people don't become *permanently* happier after strokes of good luck—such as coming into money—or unhappier after a misfortune. They instead adjust to their new circumstances and are about as happy as before. This is hedonic adaptation within a single lifetime, within a single human, not across time and space.

But this principle *can* be applied across time and space, by comparing happiness levels in different countries or centuries. I'm not arguing that Chaucer's contemporaries were happier *on average* than Eliot's—romanticizing the past is not usually a good idea. The difference in cheerfulness between the two poets was probably an individual matter, a contrast in temperaments. But the words show that fourteenth-century Englishmen could be joyful and that twentieth-century Americans could be morose, despite the latter having many times more income and creature comforts than the former.

## It's Hard to Get Happy: The Hedonic Treadmill

Hedonic adaptation within a single human life is easier to document and measure than across time and space. Psychologists originally used the concept to show that you can't get ahead: "as the example of the 'poor little rich girl' suggests, objective outcomes and happiness are not perfectly correlated," writes Shane Frederick, a marketing professor at the Yale School of Management. "For example," he continues, "winners of lotteries do not report themselves as being much happier than other people, and those who were paralyzed in an accident do not report themselves as being much less happy."[2] This particular manifestation of hedonic adaptation is called the *hedonic treadmill*, where you can try and try but still not get ahead.

This observation is surely worth further investigation, given that being paralyzed in an accident is about the worst luck most people can imagine—they might even describe it as a fate worse than death.

But Frederick then makes the link to our topic, progress over time: "Similarly, as nations get wealthier, the reported well-being of its citizens does not increase."

Really? The Norwegians are no happier than the people of Sierra Leone?

## Reconciling the Hedonic Treadmill with Reality

It's pretty clear Frederick does not believe this silliness. Having apparently been asked to write the "hedonic treadmill" article in an encyclopedia of social psychology, he obliges, but then concedes that the idea of happiness not varying with material conditions is implausible. It's also morally objectionable in that it appears to condone poverty and maltreatment. Frederick notes that studies of happiness depend on self-reported data that are hard to interpret:

> When asked "How happy are you on a scale from 0 to 100?" respondents must judge for themselves what the end points of the scale represent. Someone who has lived a tough life might interpret 0 as unrelenting torture and 100 as pleasant comfort, whereas someone who has lived an easy life might interpret 0 as the absence of joy and 100 as heavenly bliss. If these two people each declared their happiness level to be a 60

[2] Frederick (2007).

(out of 100), it would obviously be wrong to conclude that the two people really are equally happy, since one person has adopted a higher standard.

## The Satisfaction Staircase

An alternative explanation for the same self-reported happiness data is what the Nobel Prize–winning psychologist Daniel Kahneman called (I think misleadingly) a *satisfaction treadmill*, "whereby improving circumstances lead individuals to adopt successively higher aspirations for the amount of enjoyment they regard as acceptable."[3] In other words, the better your life is, the more it takes to constitute a further improvement that you'll report as an increase in happiness. The satisfaction treadmill is misnamed because it's not a treadmill but a staircase that stays in one place while you climb it. The staircase thus takes you to higher levels of accomplishment and enjoyment.

Or, as the fictional detective Sunny Randall said in Robert B. Parker's *Family Honor*, the more things you like, the happier you are. But then (she did not add) you'll be comparing yourself to a happier class of people, and you won't report a higher level of happiness in a psychological survey.

The satisfaction staircase metaphor seems consistent with ordinary experience, which is that almost everyone is continually trying to improve their lot—if it's to no avail, why bother? There are those who would say that we should not bother, that we should adapt ourselves to the circumstances in which we find ourselves and try to maximize happiness that way. However, I'd note, a little cynically, that such advice often comes out of religious traditions that developed under conditions of extreme deprivation. We do not live in those conditions.

## Happiness Economics

Based on these speculations, economists and psychologists have carved out a curious new field of investigation—happiness economics. If you told a budding economist of a half-century ago, wrestling with airline-seat-mile

[3] Frederick (2007), p. 419. See also Kahneman, Daniel, "Objective Happiness," in Kahneman et al. (1999).

data and conducting input-output analysis for heavy industry, that her successors would be asking what it means to be happy, she would have thought you were crazy. But that is a direction that economics has taken with considerable enthusiasm. Actually, you can win a Nobel Prize for it.

Sir Angus Deaton, a Scotsman who now lives in the United States and teaches at Princeton, did exactly that. He won the 2015 economics Nobel Prize for work that is closely intertwined with happiness studies. He and Daniel Kahneman write, "Emotional well-being also rises with . . . income, but there is no further progress beyond an annual income of [about] $75,000."[4] This work was done about 10 years ago and there has been some inflation, so raise the number to $90,000. The work refers to the United States; other countries presumably have different happiness functions.

Richard Easterlin, a University of Southern California economist who is one of the founders of happiness studies, did perform cross-country comparisons in the 1970s, and found a hedonic treadmill effect—that life satisfaction does not always go with economic progress—that, in the context of international comparisons, is now referred to as the Easterlin Paradox.[5] A modern example is India, where self-reported happiness declined by more than a full point (on a scale of 0 to 10) over 2006 to 2016, a period when per capita income, stated in constant dollars, rose by a startling 78%.[6]

## "I've Been Poor and I've Been Rich, and Rich Is Better"—Sophie Tucker

But it turns out, on more thorough examination, that this is much more the exception than the rule. There is a powerful correspondence between material well-being and self-reported happiness, in many different places, times, and at many different levels of economic accomplishment. It is better to be richer.

Figure 20.1, constructed by Esteban Ortiz-Ospina and Max Roser, packs a lot of data into a small space so please look at it carefully.[7]

---

[4] Kahneman and Deaton (2010).

[5] His classic paper, "The Economics of Happiness" (2004), is at www-bcf.usc.edu/~easterl/papers/Happiness.pdf. It was published in *Daedalus* 133, no. 2 (Spring 2004): 26–33.

[6] From the data underlying the chart entitled "GDP per capita vs. Self-reported Life Satisfaction, 2005 to 2016" in https://ourworldindata.org/happiness-and-life-satisfaction.

[7] Ortiz-Ospina and Roser (2013).

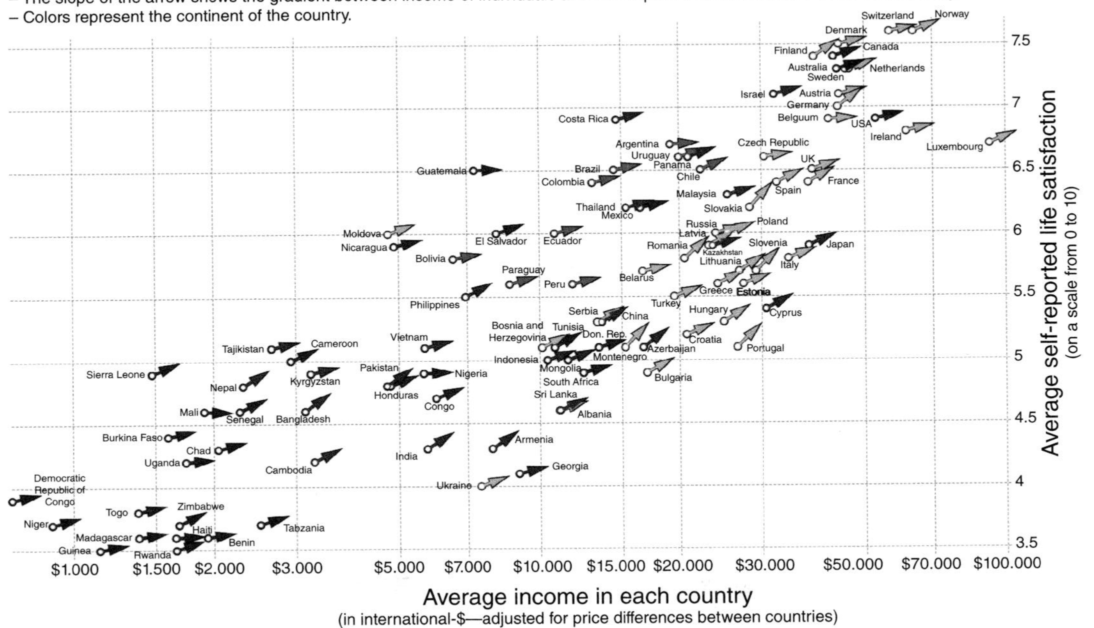


**Figure 20.1 People in richer countries tend to be happier and within all countries richer people tend to be happier.**

*Source:* Max Roser. Our World in Data https://ourworldindata.org/happiness-and-life-satisfaction. Licensed under CC BY.

Each arrow represents a country. Its horizontal (*x*-axis) position is the country's PPP GDP per capita, that is, roughly the average income in the country. Its vertical (*y*-axis) position is the country's average self-reported happiness on a scale of 0 to 10. The direction of the arrow is the extent to which, in that country, higher incomes are correlated to higher self-reported happiness. In other words, looking at the $15,000–$20,000 income range, the self-reported happiness of Mexicans and Thai is affected very little by their incomes, whereas in China the effect of increased income on happiness is dramatic.

Happiness, then, is at least partly culturally determined. Note the cluster of medium gray arrows, reporting above-average happiness per dollar earned. These are the Latin American countries. A group of light gray arrows in the middle-income ranges, reporting below-average happiness per dollar, represents eastern Europe and the former Soviet Union. (Western Europe is also shown in light gray but is richer.) Surprised? The stereotypes of the upbeat Latin and the morose eastern European may be unfair, but they show up in the data.

I'd also bet on countries with more upwardly sloped arrows, such as India, Slovakia, and Portugal, making faster economic gains than those with flatter arrows. Bangladesh, starting at a much lower level, is a good bet too, as, perhaps surprisingly, is Pakistan.

## Conclusion: How Understanding Hedonic Adaptation Helps Us to Better Measure Progress

These observations enable us to circle back to the question of how to measure progress, addressed in Chapter 16, "The Mismeasurement of Growth." If we know one thing about measuring either economic or overall human progress, it's that GDP does a very poor job of it. We are not—*cannot be*—either 30 or 100 times happier than the typical Englishman or American of 250 years ago. That's how much real GDP per capita has increased, with either the high or low figure bracketing the various estimates. But, with all the misery that has been alleviated over that period and with some new miseries that have been added (such as when my hard drive crashed), *we eat, sleep, and make merry as we always have.*

This one example suggests that increases in GDP overstate progress, as surely they may. But they can understate it as well. The change in the overall quality of human life rocketed upward on October 16, 1846, the day that the first successful surgery was performed using anesthesia. The improved experience of the one patient is trivial compared with the decline in the dread of surgery that followed, along with the lives saved because people were willing to undergo it. It was a giant leap for mankind, yet GDP increased hardly one whit on that day.

Even if you count all the paychecks received by anesthetists over the next 175 years, along with the prices paid for the drugs, they don't move GDP that much. But the innovation improved human life profoundly.

The bottom line is that we do not know how to measure progress. But we tend to know it when we see it. Perhaps the best way is to ask yourself if you would exchange your life, lived now, with that of some forebear or historical figure in similar economic circumstances who had to make do with the technological, medical, social, and environmental conditions of his or her time. Perform the thought experiment and get back to me with your answer.

I'm not saying that everyone should answer, "This is the greatest time to be alive." Honest answers will differ from person to person.

But we often hear a glamorized view of the past and the wish that one could live at some specific time (and perhaps place) in the past. Woody Allen's movie *Midnight in Paris* is a meditation on that question, as are many other literary works. In "I've Got a Little List," from *The Mikado*, Gilbert and Sullivan lampooned

> the idiot who praises, with enthusiastic tone,
> All centuries but this, and every country but his own.

No, I'm not calling you an idiot. That was W. S. Gilbert; what does he know? I'm saying that you're human and may be subject to the part of human nature that colors the past in rose and then pines for it.

"I was born in the wrong (generation, century, millennium)," think many of us. If only I lived in Cleveland in 1875, I could have easily become a millionaire. . . . Doc Brown in *Back to the Future* wanted nothing more than to return to the 1885 Wild West of his imagination, as does every urban cowboy. . . . It would have been fun to discuss philosophy with Socrates. . . . The 1960s were clearly a better time to play in a rock band.

That mode of thinking is akin to being jealous of some aspect of another person's life wherein we want a part (their wealth, physique, job) but would not be willing to swap for their whole life and take the baggage.

So, the clear-headed thinking is (usually): no, the resources necessary to succeed then were not as ample, nor were other conditions so great. You might die of a now-easily treatable disease before you realized your dreams.

But just because conditions of food, health, safety, and technology are drastically improved, that does not make the present the best time *for that person* to be alive. A musician with an interest in becoming a classical composer would probably have had a better chance in Beethoven's day—but the counterargument is that today's musicians can play Beethoven, *and* Chopin, *and* the Beatles, and on better instruments that can reach a wider audience in a fraction of the time. The trade-offs are usually in favor of the present and the future.

Just one last thought: Following the logic of the latest date being the best time to be alive, would you want to be born in the year 2054? How about 3054? I have no idea how to answer that question.

# PART V
# GREENER

# 21

# *Prologue: Why Poor Is Brown and Rich Is Green*

## Giants in the Earth

It is almost an article of faith among educated readers that nature—the environment, the Earth—is in bad shape. According to this story, the past economic development I've described in the foregoing chapters has seriously degraded the natural world. And it's widely believed that future growth on the scale I'm expecting might just finish it off, making the world uninhabitable or barely habitable for future generations.

This story reminds me of just about every religious text I've ever read. There was a Golden Age in the distant past—"there were giants in the earth in those days," says the Old Testament[1]—and an apocalypse to come.

Maybe it's an innate human characteristic to believe that. Dan Carlin, a broadcaster and lecturer on history, said, "It's one of the oldest tropes in history that people look back and pine away for a golden age that they missed that never existed."[2]

Myths usually carry some burden of truth, and it turns out that the story of environmental despoliation is half correct. Land unsettled by humans is pristine. Subsistence living, with its very high discount rates (described later), is the most environmentally destructive lifestyle

[1] Genesis 6:4

[2] Carlin (2018).

ever known, but the population may be too small for the damage to be immediately obvious. Later, rapid industrialization can cause profound environmental problems. These are legitimate and serious concerns.

## The Tipping Point

But then, when a certain level of wealth is reached, something changes. (See Figure 21.1.) After reaching a tipping point at which environmental protection becomes a desirable and affordable choice, a society becomes a better steward of its natural surroundings. Richer means greener.

**Figure 21.1 Liberty Street, Pittsburgh, in 1940 and today.**

*Source:* (Top) Pittsburgh City Photographer Collection, 1901–2002, AIS.1971.05, Archives Service Center, University of Pittsburgh. Public domain. (Bottom) Photo by Corey Grau (www.coreygrau.com); reproduced with permission.

How does this occur? Let's look at some successes in the preservation of nature. The US national parks are among the world's most beautiful, healthful, and unspoiled places. The northeastern United States is more densely forested now than it was in 1850, 1900, or 1950. Switzerland resembles one big national park, with splendidly clean air and water and unparalleled natural beauty—despite a dense population of 485 people per square mile, higher than that of China or Nigeria.[3]

These successes have one common factor: the United States and Switzerland have been rich for a long time. The US national park system was established in 1872 and was later developed into its modern form by President Theodore Roosevelt at the turn of the last century. The United States was not yet the world's richest country by 1872, but it was in the top half dozen: basic needs had been met to the extent that, in a democracy, it was not inconceivable to divert a portion of the tax revenues, as well as a good chunk of publicly owned land, to environmental protection.[4]

By Roosevelt's day, a generation later and a good bit richer than in 1872, the United States could afford to have conservation become a major government program. Switzerland and other countries at the top of the income scale tell a similar tale. Effective environmental protection cannot be achieved through private action alone; much of it requires the kind of mass cooperation that can be enforced only by government. But in a free society, taxpayers must consent to their resources being used in this way or else the environmental protection will not take place.

## The Role of Government in Preserving the Environment

Why this emphasis on government? Private agents always externalize—get other people to pay part of their costs—as best they can, and competition forces them to do so even if they are good guys and would rather not. Many, though not all, environmental effects are *externalities*. Externalities are instances of harm done to the commons (assets held in common by the people) that arise as a side effect of transactions that are arranged by, and benefit, some group of private parties. A trades with B and imposes a cost on C. Examples of externalities are pollution (air, water, noise, etc.), overgrazing,

---

[3] Amazingly, Switzerland, with an area of 15,940 square miles, is smaller than the largest US national park (Wrangell–St. Elias National Park in Alaska, with an area of 20,625 square miles).

[4] Royal parks and preserves had long been a feature of nondemocracies. Although these are now treasured elements of the landscape, they were not ceded by the people to their government voluntarily.

and overfishing. Because externalities are suffered by individuals who do not benefit from the transaction causing the harm, the problem can generally be alleviated only through some form of government action.[5]

Free-market environmentalists might object that the problem of externalities can be solved by something called a Coaseian bargain, named after the late Anglo-American Nobel Prize–winning economist Ronald Coase. In such a bargain, each resource is owned by some specific person or organization, and the owner can simply charge a would-be user for the right to use up, pollute, or otherwise despoil a resource. If the value to the resource owner is high enough, she could charge a very high price and the resource would not be used at all, but would be preserved for future generations (who might put a different price on it).

Coaseian bargaining for all natural resources is a pipedream. You can't establish enforceable property rights in resources like air and water, which blow away or flow away, or that—like wildlife—walk, fly, or swim. (An exception is revealed in Chapter 24, "We Are as Gods," where we'll see how a clever Icelandic law had the effect of privatizing the supply of fish.) If we want to preserve the environment, we're going to have to tolerate some government involvement.

## Introducing the Environmental Kuznets Curve

The environmental Kuznets curve is a fancy name for the clean-dirty-clean pattern I've been describing as a country develops. The meanings of *environmental* and *curve* are obvious. I'll get to Kuznets a little later.

In the early stages of a society's economic development, the environment becomes degraded pretty quickly. Why? If you're hungry, a meal today is worth sacrificing some indefinite benefit in the future, even if that benefit could be large. You'll do what is necessary to survive in the short run. These trade-offs are represented numerically by what economists call *a discount rate*—the rate at which one is willing to trade consumption today for consumption tomorrow.

In a poor society the discount rate is very high. For that reason, in the early stages of economic development, the environment becomes degraded pretty quickly because progress must be made on the cheap, without sacrificing immediate needs; that's what it means to be poor.

[5] Externalities can also refer to *benefits* unintentionally conveyed to a third party by the private actions of two other parties, such as when my real estate value rises because my neighbors build luxury homes on their property. But, in this discussion, we are mostly concerned with negative (harmful) externalities.

Even at moderate levels of prosperity, such as what China is now experiencing and as the United States and Western Europe did until, say, 50 or 75 years ago, environmental problems can be pretty overwhelming. Figure 21.1 contrasts the filthy air of industrial Pittsburgh in the 1940s with the clean air of that now postindustrial city today. It's a pretty dramatic change. Pittsburgh was called a two-shirt-a-day town, meaning the air was so bad you had to change your shirt in the middle of the day to avoid feeling and looking grungy.

The tipping point toward better environmental quality comes at different wealth levels for different environmental goods. Drinkable water and garbage collection are among the first environmental benefits that people demand as their standard of living rises. Air that is free of particulate matter and hydrocarbon pollution comes pretty quickly afterward, as does a desire to avoid being exposed to toxic chemicals. Limitations on carbon dioxide emissions—invisible, odorless, and harmless to the individual, although concerning when it comes to climate—come much later.

Can we formalize this relationship, clean to dirty to clean, in something resembling a theory? Figure 21.2 shows an *environmental Kuznets curve*, a stylized curve demonstrating how, over a country's development from low to middle to high income, one would expect the level of pollution (by some hypothetical substance) in that country to evolve.

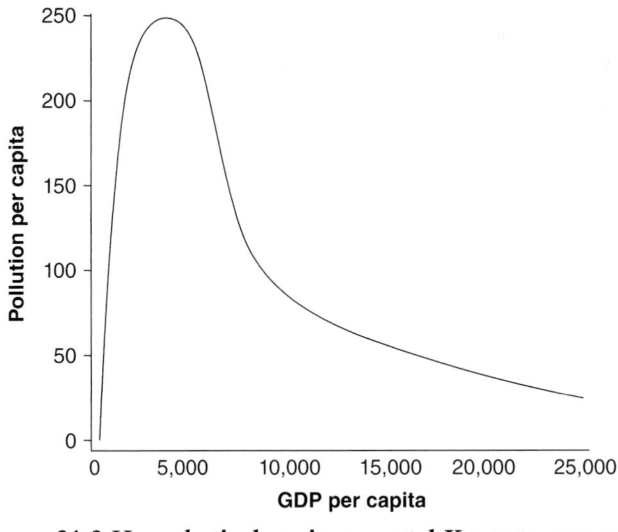


**Figure 21.2 Hypothetical environmental Kuznets curve for a generic pollutant.**

*Source:* Gallagher, Kevin P. 2009. "Economic Globalization and the Environment." *Annual Review of Environment and Resources* 34 (November): 279–304.

The environmental Kuznets curve, or EKC, is powerful in explaining differences in conditions among different countries, or in a given country at different phases of its development. The curve was first described by Gene Grossman and Alan Krueger in an influential 1991 paper.[6] So who's Kuznets?

## The Master of Measurement

Simon Kuznets, a Russian-American economist who was only the second American citizen to win the Nobel Prize in economics (he won it in 1971), thought he had discovered a similarly shaped relationship between *income inequality* and a country's level of development. At low incomes, everybody has more or less the same income: they're poor. As a country develops, landowners, entrepreneurs, merchants, and traders get ahead quickly, whereas ordinary workers languish: maximum inequality. Then, in an advanced industrial society, almost everybody is middle class and inequality is low again.

Studying the industrial countries of his time, Kuznets's observation was more or less valid. It did not work out exactly as he expected, because very advanced, postindustrial societies turn out to have a lot of inequality. But when any variable first rises and then falls as a country or society develops, the curve that describes this phenomenon has become known as a Kuznets curve. Applying it to the environment has turned out to be a very useful concept, elaborated in Chapter 22, "A Skeptical Environmentalist," which focuses on the work of the Danish statistician Bjorn Lomborg.

Kuznets had previously distinguished himself by working out the concepts of gross national product (GDP), gross national product (GNP), and national income and product accounting (NIPA), tools that every macroeconomist now uses as comfortably as a carpenter uses a hammer and saw. His contribution to economic measurement, spurred by the need to have good economic statistics to help set relief and stimulus policy in the Great Depression, is Kuznets's greatest achievement.

Simon Kuznets is the subject of a charming story. In 1971, the economics Nobel was only two years old, and the suspense and hoopla that now surrounds the awarding of the prize did not yet exist. When the 1971 prize was announced, Kuznets heard from a colleague that "some

---

[6] Grossman and Krueger (1991). A later, published version is Grossman and Krueger (1995). Grossman is a perennial Nobel candidate (for earlier work), although he has not won the prize. Krueger was chairman of President Barrack Obama's Council of Economic Advisers. Both men teach at Princeton.

guy with a Russian name" had won the prize. It was him. He did not think of himself as Russian and had not lived in Russia (now Belarus) since 1922, so he did not know he had won for some time.[7]

## An Environmental Kuznets Great Divide

You don't have to be as rich as Switzerland to engage in meaningful environmental preservation. The Dominican Republic is not exactly a rich country, but with a PPP GDP *per capita* of $16,030, close to the world average and also close to that of China, it's about nine times as wealthy as Haiti, the poorest country in the Americas. The two countries share the island of Hispaniola.

The border between them, as seen from the air, is shown in Figure 21.3. Haiti is denuded, the Dominican Republic richly forested. Partly the difference is cultural, with the Dominican Republic having absorbed far

**Figure 21.3 Border between Haiti (left) and Dominican Republic (right).**
*Source:* 2011 © United Nations Environment Programme www.flickr.com/photos/unep_dc/8578860427/.

[7] Warsh (2014).

more European influence, but mainly one country is dirt-poor and the other is an emerging middle-income economy. A starker difference in environmental quality could scarcely be imagined.[8]

## Roadmap Through This Section

The rest of the chapters in this section explore specific issues that, so far, I've raised only in the broadest terms:

- Chapter 22, "A Skeptical Environmentalist," will explore the current state of the environment, how it got that way, what its prospects are, and how all these questions relate to the environmental Kuznets curve chapter. The next chapter focuses on the work of Bjørn Lomborg, the "skeptical environmentalist."
- Chapter 23, "Dematerialization," looks at *dematerialization*, the trend in advanced economies (and even some less advanced ones) of using fewer resources instead of more, doing more with less, and seeking out experiences rather than physical possessions.
- I'll then elaborate on practical strategies for improving the future in Chapter 24, "We Are as Gods." First, I'll reflect on the extraordinary career and mind of Stewart Brand, author of the *Whole Earth Catalog*, and a founding father of environmentalism, whose proposed remedies, seemingly designed to turn conventional environmentalism on its head, deserve serious attention.
- Finally, Chapter 25, "Ecomodernism," will discuss the remedies themselves—urbanism, nuclear energy, genetic engineering, and ecosystem engineering—in the context of a multi-author document, called *The Ecomodernist Manifesto*, that calls for implementation of these approaches.

[8] In an unfortunate footnote to the mostly positive Dominican story, a massive garbage spill floated onto some of the beaches of the Dominican Republic in 2018. I do not know if this was vandalism or if, in a bizarre act of corruption, someone agreed, for a fee, to allow garbage dumping in the ocean there—but it is a reminder that the environment is fragile and threats to it may be unforeseen.

# 22

# *A Skeptical Environmentalist: The Greening World of Bjørn Lomborg*

## The Data Hound

Bjørn Lomborg (see Figure 22.1), the youthful author of *The Skeptical Environmentalist* and other books, is an enigma.[1] What can you say about a man who's written a book with 2,930 footnotes? He is obsessed with data.

And he loves controversy. Rarely presenting both sides of a story (you may have noticed that I don't either), the Danish academic argues that the environment is getting better in most ways. The reason, he says, is that we are becoming richer, more technically able, and thus continually better positioned to deal with the problems that large, technologically advanced societies face.

[1] Lomborg (2001).

**Figure 22.1 Bjørn Lomborg.**

*Source:* Photo by Emil Jupin. https://commons.wikimedia.org/wiki/File:Bj%C3%B8rn_Lomborg_1.jpg.

Essentially, he says that we're on the down side (the good side) of the environmental Kuznets curve—hereafter "EKC"—described in Chapter 21, "Prologue," although he doesn't mention the EKC by name even once.[2]

A good bit of this chapter will be about Lomborg's book and the data he presents, but first I want to place Bjørn Lomborg, the man, in context and comment briefly on his philosophy and the controversy he has engendered.

## Challenging Received Wisdom

Originally a left-leaning Greenpeace member and "vegan backpacker," Lomborg, while researching environmental science, was influenced by

---

[2] The discovery of the EKC by Grossman and Krueger was roughly contemporaneous (1991–1995) with Lomborg's early research, so it's understandable he might not have heard of it. Still, it seems like an odd omission.

Julian Simon's pro-growth book, *The Ultimate Resource* (see Chapter 18, "Simon and Ehrlich"). Lomborg writes,

> It all started in 1997, when [I] read a *Wired* Magazine interview with economist Julian Simon claiming that the environment—contrary to common understanding—was getting better, not worse. . . . [T]his had to be incorrect ("right wing, American propaganda").

I'm glad he wasn't satisfied with his first reaction! Lomborg continues,

> [I] organized a study group with some of [my] top students to prove Simon wrong. . . . [T]o everyone's surprise, much (though definitely not everything) of what Simon said was right. [The project] . . . led to the publication of a Danish book later that year and to *The Skeptical Environmentalist* in 2001.[3]

## The Critics Howled . . .

*The Skeptical Environmentalist* touched off a firestorm of criticism, resembling in emotional temperature the response to the *Ninety-Five Theses* that Martin Luther nailed to the Wittenberg church door in 1517. Lomborg's detractors fall into two categories. One group claims that he is a climate change "denier" and an anti-environmentalist.[4] He is nothing of the sort. "He has never denied that mankind's actions are making the earth hotter, merely that it makes more economic sense to adapt to higher temperatures than resist them," writes Matthew Moore in the *Telegraph*.[5] In his later work, Lomborg has become even clearer regarding his belief in the science of global warming and his view that climate change constitutes a meaningful risk.

But, pursuing economic reasoning, Lomborg wants to get good value per social or environmental dollar spent. He asks us to recognize that there are other problems that need to be solved with the resources available, which are

---

[3] www.lomborg.com/for-journalists.

[4] I don't like the word *denier* because it comes from *Holocaust denier,* a disgusting species of person who claims that the six million murdered European Jews never existed, mysteriously disappeared, or are still living. Expressing concern that climate change risk may be overstated, or that the evidence for it is incomplete, or that the proposed remedies will be ineffective and money is better spent fighting other threats, is not in the same category and people who make the connection should be ashamed of themselves. Lomborg is a climate pragmatist.

[5] Moore (2010).

always limited. Each problem needs to be balanced against every other, with regard to both importance and the likelihood that the proposed remedies will work. Lomborg's Copenhagen Consensus Center supports programs relating to education, health, hunger, inequality, population growth, and trade and migration as well as specific countries such as India, Bangladesh, and Haiti. These are not the priorities of someone who doesn't care.

The second category of critiques is that Lomborg is trained as a political scientist and statistician, not an economist or environmental biologist. He thus writes outside his field of expertise, does so with brio and confidence, and sometimes glosses over evidence contrary to the point he is trying to prove. I'd respond that an informed outsider's perspective should always be welcomed in any field, since insiders tend to reinforce each other's views and prejudices.

## Comparing Costs and Benefits

Lomborg gets one thing *exactly* right: all decisions, including environmental, personal, and political choices and any other kind of decision you might be faced with, can only be made sensibly if you apply cost-benefit analysis. *All* of the costs, including hidden ones, must be fully accounted for, and long-run costs (or benefits) must be balanced against short-run benefits (or costs).

In the same spirit, *all* the benefits, including those you can't see and that may seem completely unrelated, must be accounted for. The long-run versus short-run balance that matters for costs also matters for benefits.

This principle of economics, or of life if you'll permit me, goes back to Frédéric Bastiat, the Frenchman from two centuries ago whom we met in Chapter 2, "The Population Explosion, Malthus, and the Ghost of Christmas Present." Bastiat did not even claim originality—he acknowledged his debt to the earlier historian François-René de Chateaubriand—so the idea is old indeed. Americans may be more familiar with the brilliant restatement of Bastiat's ideas by the journalist Henry Hazlitt,[6] or with their expression by Milton Friedman in his bestselling books and public television shows.

[6] Hazlitt stated the rule in three parts: "The bad economist sees only what immediately strikes the eye; the good economist also looks beyond. The bad economist sees only the direct consequences of a proposed course; the good economist looks also at the longer and indirect consequences. The bad economist sees only what the effect of a given policy has been or will be on one particular group; the good economist inquires also what the effect of the policy will be on all groups." Hazlitt (1946).

But, wherever you have heard the idea, it is probably the most important economic concept you'll ever learn. Only "people respond to incentives" comes close.

## Prioritizing Ways of Helping the World

Even if only half of what Lomborg says is true, then, most of us are thinking about environmental remediation in the wrong way. There are only so many resources in the world, and only a fraction of them are available to be spent on social, environmental, health, and allied problems. That fraction could and should rise or fall according to need; but whatever amount is available, it should be spent as efficiently as possible. It should be spent in a way that reflects not just the severity of each problem but "how effective our solutions might be."

Lomborg's 2004 Copenhagen Consensus (a convention of experts) came to the surprising conclusion that adding micronutrients to the diets of the poor, and controlling HIV and AIDS, offered the biggest "bang for the buck," and climate change mitigation the smallest.[7] That is not because climate change is unimportant; although Lomborg had argued that past warming had mostly been beneficial, he believes warming could be very harmful in the future. It's because, according to his panel of experts, the proposed remedies are likely to have astronomical costs and will do almost nothing to solve the problem.

But what *kind* of experts did Lomborg think it was wise to assemble? *Prioritizing* the problems and solutions, given their various potential costs and benefits, is the task at hand. And, since figuring out how to allocate limited resources to satisfy unlimited needs is what economists do, Lomborg assembled his brain trust from among the community of economists. Not biologists or sociologists or doctors; *economists*. This approach provides a fresh set of eyes with which to view the problem and is to be commended.

Later versions of the Consensus have given more favorable play to climate change remediation. However, it still ranks far below a number of lower-cost, higher-impact projects such as expanding access to contraception, making it easier for workers to migrate in search of opportunity, and promoting free trade among nations.

Some of the experts in his group rebelled. The late Thomas Schelling, who won the 2005 Nobel Prize in economics, thought that if a more

[7] www.ted.com/speakers/bjorn_lomborg.

## Evidence of a Greening World: "Good but Not Good Enough"

Lomborg begins *The Skeptical Environmentalist* with what he called "the Litany," the usual recitation of ways in which the growing population and its increasing wealth are supposedly ruining the environment. Of course, there's a grain of truth to the Litany—as we saw in Chapter 21, early stages of agricultural and industrial development can be ecologically terrible.

Lomborg then demolishes most of the Litany, but since we've already heard that from Ridley, Rosling, Norberg, and Pinker,[11] we'll skip that part, except to point out that Lomborg finds a few rotten apples in the horn of plenty. For example, as of 2010, there were 680 million people in the developing world, some 9% of the world population, who were *starving*. Not malnourished—starving, the technical definition of which is having "inadequate [food] to cover even minimum needs for a sedentary lifestyle."[12] Although it's a big improvement from the recent past when many more people were starving, it's definitely "not good enough."

In the next sections of *The Skeptical Environmentalist*, Lomborg then demonstrates improvement in population control, life expectancy, health, nutrition, money incomes, consumer goods, education, leisure, and safety—concepts familiar from our earlier chapters. He then responds, mostly optimistically, to a number of concerns about depleting resources, such as forests, fisheries, energy, minerals, and water. (Water? Really? We have enough water?) This line of reasoning is not new, although Lomborg impressively squeezes it all, with mountains of supporting data, into a single volume.

Lomborg is not a Pollyanna. He says that many things about our world are good but not good enough. His Copenhagen Consensus organization has evolved from one mainly concerned with conserving the environment in an economically efficient way to a broader-based problem-solving group. The group now seeks out expert opinion in finding a balance between environmental issues, nutrition, health, social welfare, liberty, and many other questions.

---

[11] Lots of Scandinavians!

[12] Definition used by the United Nations Food and Agricultural Organization, cited in Hickel (2015). This definition gives a low estimate of the incidence of hunger because most poor people cannot survive while engaging in a sedentary lifestyle; they more typically need to perform heavy physical labor involving much higher caloric consumption.

## Cough, Cough

The part of *The Skeptical Environmentalist* that links most closely to the EKC discussed in the last chapter starts with Lomborg's chapter entitled "Pollution." Pollution is what we first think of when we contemplate environmental destruction, although it is obviously not the only serious concern.

Figure 22.2 shows the evolution of sulfur dioxide and smoke pollution in London over a *very* long period of time, from Shakespeare's day through the present.

Strictly speaking, this isn't an EKC because it plots pollution against *time*, not wealth, but London got richer so steadily over the period that pollution might as well be plotted against wealth in this diagram. Clean, dirty (peak dirty in the mid- to late 1800s), clean.

How do we know how much sulfur dioxide or smoke was in the air in London in 1585? We don't, exactly. According to Lomborg,

> On the basis of coal imports . . . the British environmental scientist Peter Brimblecombe . . . has estimated the concentrations of sulfur dioxide and smoke (particles or soot) in the air . . . from as far back as 1585. . . . [L]evels of smoke pollution increased dramatically [for] 300 years . . . only to have dropped even faster [starting in] . . . the late nineteenth century . . . such that the levels of the 1980–1990s are below the levels of the late sixteenth century.

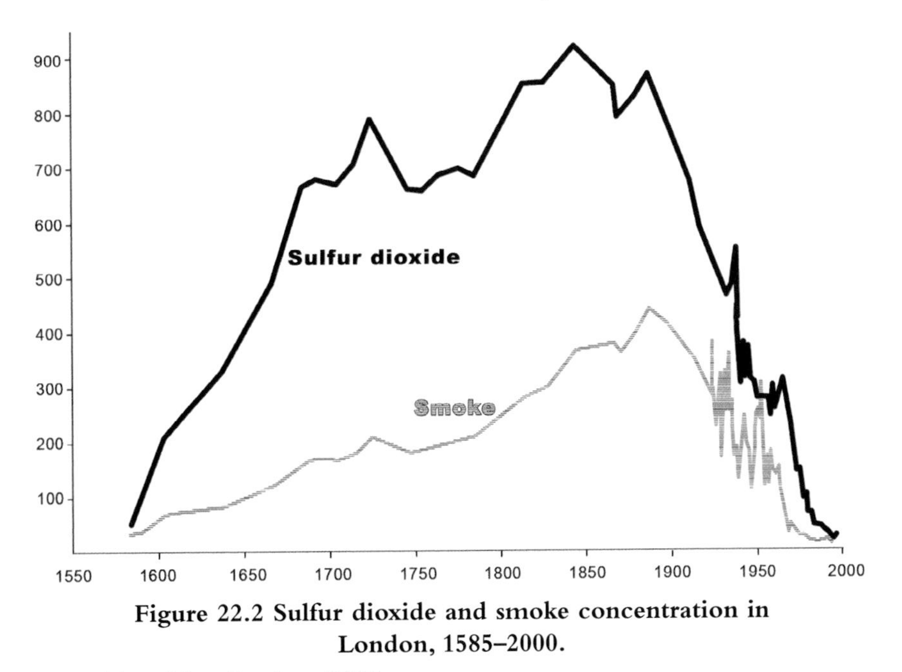


**Figure 22.2 Sulfur dioxide and smoke concentration in London, 1585–2000.**

*Source:* Adapted from Lomborg (2001).

This despite the fact that the last severe London smog, which occurred in 1952, "killed about 4,000 Londoners in just seven days," some by drowning after having fallen into the Thames because they could not see it.

Lomborg continues,

> [S]moke or particles are probably by far the most dangerous pollutant. In other words, with respect to the worst pollutant the London air has not been as clear as it is today since the Middle Ages. . . . Air pollution is not a new phenomenon that has gotten worse and worse—it is an old phenomenon, that has been getting better and better [for over a hundred years].[13]

## Incomes and Pollution

Gene Grossman and Alan Krueger, the discoverers of the EKC, whom we met in Chapter 21, hypothesized and then documented a relationship between *income* and environmental quality. We haven't quite demonstrated this, except indirectly in the case of London smog. But Lomborg's research turns up a direct connection.

Figure 22.3 shows the relation between country income (that is, PPP-adjusted *per capita* GDP) and particulate-matter air pollution in

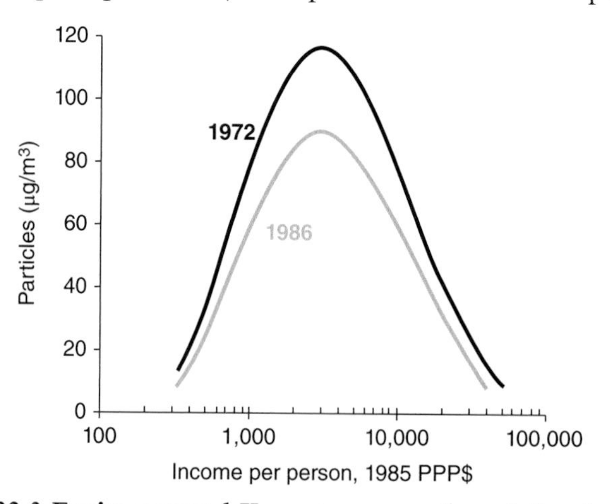


**Figure 22.3 Environmental Kuznets curves: the relation between PPP GDP *per capita* and particulate-matter air pollution in 48 cities in 31 countries in 1972 and 1986.**

*Source:* Adapted from Lomborg (2001).

[13] Lomborg (2001), p. 164. London's population also grew tremendously over the period, so if the *x*-axis represented *per capita* real income, not the total real income that is roughly captured by the time dimension, the pollution curve would have peaked earlier and come down more quickly.

1972 and 1986. In both years, pollution turned down as incomes rose above about $3,000 per year—about $7,000 in today's money, or about the income level of India or Guatemala.[14]

But the *amount* of pollution at the peak of the curve also fell by about 25% between 1972 and 1986; not only does each curve slope downward beyond a certain level of income, the *curve itself* moves lower over time, representing less pollution at *each* income level as time passes. A similar analysis of sulfur dioxide shows peak pollution at about the same income level, but with a steeper lowering of the curve over time—60% between 1972 and 1986.

## Is the Environmental Kuznets Curve a Universal Principle?

Will this upside-down U-shaped relation between incomes and pollutants *always* work? Probably not. Some pollutants continue to rise with incomes, although true believers in the curve would say that incomes are just not high enough yet for the amount of pollutant to turn down. (This appeared to be the case with carbon until recently; more on that later.) And some pollutants fall with incomes at all levels, never showing a rising part of the curve.

There are some exceptions to the EKC. Species diversity is a particularly difficult one. As income levels rise, you don't get new species to replace those that have become extinct. This is because, no matter how clean the environment, new species take a long time to "appear" (that is, to become differentiated from existing ones).[15]

Yet, if you go beyond the species to the individual animal as the unit of analysis, the curve comes back. "The Kuznets curve," writes the British researcher Alexander C. R. Hammond,

> not only applies to forest area [which is increasing in rich countries; see Chapter 11, "Cities"], but also biodiversity. [Matt] Ridley gives the example of three apex predators: wolves that live in developed

[14] Although the data in Lomborg's example are old and the specific problems represented by the example have largely been solved, the underlying principle is still valid and important.

[15] At least this is true of large, familiar animal and plant species. Bacteria and other microorganisms can evolve faster than we destroy them. What most people call *biodiversity* refers to animals and plants, which are a small fraction of the living things on earth.

> countries of Europe and North America, tigers who mainly inhabit mid-income India, Russia and Bangladesh, and lions, which live in poor Sub-Saharan Africa. Following the Kuznets curve, wolves are rapidly increasing, tiger numbers have been steady for the last 20 years (and have just begun to increase), while lion numbers continue to fall.[16]

Lomborg is sanguine about whether the general EKC principle can be relied on across time and differing circumstances: "There are no decisive reasons to assume that the same development will not happen in the Third World, which today faces serious environmental problems equivalent to those we faced 50 [to] 80 years ago."[17]

## Different Curves for Different Problems

In other work (not covered by Lomborg), the EKC is shown to move around a lot depending on what pollutant or index of environmental quality you care about. The peak is not always at $7,000. Alexander C. R. Hammond writes, "Once nations hit around $4,500 GDP per capita, forest areas begin to increase."[18] That's a pretty low income, below that of Pakistan and close to that of Bangladesh.

But, for some materials, the point at which the EKC turns down is much, much higher—and therein lies the rub. The remarkable energy density of fossil fuels makes their use very desirable if one ignores environmental considerations. For this reason, carbon emissions are among the last to decrease as incomes rise; fossil fuel consumption is a type of luxury good.

Figure 22.4 is an estimate (not from Lomborg) of the EKC for carbon, separated into emissions from production and emissions from consumption.

Note that the two curves have very different shapes. The higher one (showing less improvement) is for carbon emissions from consumption. The lower one (showing a bigger reduction in emissions) is for

---

[16] Hammond (2018).

[17] Lomborg (2001), p. 176.

[18] Hammond (2018).

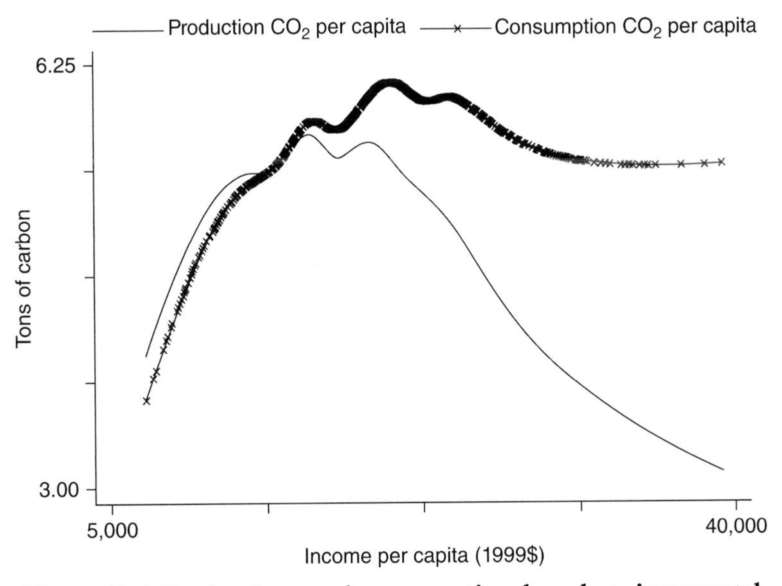


**Figure 22.4 Production- and consumption-based environmental Kuznets curves (the heavier line represents consumption).**

*Note:* The figure presents the fitted values for the 1,920 income-*per-capita* observations in the data. To convert to current (2018) dollars, multiply the 1999 dollar amounts by 151.

*Source:* Aldy (2004). Reproduced with permission.

production.[19] For example, the carbon footprint of manufacturing, say, a washing machine (production) is much lower at higher incomes, whereas that of operating that same machine (consumption) does not improve as much.

Peak carbon emissions, then, seem to be around the income levels of the world's newly industrialized or moderately rich countries. Rich-country $CO_2$ emissions are beginning to turn down sharply, as Western European, the United States, and Japanese incomes pass through the $40,000–$60,000 range. The change is due to several factors: more efficient vehicles and power plants, the growth of alternative energy sources, and social developments such as the downturn in miles driven per capita in the United States after 2005.[20] These trends bode well for the long-term future of the Earth, when today's middle-income countries are rich and today's poor countries are middle-income—if we can get there in one piece.

[19] Aldy calls the two kinds of activities *pre-trade* and *post-trade*. I have reinterpreted these terms as production and consumption.

[20] Federal Reserve Bank of Saint Louis (2014).

By the way, the EKC doesn't work "all by itself." Public policy is instrumental in limiting pollution or emissions of any kind, for the reason stated earlier: private parties always try to externalize. The reason a country with a *per capita* GDP higher than, say, $7,000 begins to reduce particulate-matter air pollution is that the people, expressing their will through the political system, are at that point able and willing to make the *sacrifice* necessary to reduce pollution. The same principle works with carbon, at a different income level. That is one reason why we should encourage as much economic development as possible.

## Conclusion: Some Ways to Spend a Trillion Dollars

Lomborg's economic-efficiency approach to thinking about the environment leads me to propose a thought experiment. Suppose a trillion-dollar expenditure, now, devoted to decarbonizing the global energy supply, would (with certainty) lead to a 0.5°C (about 1°F) decrease in the global average temperature in 2100, relative to what it otherwise would be.

It is very unlikely to do anything of the kind: the fully allocated cost, including losses in transportation efficiency and so forth, would be many times larger. But let's say for the sake of argument that the strategy will work.

Assuming that the point of spending the trillion in the first place is to *help people*, what else could we do with the money? If, as I said earlier in this chapter, 680 million people are starving, we could buy each of them 1,470 bowls of golden rice at a dollar a piece—now *that* would make a difference! We would almost certainly be able to eradicate malaria—about 10 times over, according to the Gates Foundation. That would make hundreds of millions of tropical residents dramatically more productive. They would feel much better too.

A trillion dollars would probably suffice to build a seawall that would protect Bangladesh from flooding due to rising sea levels for centuries. With the leftover money, we could protect Venice, New Orleans, and (if you care) Miami Beach. We could educate a hundred million girls in Africa and the Middle East. The possibilities are endless unless you consider the trillion dollars to be spoken for by the One True Cause.

Of course, given enough time, we can come up with more than just one trillion dollars to spend. But resources are not unlimited and never will be. To paraphrase Senator Everett Dirksen, a trillion here and a trillion there and pretty soon you're talking about real money. Trying to control the earth's climate fluctuations by removing the part of them that is manmade is one strategy, but trying to efficiently adapt to the fluctuations—and helping people in a myriad of other ways—is another, more consistent with the principles of economic reasoning.

There is no single answer. Economic growth will help—a lot.

# 23

# *Dematerialization: Where Did My Record Collection Go?*

So how does the idea that economic growth helps the environment actually work? It's quite a claim, and extraordinary claims require extraordinary evidence.

Let's delve a little deeper, into the idea of *economizing*, which is doing more with less—not *doing less*, which some people advocate, but *doing more* while *using* less. Less labor, less energy, fewer natural resources, and less pollution.

All creatures, even primitive single-celled organisms, economize: they do the best they can with resources that are costly and difficult to acquire. But human beings have found a new way to economize, which is to *dematerialize*—to substitute experiences, which in some sense are made out of nothing, for possessions. We also use technology to replace large and heavy things with small and light ones that provide the same or greater utility.

Almost a century ago, in his book, *Nine Chains to the Moon*, the visionary inventor Buckminster Fuller called this process *ephemeralization*.[1] Since Latin is often clearer than Greek, we'll call it *dematerialization*. It's part of the larger theme of greater efficiency in the use of resources.

[1] Fuller (1938).

## The World's Largest Vinyl Mine

In the 1960s and 1970s, just about every music fan amassed an impressive record collection, often weighing hundreds of pounds and taking up the better part of a room. In an advertisement, Warner Brothers joked about this fantastically prolific use of plastic, saying that Warner made the best records because its plant was "located atop the world's largest vinyl mine" in Burbank, California. After a number of claims about the company's virtues, the ad concluded, "It's the plastic!"

Half a lifetime later, which is an instant in the context of human history, the records are gone, except in museums and the homes of specialized collectors. They've been replaced, first by tapes, then by sequences of zeroes and ones stored on little records (CDs), and later on hard drives or in solid state memory (MP3s). Now, MP3s have a slightly antique feel about them. Most music services store 25 million "songs" (including nonsongs such as classical pieces and spoken word) on their own servers and stream them on demand to subscribers (see Figure 23.1).

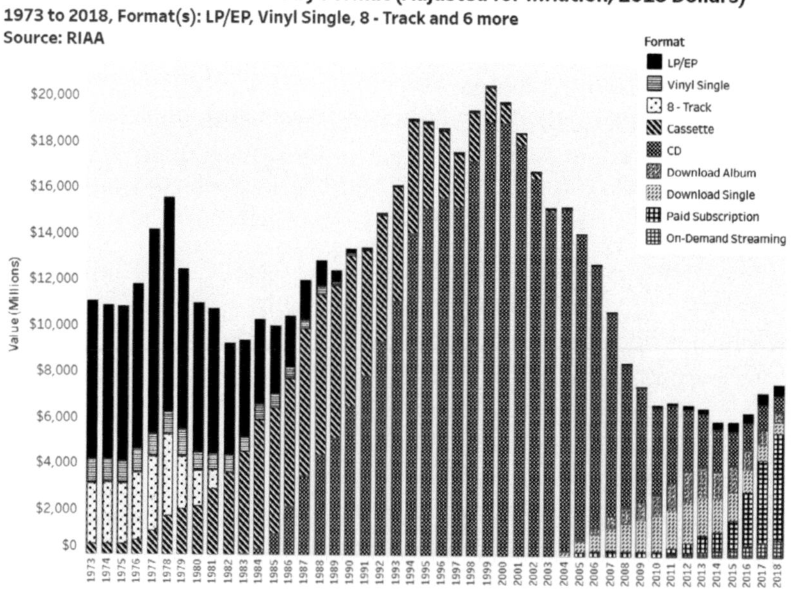


**Figure 23.1 US recorded music formats by sales volume, constant dollars, 1973–2018.**

*Source:* Recording Industry Association of America US Sales Database, https://www.riaa.com/u-s-sales-database/.

## Degrowth and the Religion of Radical Environmentalism

In the 1970s, environmentalists, noting that economic and population growth were causing us to use up resources quickly and pollute the air and water, began to rally around "degrowth" as the remedy. "Growth for the sake of growth is the ideology of the cancer cell," intoned the influential writer Edward Abbey.

But "degrowth didn't happen," writes Brian Eastwood, an MIT Sloan School *Newsroom* reporter. He continues,

> In the decades since Earth Day [April 22, 1970], population growth in the United States continued at the same pace, gross domestic product growth has increased, and our desire for material goods has not abated—yet the planet remains intact. That's because this growth coincided with reductions in the use of water, metals such as iron and steel, and minerals such as phosphate and potash, . . . said [Andrew] McAfee [an MIT professor and co-author of *The Second Machine Age*].[2]

Yet, in the decades since modern environmentalism began, the movement has taken on some of the characteristics of a religion: claims not backed by evidence, self-denying behavior to assert goodness, focus on the supposed end times. The celebrated *New Yorker* writer John McPhee, in *Encounters with the Archdruid*, a series of dialogues with the Sierra Club leader David Brower, captures the religious fervor perfectly:

> The outermost circle of the Devil's world seems to be a moat filled mainly with DDT. Next to it is a moat of burning gasoline. Within that is a ring of pinheads each covered with a million people—and so on past phalanxed bulldozers and bicuspid chain saws into the absolute epicenter of Hell on earth, where stands a dam. Conservationists who can hold themselves in reasonable check before new oil spills and fresh megalopolises mysteriously go insane at even the thought of a dam . . . because rivers are the ultimate metaphors of existence, and dams destroy rivers.[3]

Holy smoke!

The illusion that degrowth, better understood as a return to widespread extreme poverty, is somehow desirable has not gone away since its first mass flowering in the 1970s. In 2012 Erik Assadourian, a senior

[2] Eastwood (2018).

[3] McPhee (1971).

fellow at the Worldwatch Institute, published *The Path to Degrowth in Overdeveloped Countries,* the gist of which is obvious from the title—Overdeveloped. The pessimism that is still fashionable in some quarters, after decades if not centuries of countervailing evidence, oozes from the front page of Assadourian's website:

> You can . . . explore the mind of someone obsessed with if/how the human species is going to stop its rapid slide into civilizational collapse. . . . [W]e probably won't, though the survivors might be able to build a more sustainable civilization from its ashes in a few thousand years.

Assadourian is not considered a crank—his organization is supported by, among others, the Ford and Hewlett Foundations and the United Nations—and his words are fairly mild. Some other environmentalists still sound like they are rewriting Dante's *Inferno*, as McPhee did so vividly in his report on Brower.

It's embarrassing. Instead of an environmentalist, may I call myself a conservationist instead?

## Environmental Efficiency: The Path Forward

The degrowth advocates of earlier times and of our own could not be more wrong: as we saw in Chapter 22, A "Skeptical Environmentalist," on the environmental Kuznets curve, simple living is dirty living. The remedy is not degrowth, but a *different kind* of growth: an increase in efficiency, fewer inputs being converted to more outputs.

### A Basket of Goodies from Radio Shack

In a TEDx talk,[4] Andrew McAfee, the MIT professor whom we met earlier, pointed out that, 20 years ago, most well-to-do people owned a personal computer, camcorder, calculator, radio, clock, CD player, answering machine, and many of the other devices advertised in a 1991 Radio Shack ad.[5] Not only did these gadgets cost a total of $3,055 according to the ad, or about $5,225 in today's money, they used a lot of steel, aluminum, copper, silicon, electricity, and space. Today the functionality of 13 of the 15 devices in the ad is achieved by a $200 smartphone that uses much less of each of these materials, weighs four

[4] https://youtu.be/uEQDLUdMb5M.

[5] Cichon (2014).

ounces, and can fit in a shirt pocket.[6] Figure 23.2 is a visualization of this compression, but with 24 devices instead of 13.

**Figure 23.2 Devices fully or partially replaced by smartphone.**

*Source:* Constructed by Dave Stanwick.

[6] The $200 is for a slightly outdated but perfectly good iPhone 5s. Perry (2016).

## Delving Deeper: The Dematerializing World of Vaclav Smil

The modern guru of dematerialization, superior in talent if not quite equal in visibility to Buckminster Fuller, is Vaclav Smil, the Czech-Canadian scientist whom we met in Chapter 10, "Energy." (Bill Gates says he's read all 36 of Smil's books.) In *Making the Modern World: Materials and Dematerialization,*[7] Smil argues that we are doing more with less in almost every aspect of production, at an accelerating pace—and, yet, in the aggregate, increases in demand for material goods in the developing world will overshadow the effect of dematerialization. Until they don't. As much of the world becomes developed and economic growth (as well as population growth) slows, our use of materials will eventually fall in absolute terms.

Smil writes in a complicated style that is not particularly quotable—but Ronald Bailey, a columnist for the libertarian magazine *Reason*, summarizes Smil's work concisely. "Today," Bailey writes, "it takes only 20% of the energy it took in 1900 to produce a ton of steel. Similarly, it now takes 70% less energy to make a ton of aluminum or cement and 80% less to synthesize nitrogen fertilizer."[8]

## Planes, Trains, and Automobiles

The examples go on and on, embracing not only agriculture and basic industries and but also automobiles, airplanes, housing, mining, technology, and many other aspects of life. For example, a good car now lasts 200,000 robust miles instead of 100,000 wobbly ones. Car sharing nudges the utilization rate of a particular car (the percentage of time it is not parked) toward 100% from the current 5%.[9] If self-driving cars become widely used, they can be packed much more tightly, either on existing roads or on super-roads yet to be built. (A possible design for a system of super-roads is one that collects cars from local streets and delivers them onto a long-distance conveyance that's as fast as a high-speed train is now, or funnels them into a convoy in which the cars become self-driving at high speed, but only within the controlled confines of the super-road.)

---

[7] Smil (2014). Another useful resource, carefully covering the UK in terms of inputs use, is Goodall (2011).

[8] Bailey (2014).

[9] Morris (2016).

Passenger aircraft have improved their fuel efficiency per seat by 55% in four decades.[10] Housing is lighter in weight, and more fuel-efficient with regard to heating and cooling, than ever before. As we saw earlier, a smartphone weighing a few ounces displaces a massive pile of electrical gizmos. Your photographs, newspapers, letters, bills and checks, and to some extent books went the same place as your phonograph records.

## Dark as a Dungeon Down in the Mine?

Even coal mining, the poster child for environmental awfulness, has become more efficient, cleaner, and radically safer. It can now be done (surprise!) by computer.

In Australia, about 90% of underground coal production traditionally comes from a technique called longwall mining. A mechanical shearer, controlled by the miners from inside the mine, slices off a coal-rich wall of the mine's interior. Although the technique is much better than traditional hand digging, it's still dangerous and labor-intensive.

To make longwall mining radically safer and more efficient, an Australian government agency called CSIRO has developed an "underground automation system that both isolates people from mining hazards and improves productivity. Specialised remote guidance technology continuously steers longwall equipment, automatically plotting its position in three dimensions." About 60% of Australia's longwall coal mines have adopted the new system, which "removes personnel from direct hazards and increases safety. Real-time progress can be monitored *from anywhere in the world*, leading to further efficiency gains."[11]

Although this sounds more like a story about improved safety and labor efficiency than one about economizing on physical inputs, it fits the general theme of doing more with less. Less labor, less breathing of polluted air, less heartache from the deaths of miners.

## The Limits to Dematerialization

But massive economic growth outside the developed world means not less stuff but more, at least for a while—even as dematerialization proceeds. "Too many people still live in conditions of degrading and

[10] Peeters et al. (2016). Much of the improvement was between 1971 and 1998, when efficiency improved at 2.4% per year. Despite appearances, not all of the increase in efficiency comes from packing more seats into a plane.

[11] CSIRO (undated). My italics.

unacceptable material poverty," writes Vaclav Smil, and "all of those people . . . need to consume more materials *per capita* in order to enjoy a decent life." Thus, the aggregate use of materials will increase for a while, even while we achieve a fall in the volume, weight, and cost of materials needed to produce a given unit of output across many industries.

## It's Really Happening: Dematerialization in Commodity Usage Data

When a trend shows up in the media "buzz," but you can't find it in the data, it's proper to be skeptical. That's why I was pleasantly surprised to find that Jesse Ausubel, an environmental scientist of the first rank, has documented a major downturn in raw materials consumption in the United States. Figure 23.3 shows the trend for some commodities that have just recently begun to be used less:

Only one of the commodities, potash (used in fertilizer), faced increased usage after 2000. The biggest declines were in aluminum and copper, and the reduction in the use of plastics (which have a long half-life in the environment) was particularly impressive.

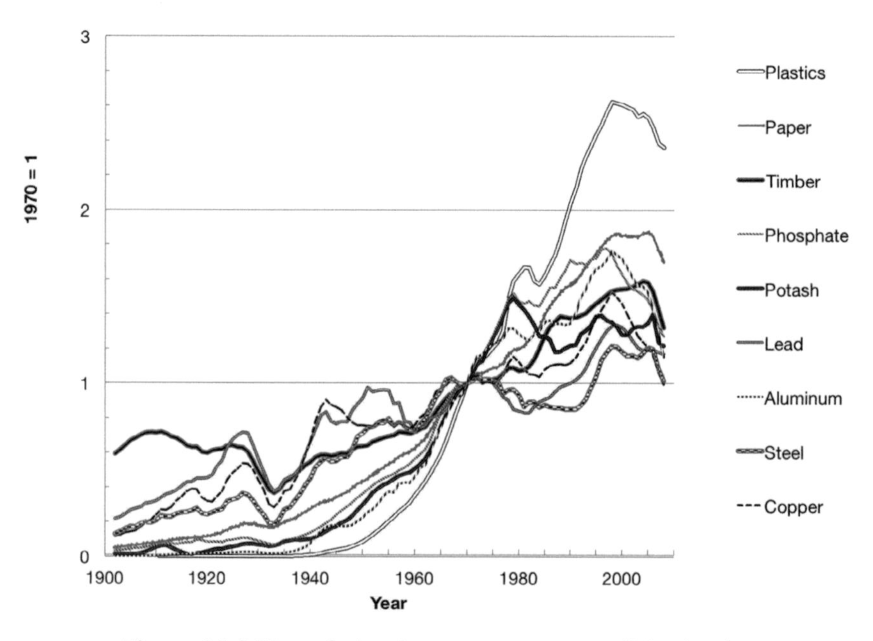


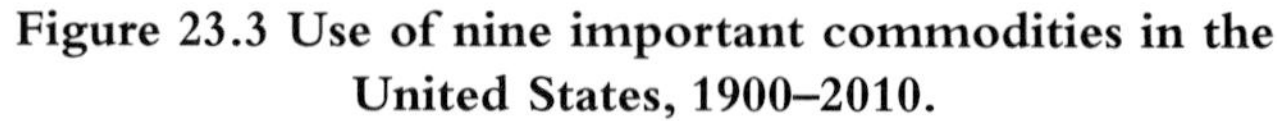
**Figure 23.3 Use of nine important commodities in the United States, 1900–2010.**

*Source:* Ausubel (2015). Reproduced with permission.

"America has started to dematerialize," writes Ausubel. In case you think he cherry picked, he discloses:

> The reversal in use of some of the materials so surprised me that Iddo Wernick, Paul Waggoner, and I undertook a detailed study of the use of 100 commodities in the United States from 1900 to 2010. One hundred commodities span just about everything from arsenic and asbestos to water and zinc. The soaring use of many resources up to about 1970 makes it easy to understand why Americans started Earth Day that year. Of the 100 commodities, we found that 36 have peaked in absolute use, including the villainous arsenic and asbestos.[12]

A number of the commodities, shown in Figure 23.4, peaked around 1970, contemporaneous with Earth Day. Some, not shown, peaked even earlier (whale oil, thank goodness, peaked in 1846).[13]

Ausubel's analysis is too detailed to show graphically in its entirety here, but you get the point:

> Another 53 commodities have peaked relative to the size of the economy, though not yet absolutely. Most of them now seem poised

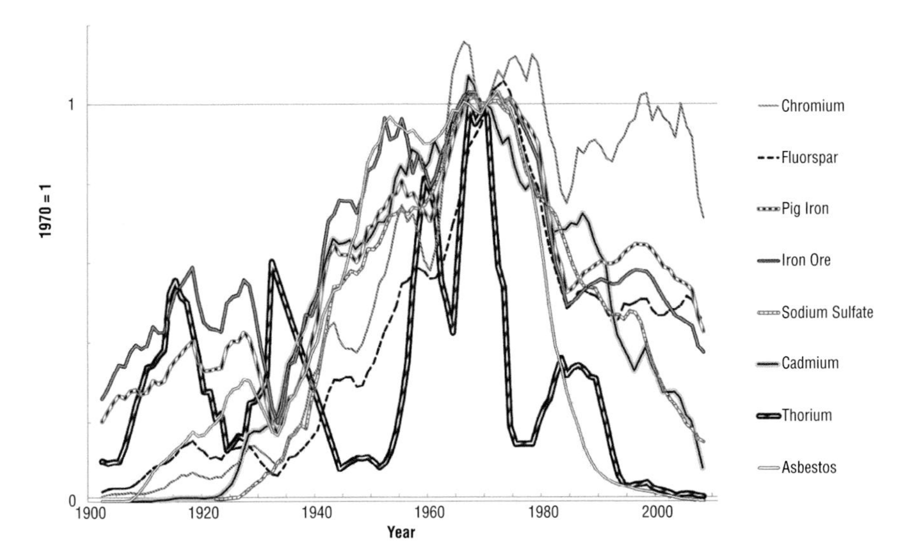


**Figure 23.4 Use of eight basic materials in the United States, 1900–2010.**
*Source:* Ausubel (2015). Reproduced with permission.

[12] Ausubel (2015).
[13] Dvorsky (2012).

> to fall. . . . They include not only cropland and nitrogen, but even electricity and water. Only 11 of the 100 commodities are still growing in both relative and absolute use in America.[14]

## Dematerializing Water . . . Really?

Ausubel is not alone in noticing the more obvious aspects of dematerialization, in particular of electronics, as I noted earlier. He writes: "Much dematerialization does not surprise us, when a single pocket-size smartphone replaces an alarm clock, flashlight, and various media players, along with all the CDs and DVDs." But he is surprised by the declining use of *water*:

> [E]ven Californians economizing on water in the midst of a drought may be surprised at what has happened to water withdrawals in America since 1970. Expert projections made in the 1970s showed rising water use to the year 2000, but what actually happened was a leveling off. While America added 80 million people—the population of Turkey—American water use stayed flat. In fact, U.S. Geological Survey data through 2010 shows that *water use has now declined below the level of 1970*, while production of corn, for example, has tripled. More efficient water use in farming and power generation contribute the most to the reduction. [italics mine]

Somehow, we even got through the great California drought of 2014–2017. I was there when it ended, with massive flooding, deadly mudslides, and an avalanche that closed part of the Big Sur stretch of Highway 1 for a year and a half. But droughts will come and go, and economizing on water is still important. Hurrah for dematerialization.

## Experiences, Not Stuff

A related trend, one that we can expect will be long-lasting in developed countries, is the substitution of experiences for "stuff" as pleasurable ways of spending one's money.[15] Although, in the United States at least, there was a fad for massive houses that has not completely faded, many people with the means to choose either more stuff or more

[14] Ausubel (2015).

[15] See, for example, Pine and Gilmore (2011) and Pine and Gilmore (1998).

experiences are opting for the latter. Young people, awkwardly called *millennials*, "aren't spending our money on cars, TVs and watches," says an industry executive. "We're renting scooters and touring Vietnam, rocking out at music festivals, or hiking Machu Picchu."[16]

Trends like this one are sometimes widely reported well before they show up in the data. However, data from the US Bureau of Economic Analysis verify that these trends are in operation, although not as dramatically as qualitative reports might suggest. Figure 23.5 compares spending on experiences to spending on other services, goods, and total consumption during the period of 2010 to 2017.[17]

We tend to overestimate the impact of change that we see occurring on the margin, so we should be careful not to extrapolate too far the meaning of failing furniture stores and booming music festivals. But

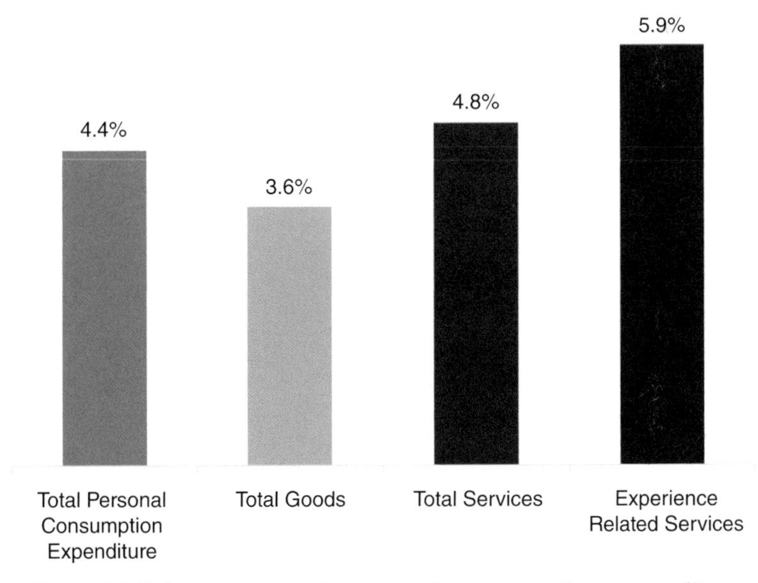


**Figure 23.5 Average annual personal consumption expenditure growth, 2010–2017.**

*Source:* Constructed by the author using data from the US Bureau of Economic Analysis.

[16] As told by Taylor Smith, CEO and co-founder of Blueboard, to CNBC reporter Uptin Saiidi; see Saiidi (2016). A study by Harris Group found that 72% of millennials prefer to spend more money on experiences than on material things: "More than 3 in 4 millennials (78%) would choose to spend money on a desirable experience or event over buying something desirable" (http://eventbrite-s3.s3.amazonaws.com/marketing/Millennials_Research/Gen_PR_Final.pdf). "Only" 59% of Boomers say this, says Dan Kadlec (2015).

[17] "Experiences," in this figure, consist of membership clubs, sports centers, parks, theaters, and museums; gambling; other recreational services; foreign travel; personal care and clothing services; and air transportation.

one can be pretty confident that, compared to our own history, experiences will continue to receive a disproportionate share of the consumer dollar, contributing to the general trend of dematerialization.

## Rematerialization: It's Not Easy Being Green

Even as dematerialization proceeds, there are countercurrents that involve using more stuff. I'll call these *rematerialization*. First, massive economic growth outside the developed world means not less stuff but more, at least for a while; we can't and shouldn't deprive the less fortunate of what we have so much of. Second, renewable energy, which is something of a First World problem (although maybe it shouldn't be), uses a lot of materials. Third, short of extreme repression imposed from outside (a morally unacceptable choice), we don't really have much say over how other countries develop. The will to pursue a better life for oneself and one's children—a life that includes material possessions—is not a bug but a great feature of human nature.

### Blowing in the Wind

If we're going to replace much of the fossil-fuel economy with something else, a lot of new infrastructure will have to be built. Everyone has seen the huge wind turbines that are now used for electric power generation. Although wind is free, wind turbines are not; they're neither cost-free nor are they free in terms of environmental impact.

A large wind turbine—the General Electric 1.5-megawatt model—weighs 164 tons. An organization skeptical of wind power reports, "The steel tower is anchored in a platform of more than a thousand tons of concrete and steel rebar, 30 to 50 feet across and anywhere from 6 to 30 feet deep. . . . Mountain tops must be blasted to create a level area of at least 3 acres."[18] Mining, processing, and transporting these materials consume a fantastic amount of energy (that has to come from currently existing sources, mostly fossil fuels) as well as metals and concrete.

Regarding wind turbines, Vaclav Smil cautions:

> If wind-generated electricity were to supply 25 percent of global demand by 2030, [the build-out] would require roughly 450 million metric tons of steel . . . [plus] the metal for towers, wires, and transformers . . . that would be needed to connect it all to the grid.

[18] www.wind-watch.org.

> A lot of energy goes into making steel. . . . Steel used in turbine construction embodies typically about 35 gigajoules per metric ton. . . . [Thus,] to make the steel required for wind turbines that might operate by 2030, you'd need fossil fuels equivalent to more than 600 million metric tons of coal.[19]

## Blinded by the Sun

Solar energy faces similar obstacles. Michael Shellenberger, a pro-nuclear environmental activist and, in 2018, a centrist Democratic candidate for California governor, cautions: "The centuries-long process of energy dematerialization could reverse if wind and solar power catch on. They are very materials-intensive and involve a lot of solid waste. . . . The International Renewable Energy Agency (IRENA) . . . projected that . . . [the] amount . . . of solar panel waste in the world . . . could reach 78 *million* metric tonnes by 2050."[20]

## A Giant Oil Farm?

Craig Venter, the man who sequenced the human genome, has set forth a proposal even more daring than the now-accepted drive to replace fossil-fuel power with wind, solar, tide, and geothermal energy. Backed by ExxonMobil, Venter is designing algae that are genetically engineered to secrete biofuels—oils—at a contemplated rate of 10,000 gallons of fuel per acre per year. If the rate is achieved, Diamandis and Kotler ask, "So how many acres does it take to power America's entire fleet [of cars]? With roughly 250 million automobiles . . . that translates to about 18,750 square miles, or about . . . 17 percent of Nevada."[21]

Cool stuff, and I hope he succeeds. But Nevada is very big, and covering 17% of it with anything, no matter how thin, light, or clean, involves an immense use of resources. It's also not clear that a giant algae farm will be thin, light, clean, odor-free, or disease-free. It might make the problem of solar panel waste seem trivial.

---

[19] Smil (2016).

[20] Shellenberger (2018). More detail is at http://environmentalprogress.org/big-news/2017/6/21/are-we-headed-for-a-solar-waste-crisis.

[21] Diamandis and Kotler (2012), p. 162.

## The Environmental Costs of Saving the Environment: Summary

All this does not mean we won't gradually reduce our dependence on fossil fuels and generate energy in other ways. We will. (Nuclear power is the easy answer and can be made safe; see Chapter 25, "Ecomodernism.") Decision-making can only be rational if we account for all the costs and benefits of a given energy source, not just those that are easily visible. And we need to be realistic about the amount of rematerialization that will be involved, in an era when dematerialization is making the world a cleaner and more resource-rich place.

## Lost in Space

Finally, space travel, about which there is renewed interest, is also very materials-intensive. The inventor-entrepreneur Elon Musk has stated his intention to place a colony of a million people on Mars within 40 to 100 years. I have no idea how he's going to do it, but unless he invents teleportation ("beam me up, Scotty"), the plan *must* involve moving a million people a minimum of 35 million miles each, using life-supporting spacecraft, or else moving a smaller number and breeding them once they get there. Lots of materials will be involved. Despite Musk's many successes, I wouldn't count on it happening. There is plenty of work to do here on Earth.

# 24

# *"We Are as Gods": The Fertile Mind of Stewart Brand*

"We are as gods," wrote the *Whole Earth Catalog* author Stewart Brand (see Figure 24.1) in 1968, "and we might as well get good at it."[1] What in heavens was he talking about?

Living fruitfully on the Earth, using technology responsibly, and harnessing our collective brainpower to improve our lives, that's what. Important stuff. To better understand Brand's unique contribution to thinking about humans and the environment, it's helpful to first get to know the man.

Unlike many of the thinkers and doers cited in this book, Stewart Brand, Renaissance man, author, businessman, and futurist brims with left-liberal credentials: he is one of the founding fathers of modern environmentalism. It was Brand who printed the NASA photo of the whole Earth, taken from the Moon, on the cover of the first *Whole Earth Catalog*, thereby directly inspiring Earth Day (April 22, 1970). Brand was one of the few hippie radicals who did not disrespect the space program as a costly and militaristic misadventure; he thought it would yield a cornucopia of benefits.[2]

---

[1] Brand (1968).

[2] He was also one of the older and better educated of the hippie patriarchs, descended from engineers; perhaps that explains his more rationalistic outlook.

**Figure 24.1 Stewart Brand.**

*Source:* Photo credit: Ryan Phelan (his wife). http://sb.longnow.org/SB_homepage/SB_photos_big.html.

A computer and technology enthusiast from the start,[3] and educated as a scientist (Paul Ehrlich was his ecology professor), Brand helped build the bridge between the communitarian but venturesome hippie culture of the Bay Area in the 1960s and the tech culture of quirky individualism that emerged in Silicon Valley a few years later.[4] In 1985 Brand and Larry Brilliant founded the WELL, the Whole Earth 'Lectronic Link (I was an early member), which describes itself as "the primordial ooze where the online community movement was born."[5]

## A Merry Prankster

Brand was more than a literary man and a tech visionary. He was a do-bee, in fact a member of the Merry Pranksters, the drug-infused band of provocateurs that promoted be-ins, love-ins, and even die-ins in the

[3] Brand's interest in computers started early in life. On December 9, 1968, Brand assisted the late inventor Douglas Engelbart, of the Stanford Research Institute, with a presentation of then-revolutionary computer technologies, including hypertext, e-mail, and the mouse, at the Fall Joint Computer Conference in San Francisco. The conference would become known as "the mother of all demos."

[4] It's not really called "Silicon Valley." It's the Santa Clara Valley.

[5] www.well.com/about-2/.

streets of San Francisco in the late 1960s. Tom Wolfe wrote *The Electric Kool-Aid Acid Test* about this scene; Brand is a character in it.

Brand does not disavow those days, and his current iconoclasm shares much with the unconventional spirit of that olden time. He might be a little embarrassed about the die-in, a piece of street theater in which the participants tried to stir up sympathy for the starving in other countries by subsisting only on water for a week (everyone was free to leave and most did), but there is nothing to be embarrassed about in the *Whole Earth Catalog*.[6] "Access to Tools," he subtitled it, and that's what it provided for the mostly young back-to-the-landers who imagined themselves reinventing civilization with new, more humane rules and more ecologically sound practices.

Mostly that movement didn't work, but the freedom to think unconventional thoughts and experiment with new lifestyles has had a profound effect on our culture, which has never been the same since.

## "We Are as Gods"

The epigraph to the *Whole Earth Catalog*, which also began this chapter, is a stunning boast: "We are as gods, and we might as well get good at it." The Greek gods proved their divinity through their technological superiority: they could make it rain or stop a flood, create Athena from the brow of Zeus, and perform what we would call miracles that mortals could not. (In an interesting twist, the Greek gods bred with mortals to form half-mortals, an idea that should be fodder for someone's science fiction story, set in the near future, involving robot-human hybrids.)

Brand implied that, having developed technology that would convince any ancient Greek that we were in fact gods, we are now responsible for what that technology can accomplish. We might as well get good at it, he gently suggested in 1968. Today, an older Brand has revised the boast to a moral imperative: "We *have* to get good at it." The change in his tone suggests that our responsibility to do good and not evil with our godlike powers has increased because our technology has grown so much more powerful—and because, in his view, we face a small but meaningful risk of ecological catastrophe.

[6] About the die-in, he wrote, "I was so moved by his book *The Population Bomb*, I organized a public fast to dramatize the famines he said were coming soon. They didn't come and weren't going to. That was clear by the end of the decade the book was written in. . . . His prophecies about overpopulation were highly important in drawing attention to the issue and may rank as one of the great self-defeating prophecies. World saved: Ehrlich embarrassed. Pretty good deal. He was right to be wrong." In "Environmental Heresies" (2008).

## How Buildings Learn

When Brand wasn't busy saving the earth, creating a new literary genre (the *Whole Earth Catalog* is a sort of print-format prototype of the World Wide Web), or starting businesses (Global Business Network) or foundations (the Long Now Foundation), he was moonlighting as an architectural critic and historian. His *How Buildings Learn* is one of the more intriguing amateur literary efforts of our time, reminiscent of the works of the great Victorian polymaths. Figure 24.2 shows how, in Brand's imagination, two particular buildings *learned from their users* over more than a century.

The buildings looked much plainer and almost identical to each other in the 1860s when they were built. Since then, both acquired cast-iron balconies, called *galleries* in New Orleans, to accommodate the changing tastes of the owners. The balcony on the left is quite a work of art.

Facing a population boom, the building on the left grew by a whole floor. The one on the right grew forward toward the street and well to the right, adding retail space (as of 1994 when Brand wrote his book) as well as living space; it also sprouted fancy shutters by that time, then

**Figure 24.2 How two buildings in New Orleans learned from their occupants, 1860s–2019.**

*Source:* Photo by Todd Crane. Reproduced with permission.

lost them (the photo shown was taken in 2019). The buildings also joined hands, growing a beautiful archway that connects them. What both buildings have in common is that they grew, modernized, acquired luxurious features, and *became more differentiated* as time passed by. They adapted because they had to, or else be torn down.[7]

What is the point of this digression, other than fooling around and having fun? One point is simply that, as Yogi Berra is reputed to have said, you can observe a lot by just watching. A careful eye is the key to discovery. Travel slowly.

But another point is that, by turning conventional thinking (the occupants conformed the buildings to their changing needs) on its head (the buildings learned what their users needed and they adapted accordingly), you can see the world in a different light. The built environment is very important—"we shape our buildings and afterwards our buildings shape us," said Churchill—and understanding it in new ways makes us more whole, better educated, better able to function in an always-changing environment.

## The Romance of Maintenance

My favorite chapter in *How Buildings Learn* is called "The Romance of Maintenance." Although my book emphasizes ingenuity and dynamism, much of our job as human beings on this Earth is stewardship: taking care of what has been created by our forbears, so that others can enjoy it and benefit from it in the future. Brand writes: "The romance of maintenance is that it has none. Its joys are quiet ones. There is a certain high calling in the steady tending to a ship, a garden or a building. One is participating physically in a deep long life."[8]

Brand tells a story in which his friend, the anthropologist Gregory Bateson, visited New College at Oxford University and asked about a particularly old stand of trees. He is told that they were planted in the late fourteenth century, not that long after the school was founded, so that they could be cut down some 500 years later to replace the beams in the College Hall (Figure 24.3) when the originals wore out.

---

[7] Brand's 1994 book contains a then-current photo of the buildings by Robert Brantley. A comparison of that photo (from the unfortunately out-of-print book; try used-book sites) shows that the buildings continued to learn enthusiastically between that time, which is not so long ago, and the present.

[8] Brand (1994), p. 130.

**Figure 24.3 College Hall, New College, Oxford. The late-fourteenth-century oak beams are, unfortunately, not quite visible at the top.**

*Source:* https://commons.wikimedia.org/wiki/File:New_College,_Oxford_(3915166725).jpg. Licensed under CC 2.0.

"That's the way to run a culture," Brand exclaimed. There's an old Greek proverb that says a society grows great when old men plant trees whose shade they know they shall never sit in.

## Whole Earth Discipline

So, why—in my book—all this fuss about an octogenarian do-gooder with a lot of energy and a high IQ? The reason is that Brand's latest book, *Whole Earth Discipline: An Ecopragmatist Manifesto*, sets forth a distinctively modernist, iconoclastic, and technophilic program to "save" the environment—or, more precisely, to build a world where we are in harmony with our environment instead of being opposed to it.[9] We are a part of nature. By saving nature, we save ourselves.

Brand is in good company with this new view of his. There's an emerging and very hopeful literature on the use of technology to improve relations between man and his environment. I cover this literature

[9] Brand (2009).

in Chapter 25, "Ecomodernism," which sets forth practical ways of solving the serious environmental problems we face, as outlined in a remarkable multi-author document called *An Ecomodernist Manifesto.* Its authors and signers, of whom Brand is one, outline a way forward that helps people (rather than treating them as the enemy), preserves and enhances the Earth, is consistent with economic growth way out into the future and for everyone in the world, and puts technology to its best use.

In this chapter I stick with Brand, the pioneer.

## Laissez Faire, Property Rights, and a Cleaner Environment

But first, a diversion: How do markets and the environment interrelate? The answer is key to understanding Brand's new positions.

In Chapter 22, "A Skeptical Environmentalist," we made the case that the world is already starting to get greener: population stabilization, economic growth, and technological improvement are the ingredients that bake the green cake. But that is not the whole story. *Laissez faire* may be a productive and moral way to organize an economy but it does not take account of externalities.

In the economist's jargon, externalities are the effects that A's decision to transact with B have on C, without C's consent or even knowledge. A builds a coal-fired power plant; B consumes electricity; both are happy. But C breathes the polluted air. That is why we must be environmentalists, or, if you prefer, conservationists (which I claimed in Chapter 23, "Dematerialization," was a more fitting name for that philosophy).

One conservationist strategy that, at least theoretically, compensates C for the damage done by A and B (or discourages A and B from polluting in the first place) is *strictly defined and enforced property rights* in environmental goods. Such goods can include air, water, quiet, open space, and even a dark sky.[10] Such a policy meets the classical liberal test of avoiding harsh regulation of business and personal behavior, when fair pricing of goods held in common can achieve the same result.[11] But we don't quite have the technology to do it yet.

---

[10] The International Dark Sky Association, a group of astronomers, promotes policies that avoid light pollution of the night sky. It's a good cause, worth investigating and supporting.

[11] Classical liberalism, the liberalism of John Stuart Mill, combines the philosophies of self-ownership and the primacy of the individual with the freedom inherent in a capitalist or market economy. According to https://mises.org, named after classical liberal author Ludwig

Carbon taxation could be a step in that direction; so are other "rights to pollute" that are issued by governments and that can be bought by whoever derives the greatest advantage from polluting. Such strategies really do limit environmental damage. But we can't create conventional property rights in cubic meters of air and liters of water in ways that could make these strategies function optimally; the air and water move around too quickly. So we need some explicit regulation. Don't pollute, dammit.

## How Do You Privatize a Fish?

Occasionally, however, it's easier and more satisfying to define environmental property rights than you might think. Michael Lewis, that elegant chronicler of all things financial,[12] noted that although a fishing ground is self-renewing in the long run, the incentive for overfishing is very powerful, so fisheries typically degrade over time. A system of enforceable property rights would probably eliminate this problem, but it is hard to establish property rights in fish (they swim away). But it's not impossible! In Iceland, faced with overfishing,

> [T]hey privatized the fish. Each fisherman was assigned a quota, based roughly on his historical catches . . . [entitling him] to, say, 1 percent of the total catch allowed to be pulled from Iceland's waters that season. Before each season . . . scientists . . . would determine the total number of cod or haddock that could be caught without damaging the fish population; from year to year, the numbers of fish you could catch changed. But your percentage of the annual haul was fixed . . . in perpetuity [and], even better, if you didn't want to fish you could sell your quota to someone who did. The quotas thus drifted into the hands of . . . the best fishermen, who could extract the fish from the sea with maximum efficiency.[13]

---

von Mises, classical liberalism "is the term used to designate the ideology advocating private property, an unhampered market economy, the rule of law, constitutional guarantees of freedom of religion and of the press, and international peace based on free trade. Up until around 1900, this ideology was generally known simply as liberalism. The qualifying "classical" is now usually necessary, in English-speaking countries at least (but not, for instance, in France), because liberalism has come to be associated with wide-ranging interferences with private property and the market on behalf of egalitarian goals."

[12] Who also wrote one of the finest modern morality plays, *The Blind Side: Evolution of a Game* (2006), and (it needs no description) *Moneyball: The Art of Winning an Unfair Game* (2003).

[13] Lewis (2011). The surrounding discussion is adapted from Siegel (2012).

Creative thinking along these lines will be required if the potential environmental benefits of greater wealth are to be realized.

## Some Uncomfortable Proposals

And creative thinking is what Stewart Brand is best at. His *Whole Earth Discipline* is subtitled "An Ecopragmatist Manifesto," ecopragmatism being the use of existing and soon-to-be developed technology, rather than unacceptable self-sacrifice or pie-in-the-sky dreaming, to meet environmental challenges. In Chapter 25, I say why I prefer the similar term *ecomodernism*, used in a manifesto signed by Brand and many others, but let's stick with ecopragmatism and *Whole Earth Discipline* for now.

Here are some of the main points of *Whole Earth Discipline*:

1. Having tried to save the Earth from civilization, we are now in the unfamiliar position of saving civilization from the Earth (which may be warming catastrophically—it's not my forecast and Brand doesn't treat it as the most likely scenario, but it is a low probability event with very bad consequences if it happens).
2. Technology, which is at the heart of civilization, is the tool we're going to need to accomplish that.
3. Brand writes, "My opinions are strongly stated and loosely held." Modesty is an attractive but uncommon trait in visionaries. Brand once tweeted, "My own scenario on climate change is more alarmist than @mattwridley's [which is mostly benign], but his could be right."[14] Very unusual.
4. The solutions, as best Brand can discern them, are
   - Urbanism
   - Nuclear power
   - Genetic engineering
   - Ecosystem engineering

We'll go into each of these four "uncomfortable" solutions in some detail in Chapter 25. Ecoengineering blends into geoengineering (what I call "terraforming the Earth"), a topic that Brand also addresses in *Whole Earth Discipline*.

[14] https://twitter.com/stewartbrand/status/789867975104827392?lang=en.

Has Brand gone crazy? Betrayed his past as an eco-warrior? Become an anti-environmentalist?

No—he is doing what he has always done, thinking unconventionally and searching for pragmatic solutions to problems so big they seem to require cosmic solutions instead. But Brand knows there are no cosmic solutions. There is only the art of the possible. Due to human ingenuity, the possible usually grows with time, in surprising and unforeseen ways.

# 25

# Ecomodernism: A Way Forward

Although it sounds new and edgy, ecomodernism just means using technology (instead of "degrowth") to improve the environment. It's what we've been doing since we first noticed that the environment was not just a provider of bounty, but also a resource that needed to be preserved and nurtured. The technologies change with the times, but the ongoing effort to develop technologies for improving our surroundings is one of the attributes that make humans unique among species.

As used by the authors of *An Ecomodernist Manifesto*, a short essay signed by 18 authors, *ecomodernism* also means the use of economic reasoning, knowledge, and data to prioritize environmental efforts. The authors write:

> We offer this statement in the belief that both human prosperity and an ecologically vibrant planet are not only possible, but also inseparable. By committing to the real processes, already underway, that have begun to decouple human well-being from environmental destruction, we believe that such a future might be achieved. As such, we embrace an optimistic view toward human capacities and the future.[1]

This statement captures the essence of the "Greener" section of my book. In this chapter, I will summarize their *Manifesto* and go into detail

[1] www.ecomodernism.org.

regarding the chief elements of ecomodernism that I first brought up in Chapter 24, "We Are as Gods": urbanism, nuclear energy, genetic engineering, and eco- and geoengineering. These are the keys to the decoupling that the *Manifesto* advocates and are the ways we will leave a better future, not just economically but ecologically, to our children and grandchildren.

## An Ecomodernist Manifesto[2]

We've met several of the authors or signatories of the *Manifesto* already: Stewart Brand, Michael Shellenberger, Ted Nordhaus. The others are equally prominent. The philosophy of the *Manifesto* authors has much in common with that of Bjørn Lomborg (see Chapter 22, "A Skeptical Environmentalist"), although Lomborg did not contribute to it.

### Identifying the Problem

The authors begin by asserting that, while catastrophic forecasts about the environment have turned out to be wrong (they cite 1970s "Limits to Growth" thinking as the best example),

> There remain . . . serious long-term environmental threats to human well-being, such as anthropogenic climate change, stratospheric ozone depletion, and ocean acidification. . . . [T]hey could cause significant risk of catastrophic impacts. . . . Even gradual, non-catastrophic outcomes associated with these threats are likely to result in significant human and economic costs as well as rising ecological losses.

The authors then add that "[m]uch of the world's population still suffers from more-immediate local environmental health risks," such as air and water pollution. We can fix this. The air and water in the United States, Europe, Australia, Japan, and many other places is now generally cleaner than at any time since before the Industrial Revolution, and developing countries, led by China, have started to move in the right direction although they have a long way to go.

### Decoupling Human Betterment From Environmental Degradation

The *Manifesto* authors then address decoupling, which can be relative or absolute. Relative decoupling occurs when "human environmental

---

[2] All of the quotes in this section are from *An Ecomodernist Manifesto*, www.ecomodernism.org.

impacts rise at a slower rate than overall economic growth. Thus, for each unit of economic output, less environmental impact (e.g., deforestation, defaunation, pollution) results. Overall impacts may still increase, just at a slower rate than would otherwise be the case."

Absolute decoupling occurs when human welfare and environmental degradation move in the opposite direction: people become better off and so does the environment. The latter is the real goal, and it will not take forever (nor do we have forever). "Decoupling," they point out, "can be driven by both technological and demographic trends and usually results from a combination of the two." The demographic trends (see Part I, "Fewer") have begun to work in favor of absolute decoupling.

Now, how will all this happen?

The *Manifesto* authors start with urbanism (a more efficient use of space for housing people), agricultural intensification (growing more food on less land), and reforestation (which helps with atmospheric carbon concentration too), nuclear power, aquaculture, and desalination. They also mention dematerialization (my Chapter 22): "[D]emand for many material goods may be saturating as societies grow wealthier."

## Sparing Nature

They conclude: "[H]umans are as likely to spare nature because it is not needed to meet their needs as they are to spare it for explicit aesthetic and spiritual reasons. The parts of the planet that people have not yet profoundly transformed have mostly been spared because they have not yet found an economic use for them—mountains, deserts, boreal forests, and other 'marginal' lands. . . . Nature unused is nature spared."

Their program is intended not just to spare what remains of nature, by making it economically more efficient to leave it alone, but to roll back human despoliation of nature. Fewer, richer, greener.

## Energy and Carbon

The *Manifesto* authors begin by noting the tremendously beneficial effects of abundant, cheap energy: plentiful food grown on less land, the reduced or eliminated need to burn forests for fuel, the ability to desalinate or decontaminate water. "Looking forward," they add, "modern energy may allow the capture of carbon from the atmosphere to reduce the accumulated carbon that drives global warming."

But, they warn,

> In developing countries, rising energy consumption is tightly correlated with rising incomes and improving living standards. . . . [E]nergy consumption will continue to rise through much if not all of the 21st century. . . . [A]ny conflict between climate mitigation and the continuing development process through which billions of people around the world are achieving modern living standards will continue to be resolved resoundingly in favor of the latter.

This means that we cannot achieve climate or carbon goals simply using less energy, when we need the opposite. It will take major changes in the way we generate energy. Although the authors believe that solar power "produced from earth-abundant materials" has the potential to "provide many tens of terawatts," what we need is petawatts. A petawatt is a thousand terawatts.[3] They conclude that "[n]uclear fission today represents the only present-day zero-carbon technology with the demonstrated ability to meet most, if not all, of the energy demands of a modern economy." In the future, nuclear fusion, advanced fission, and next-generation solar sources of energy will contribute to the solution.

## Decoupling Won't Happen by Itself

The authors conclude by cautioning,

> Technological progress is not inevitable. Decoupling environmental impacts from economic outputs is not simply a function of market-driven innovation and efficient response to scarcity. The long arc of human transformation of natural environments through technologies began well before there existed anything resembling a market or a price signal. Thanks to rising demand, scarcity, inspiration, and serendipity, humans have remade the world for millennia.

Because markets don't provide the exact price signals that are needed, that is, because there are external costs that can be foisted off on nature, as they always have been, "Accelerated technological progress will require the active, assertive, and aggressive participation of private sector entrepreneurs, markets, civil society, and the state." It's a team effort.

We now explore some of the technologies favored by the authors of the *Ecomodernist Manifesto* and introduced in the previous chapter in connection with Stewart Brand's book, *Whole Earth Discipline*.

---

[3] A terawatt is a trillion or $10^{12}$ watts, and a petawatt is a quadrillion or $10^{15}$ watts.

## Urbanism

Why is urban living good for the environment, and why is more of it needed? Many people think of dirty, smoky, crowded cities as the antithesis of environmental goodness, which we associate with wild nature.

The reason is related to what I said earlier about self-sufficiency being the surest road to poverty and deprivation. Specialization and trade are the creators of wealth, and cities are the mechanism that best enables people to find their areas of comparative advantage and trade easily and cheaply with others.[4]

### The Virtues of Density

But there's another reason. There is nothing efficient about having a vast spread of land around one's dwelling and having to transport oneself (at great energy cost) a long way to do anything involving another person or physical object. Housing seven billion people, which will eventually be nine billion or more, must be done efficiently if we are to preserve any sort of nature at all, while also making room for agriculture and other uses of land.

We've already praised the contribution of cities to the economy and of human development in Chapter 11, "Cities." But what about the environment? Stewart Brand writes,

> In his 1985 article that introduces the concept of walkability, Peter Calthorpe made a statement that still jars most people: "The city is the most environmentally benign form of human settlement. Each city-dweller consumes less land, less energy, less water, and produces less pollution than his counterpart in settlements of lower densities."

"Green Manhattan," Brand continues, "was the inflammatory title of a 2004 New Yorker article by David Owen." Owen wrote,

> New York is the greenest community in the United States, and one of the greenest cities in the world. . . . The key to New York's relative environmental benignity is its extreme compactness. . . . Placing one and a half million people on a twenty-three-square-mile island . . . forced the majority to live in some of the most inherently energy-efficient residential structures in the world.

---

[4] All the quotes in this subsection are from Brand (2009), pp. 67–73. Italics are his. Subquote sources: Calthorpe (1985); Owen (2004); London and Kelly (2007).

## How Cities Help Save the Countryside

"But what about the ecological footprint?" asks Brand, understandably. Citing a 2006 United Nations report, he notes, "in the developed part of the world, cities are Green mainly because they reduce energy use, but in the developing world, the primary Greenness of cities lies in their ability to draw people in and take the pressure off rural natural systems."

So that we can better understand this principle, Brand points us to Mark London and Brian Kelly's 2007 book on the Amazon rainforest, *The Last Forest*. Brand suggests that "the nationally subsidized city of Manaus in northern Brazil" answers the question, in London and Kelly's words, "'How do you stop deforestation?' Give people decent jobs . . . [and] they can afford houses . . . their family has security . . . and their vision shifts to the future."

"One hundred thousand people," Brand concludes, "who would otherwise be deforesting the jungle around Manaus [to survive] are prospering in town making such things as cellphones and TVs."

But urbanism does not achieve its maximum greening impact all by itself. The abandoned countryside must be protected—or, potentially, "gardened" with the goal of reducing $CO_2$ in the atmosphere. (Plants are geniuses at it.) And cities can be greened through approaches such as urban farming, which reduces food transportation costs and may increase food quality and availability.

## Urbanism: Conclusion

Urbanism is not all environmental sweetness and light. Urban sprawl is better than distributing large, ecologically destructive populations across the countryside, but is still not ideal. Cities require a lot of infrastructure investment. And "[t]hey concentrate crime, pollution, and injustice as much as they concentrate business, innovation, education, and entertainment," Brand warns. But he sums up by saying that, on balance, cities make for a culture that is "multicultural, multiracial, global, worldly wise, well traveled . . . cultured, sophisticated . . . *urbane*." And many times richer, better fed, and healthier than they would ever be as self-sufficient rural farmers. This sounds good to me.[4]

## Nuclear Power

Why do we drive around with what amounts to a bomb mounted underneath our cars? (That's what a gas tank really is.) Because the *energy density* of gasoline is unsurpassed by any other easily obtained and stored liquid, that's why. The energy density of a fuel is the amount of energy you can extract from that fuel per unit (weight or volume) of the fuel. When you have to drag around your energy source with you, energy density becomes a critical variable. Gasoline is also pourable, storable (it doesn't spoil), and, where I live, cheaper than milk.

So far, so good. Gasoline sounds like an ideal fuel for transportation—except for carbon emissions, uncertainty about the long-run future supply of it, and the fact that a much more energy-dense material is sitting right under our noses.

That material is uranium-235, the radioisotope of uranium used in nuclear fission reactors.[5] It has an energy density of 24,513,889 kilowatt-hours per kilogram (kWh/kg), compared to gasoline's energy density of 9 kWh/kg, or 1/1,930,227 as much as U-235. It would be impractical to put nuclear fission reactors in cars (although we've put them in submarines). But electric power plants can run beautifully on the stuff, and the electricity can be used to charge the batteries in electric cars, which are multiplying like rabbits on our highways.

So why don't we have more nuclear power plants? Fear. As a result of:

- *Three Mile Island* (no fatalities).
- *Chernobyl* (an inexcusable failure of the Soviet Union to use basic safety precautions in building and maintaining the reactor, and deliberate lying about the danger, greatly slowing evacuation).
- *Fukushima.* One commentator said that while he was skeptical of the safety of nuclear power before Fukushima, he is convinced of it now. It was a perfect storm—earthquake, tsunami, and old equipment—and there has been one fatality from radiation so far. (Evacuation of the sick and elderly to substandard quarters caused their death rate to increase substantially, but none have died from radiation.).[6]

[5] U-235 makes up 0.7% of natural uranium, most of the remainder being U-238, which is not fissionable. Although U-235 is amazingly energy-dense, *antimatter* is actually the most energy-dense substance because *all* of its mass is converted to energy in a matter-antimatter collision, compared to about 0.1% of the mass in a U-235 nuclear fission reactor. However, the use of antimatter to generate energy is purely theoretical—we don't have any (other than in subatomic amounts); if we did it would annihilate the container it was stored in, and using it presents hazards we can't anticipate. This does not stop some dreamers from contemplating antimatter reactors: see Millis (2018).

[6] See Shellenberger (2019). (Shellenberger is not the source of the anonymous Fukushima comment in the text.)

Meanwhile, between 1900 and 2006, there were 11,606 coal miners who died in underground coal mining disasters in the United States alone.[7] This doesn't count deaths from black lung disease, accidents involving railroads carrying coal, or nonunderground (strip mining) deaths. Nonworker deaths also occur due to coal emissions, as evidenced by the 4,000, or more, who died in the great London smog of 1952. The gradual phasing out of coal mining, along with greatly enhanced safety provisions, has reduced the annual death rate to a low level: 17 dead from direct coal-mining causes in 2017.[8] But that is still 17 times as many as died at Fukushima.

Oil and gas extraction looks and feels safer, and on a per-worker basis it is. But the oil and gas industry lost 101 workers to accidents (many of them motor vehicle-related) in 2014, a representative recent year.[9]

Uranium mining isn't completely safe either. Because we don't do much of it, the safety statistics aren't directly comparable with those for coal, oil, and gas. But the safety of an energy source must be evaluated using all of the steps: exploration, extraction, transportation, refining, and final use for energy production (including externalities like air and water pollution and radioactive waste).

## Petr Beckmann and the Moral Case for Nuclear Power

On this basis, Petr Beckmann, the late University of Colorado engineering professor who unfortunately spent the last years of his life in a futile attempt to disprove Einstein's theory of relativity, made what he called the "moral case for nuclear power" in a memorable lecture that I attended about 40 years ago.[10] The book on which his lecture was based is out of print, and it's hard to remember details for 40 years. Fortunately, two reviews of the book are still available, and one (by R. B. Murray, a Northern Irish professor) summarizes one of Beckmann's many arguments:

> [Nuclear] waste is . . . small in volume. Beckmann claims that, if all the electricity in the United States were generated by nuclear reactors, the waste would amount to the volume of an aspirin per head of population per year. Compare this with the 320 [pounds] of obnoxious waste produced today for each American as a result of

[7] Centers for Disease Control (2009).

[8] Raby (2018).

[9] And the most recent year for which we have the data. Ridl et al. (2017).

[10] The lecture summarized ideas in Beckmann (1976).

> coal-fired electricity! Reactor safety is [also] under extensive scrutiny, but . . . there is little to indicate that a sufficiently long chain of human errors could occur in such a way as to overcome the intricate failsafe systems.[11]

Beckmann also projected the decline in the number of deaths from railroad and truck accidents that would occur if nuclear fuel replaced coal entirely, and he makes other points that support the idea that nuclear power is not only the most efficient but the safest form of electrical power generation currently available to us.

And, although Beckmann didn't seem too concerned about the carbon footprint of nuclear power at the time, he might be now. It is zero.[12]

No method of generating electricity is completely safe, but Beckmann's 40-year-old arguments are truer now than they were when he made them. The four decades of additional technological advance, plus our much greater need for energy, make them so.

## The Molten Salt Reactor: Just One More Wild Idea for Improving Nuclear Power Generation

Today's ecomodernists are exploring new ways of making nuclear power efficient, safe, affordable, and publicly acceptable. Although existing reactors can be used—the design currently preferred by the cognoscenti seems to be the French one, small and standardized instead of large and customized—some of the technology being developed sounds like science fiction (but so did nuclear power itself 90 years ago). Here's a sample:

> [T]he beer-and-nuclear group [of physicists who met socially] found that one such design, the molten salt reactor, had a simplicity, elegance and, well, weirdness that especially appealed.

---

[11] Murray (undated). Professor Murray, of the New University of Ulster, Coleraine, Northern Ireland, was an executive editor of the *International Journal of Energy Research*, which was published by John Wiley & Sons, Hoboken, NJ. I do not have the name or date of the journal in which the review appeared. See also the review by Mark K. Enns, of the Power Systems Laboratory at the University of Michigan, at https://ieeexplore.ieee.org/stamp/stamp.jsp?arnumber=6501625.

[12] The *operating* carbon footprint (the amount of $CO_2$ generated by the nuclear reaction) is zero. $CO_2$ may be released during the extraction of materials used to build the plant, construction of the plant, mining and transportation of the fuel, and waste disposal. This ancillary footprint should not be ignored (with any energy source).

> The weird bit was that word "molten," says [entrepreneur Troels] Schönfeldt: Every other reactor design in history had used fuel that's solid, not liquid. This thing was basically a pot of hot nuclear soup. The recipe called for taking a mix of salts—compounds whose molecules are held together electrostatically, the way sodium and chloride ions are in table salt—and heating them up until they melted. This gave you a clear, hot liquid that was about the consistency of water. Then you stirred in a salt such as uranium tetrafluoride, which produced a lovely green tint, and let the uranium undergo nuclear fission right there in the melt—a reaction that would not only keep the salts nice and hot, but could power a city or two besides.
>
> ...And more to the point, the beer-and-nuclear group realized, the liquid nature of the fuel meant that they could potentially build molten salt reactors that were cheap enough for poor countries to buy; compact enough to deliver on a flatbed truck; green enough to burn our existing stockpiles of nuclear waste instead of generating more—and safe enough to put in cities and factories. That's because Fukushima-style meltdowns would be physically impossible in a mix that's molten already. Better still, these reactors would be proliferation resistant, because their hot, liquid contents would be very hard for rogue states or terrorists to hijack for making nuclear weapons.
>
> Molten salt reactors might just turn nuclear power into the greenest energy source on the planet.[13]

The irony here is that the molten salt reactor (see Figure 25.1) was conceptualized in 1954 and a successful prototype operated from 1965 to 1969. It was brought out of the junk heap of discarded reactor designs in 2009 at the beer-and-nuclear meet-up of young physicists in Copenhagen. The design's challenges had gotten the original inventor, Alvin Weinberg, fired from Oak Ridge National Laboratory in 1973.

Scientists are both creating entirely new designs and reimagining old ones like the molten salt reactor. Despite past disappointments, they are also continuing to research nuclear fusion. One, or several, of these innovations will fulfill the decades-old promise of nuclear power being "too cheap to meter," and will energize the enrichment of the world's poor, as well as the rest of us.

---

[13] Waldrop (2019). Waldrop is an elementary particle physicist and author and was the editorial page editor at *Nature*.

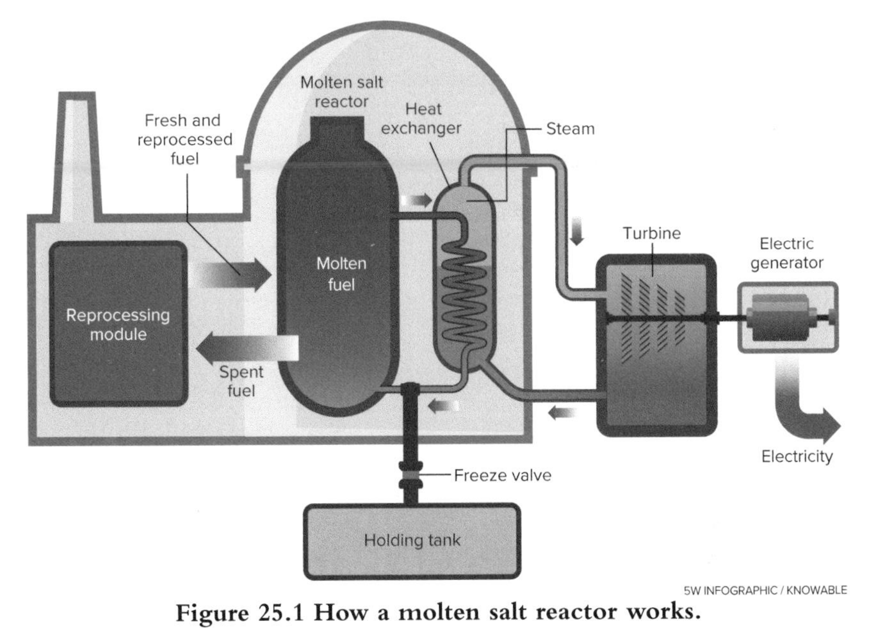


**Figure 25.1 How a molten salt reactor works.**

*Source:* Waldrop (2019), © Knowable Magazine. Reproduced with permission.

## Nuclear Power: Conclusion

For most applications, nuclear power dominates both fossil fuels and renewables in almost every aspect: efficiency, safety, reliability, carbon neutrality, fuel abundance, and eventually price. Although renewables make people feel good and have a place in the energy spectrum (we're going to need as much energy as we can get), "wind and solar power by themselves couldn't offer nearly enough energy, not with billions of poor people trying to join the global middle class," writes Waldrop.

Renouncing nuclear power just when we need it most is almost too much for me to bear. Have we all gone crazy? But enough of my kvetching about nukes. Let's turn to bioengineering, the most interesting and complex of the ecomodernists' proposals.

## Bioengineering

There are many forms of bioengineering, ranging from the development of prosthetic medical devices to growing of replacement organs to the breeding of a new variety of tulip. Here, however, we'll be mostly concerned with genetic engineering, the best-known and most controversial branch of bioengineering. At present, the most important uses of genetic engineering are agricultural and medical, although that list will expand in the future.

## Genetic Engineering for Agriculture

Genetic engineering or genetic modification of organisms (GMO) is principally used to grow more and (usually) better food with fewer resources. It is a key ingredient of ecomodernism in that we need to use land more efficiently, because the population is still growing (although it will not do so forever), some people still do not have enough to eat, and we need to preserve and enlarge wild lands both for their own sake and for their possible benefit in controlling the level of atmospheric $CO_2$.

Many people oppose genetic engineering of foods because it's "unnatural" and they are afraid it will cause cancer or make you sick in some other way. It also just sounds yucky, putting an oyster gene in broccoli. However, almost everything we eat is, technically speaking, unnatural: it has been genetically modified over the millennia through selective breeding and hybridization. In fact, the wild version of most of our plant-based foods is inedible or almost inedible. We saw this with teosinte, the precursor to corn, in Chapter 8, "Food."

But genetic engineering (GE) in the modern sense does differ from selective breeding and hybridization. It's a faster pathway to the genetic design one is seeking. And, more importantly, GE only brings along the genetic characteristics you want, while hybridization and selective breading drag all kinds of unwanted characteristics along with the wanted ones.

## "Please Give Us the Food"

One of the tragedies of modern life is that the increase in agricultural yields made possible by GE has sometimes been denied to those who need it the most—sub-Saharan Africans—for political reasons verging on the homicidal. And these political pressures appear to originate with the "hypercivilized" Europeans. Brand writes:

> Outside a locked warehouse in Shimabala, Zambia, where [GMO] corn was stored for free distribution . . . an elderly man pleaded with officials to release the corn: "Please give us the food. We can't care if it is poisonous because we are dying anyway." . . . The World Health Organization estimated that 35,000 Zambians would starve to death in the coming months.[14]

[14] Brand (2009), p. 155.

Who was stopping the starving Zambians from eating? The direct cause was the policy of Zambian president Levy Mwanawasa, who said, "Simply because my people are hungry, that is no reason to give them poison."[15] Maybe he was just ill-informed; the corn was perfectly safe. I'm willing to bet, however, that Mr. Mwanawasa himself had plenty to eat, or he would have changed his tune.

But the story behind the story is that that Zambian farmers are highly dependent on grain sales to Europe, which would not accept GMO foods. They would be unable to sell GMO corn in their most important market. Testifying to the US Congress in 2003, Kenyan biologist Florence Wambugu said, "The primary accomplishment of the mainly European antibiotech lobby . . . was . . . to keep safe and nutritious food out of the hands of starving people"[16]—not in well-fed Europe, but in Africa, *where the food is grown.*

Fortunately, this situation is improving. Many sub-Saharan African countries now produce GMO foods, at great benefit to themselves and their customers. There is no longer a famine in Zambia,[17] South Africa and Sudan have burgeoning GMO food export industries,[18] and "11 African countries . . . [engage in] biotech crop research, with 14 traits on 12 crops under various stages of development," writes the Ugandan journalist John Agaba. But the battle is not over. "[I]ndecision by political leaders in these countries," says Agaba, "due partly to a rise in anti-GMO activism, is threatening to ruin these strides."[19]

And, whether you find GMO foods aesthetically acceptable or not, we're going to need them. We all need to eat, we will have an additional two or three billion people on the Earth in this century, and the quality and quantity of food demanded by an ever-richer global population will also increase. GMO foods are here to stay.

## Genetic Engineering and Urbanism

I said earlier that urbanism is good for the environment, but where does the benefit come from? Mostly from the fact that, as I said, a large number

---

[15] Ibid., p. 156.

[16] Ibid.

[17] Although it's hard to tell from the Famine Early Warning System (fews.net) headline, "Minimal acute food insecurity outcomes expected to continue." Decoded, that means there is enough food in Zambia.

[18] Agaba (2019).

[19] Ibid.

of people spread out across the countryside, engaged in subsistence or traditional farming, tend to despoil it. In contrast, large, technologically advanced farms, especially those that grow GMO crops, are many times more efficient. They leave more land for either wild nature or carbon-sink plantations (consisting, I imagine, of GMO plants that suck up $CO_2$ the way I suck up coffee). And if fewer people live in the country, they have to live in towns and cities by default.

Transportation costs are also lower in cities, as anyone who has lived in the country knows from his or her motor fuel bill—country people often drive 50 miles each way to work and 15 miles to buy groceries. But if we really want to reduce the $CO_2$ footprint of the cities, we should grow food in them! In Chapter 8 I noted that vertical farming (which is most likely to take place in or near cities) may not be needed to provide sufficient calories to the world's population; we may get enough food just by using GE, irrigation, and other existing technologies. But greatly reduced transportation costs *for food* could make a meaningful contribution to climate change abatement.

## Genetic Engineering and Better Health

Stewart Brand mused that, if Americans ever embrace genetically modified foods, it will be because they "offer benefits seen as somehow medical. Coming soon is a GE pig whose pork has heart-healthy, omega-3 fatty acids as good as those found in fish. . . . [And] [n]ow that we know that resveratrol in red wine is what keeps the French living longer despite all that butter, there's a GE wine on the way from China that has six times as much resveratrol."[20]

But there are GE applications with medical benefits much more profound than simply enabling us to eat rich foods more safely. We can almost certainly eliminate malaria.

## Patenting a Mosquito (Seriously)

Dengue fever is one of the most painful communicable diseases known to man. That's why it's also called breakbone fever; it doesn't break your bones, it just feels that way. The Aedes mosquito, widespread in sub-Saharan Africa, transmits this disease along with the horribly teratogenic, or birth defect-causing, Zika virus. Researchers have developed a genetically modified male Aedes mosquito that spreads sterility in the wild

[20] Brand (2009), p. 196.

Aedes population by mating with the females and producing offspring that do not live to adulthood, so they're effectively sterile.

A patent on the self-eliminating mosquito is held by a British company called Oxitec, and the little bug is called Friendly Aedes.[21] According to the company's web site,

> Friendly™ Aedes males . . . do not bite and do not transmit diseases. When released, these males search for wild females to mate [with], and their offspring inherit a self-limiting gene that causes them to die before reaching functional adulthood. Friendly™ Aedes' offspring also inherit a fluorescent marker that allows tracking and monitoring. . . . Unlike other approaches, Friendly™ Aedes mosquitoes die along with their along with their offspring, and therefore do not persist in the environment or leave any ecological footprint.

Begone, loathsome bugs! But you probably paused when I said that Oxitec, in cooperation with the Gates Foundation, had patented a mosquito. How can it be legal or moral to do that?

"Inventing" a new mosquito that can breed its disease-bearing cousins out of existence is hard work that has to be done by highly skilled biologists. They could do the work out of the goodness of their hearts, but they have families and mortgages and a retirement plan to contribute to and need an incentive to do this work instead of something else. Intellectual property rights in the genetic composition of a new breed of mosquito are the right way to provide that incentive. (Ask someone who's had dengue whether they mind.) Bravo for patents, and bravo for Oxitec and the Gates Foundation!

But there's a fly in the ointment. Oxitec mosquitoes die out after two generations—the genetically altered male mosquitoes and their resulting "zombie" offspring—allowing regeneration of the wild, fertile type of Aedes from mating of unmodified mosquitoes. So the eradication of the disease vector is not forever.

But the genetically modified mosquito story is not over—far from it.

## Eradicating Malaria

Malaria, a parasitic disease, is an even greater plague on sub-Saharan Africa and other tropical regions. Some 200 million (!) new cases occur annually, amounting to almost 3% of the world's population—each year.

---

[21] Oxitec has been acquired by U.S.-based Intrexon Corporation.

More often than not it can be treated or even cured, but the sufferer loses productivity—you can't work—and recovery is often not complete.

Efforts to eradicate malaria have been partially successful using traditional methods, such as vaccines, bed nets to keep malaria-bearing mosquitoes away from people, and draining the swamps in which the mosquitoes live. However, the most promising approach involves genetically modifying the malaria-bearing Anopheles mosquito. Lee Kaplan, a retired clinical professor at University of California, San Diego (UCSD), writes,

> Gantz *et al.* created a genetically modified version of the malaria-bearing Anopheles mosquito in which two anti-malarial genes are inserted into the mosquito by CRISPR-Cas9 technology, and then Gene Drive technology allows passage of these genes to all subsequent offspring. The result is astounding! In ten (mosquito) generations—less than 1 year—these genes will be permanently spread through 100% of the targeted mosquito population.[22]

Figure 25.2 illustrates the basics of this technology. Using CRISPR-Cas9, "an enzyme that acts like a pair of molecular scissors, capable of cutting strands of DNA,"[23] and an RNA guide (gRNA in the diagram), an antimalarial gene is embedded in one strand of the mosquito's DNA, along with a gene to cause a similar break in the other strand—a

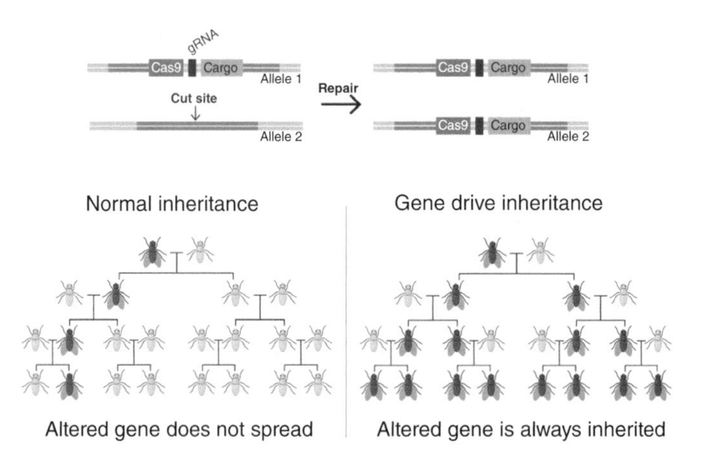


**Figure 25.2 Modifying a mosquito population so that it does not transmit malaria.**

*Source:* Mariuswalter, 2017, https://commons.wikimedia.org/wiki/File:Gene_Drive.png. Licensed under CC BY-SA 4.0.

[22] Personal communication, March 2019.

[23] Vidyasagar (2018).

gene drive—which induces a repair mechanism by which the modified, antimalarial and "drive" genes are copied over to the other strand of the "double helix." The presence of the modification in both strands means that, when the mosquito mates with a wild, unmodified mosquito, the modification will be transmitted to all offspring (instead of half of them, which would be the normal outcome). Thus, the modification quickly spreads throughout the mosquito population, as shown on the right side of the diagram under "Gene Drive Inheritance." Miraculous!

"But this technology raises an ethical issue," continues Kaplan:

> Should we eliminate one genetically defined species to replace it with another? In the case of Anopheles mosquitoes capable of transmitting malaria, the answer should be yes, because malaria is so serious and widespread. And, given the economic and epidemiologic pressures, their development and release will almost certainly occur in the next decade. Tata Trust just gave UCSD $70 million to research and develop this technology.[24]

## More Medical Uses for Biotechnology

Bioengineering isn't just used to combat dengue and malaria; it's the central weapon in a fight against many diseases, including cancer and Parkinson's disease. A whole class of medications called biologics is grounded in GE. Gene therapy is the application of GE to human beings, typically to combat genetic defects; it replaces defective genes with good ones. *Somatic* gene therapy does this only for the affected individual; *germline* gene therapy allows the genetic change to be inherited and, potentially, to spread through the population. As such, it is widely considered dangerous and wrong, but the technology is in its infancy and, regardless of present concerns, we can expect further advances.

## Bioengineering: Conclusion

Genetic engineering is probably the most important general-purpose technology to appear in our lifetime. In the very long run—say, a few hundred or a thousand years—it will be more important than the microprocessor or the Internet, which will have been supplanted by something else.

[24] Personal communication, May 2019. See also Gantz et al. (2015).

Unlike information technology, the benefits of GE will emerge slowly over time. In a thousand years, and possibly much sooner, the ability to change *who we are* and to change the makeup of other organisms (such as microbes) that profoundly affect our well-being will seem natural and we will take it for granted, just as we now take antibiotics and anesthetics for granted. There is always the risk of unwanted effects (all technologies, including clubs and knives, have a potential downside), so we have a responsibility to understand and limit that downside. But the power to modify the genetic composition of living things is a revolution in human capability like no other.

## Terraforming Earth: Ecosystem Engineering and Geoengineering

Birds do it; bees do it; even sophisticated fleas do it. More familiarly, earthworms and beavers do it with great abandon. Do what? *Ecosystem engineering* (henceforth, *ecoengineering*), modifying their environment to suit their needs. Although I've claimed in a few places that we are the only species that modifies its environment on a massive scale, this turns out to be an exaggeration; we all live in a world genetically modified by microbes and grow food in dirt manufactured by worms.

Like earthworms and beavers, Stewart Brand notes, we can "tweak our niche (the planet) toward a continuing life-friendly climate, using methods such a cloud-brightening with atomized seawater and recreating what volcanoes do when they pump sulfur dioxide into the stratosphere, cooling the whole world."[25] This is *geoengineering*, a form of ecosystem engineering that Oxford University described as involving "deliberate large-scale intervention in the Earth's natural systems" for any purpose, not limited to preventing or counteracting climate change but most often discussed in connection with that goal.[26]

The two forms of engineering, ecosystem and geoengineering, are closely related and there is no bright line between them. If we want to make a distinction for pedagogic purposes, I'd say that ecosystem engineering uses living things as tools and geoengineering uses nonliving things.

[25] Brand (2009).

[26] Oxford Geoengineering Program (2018).

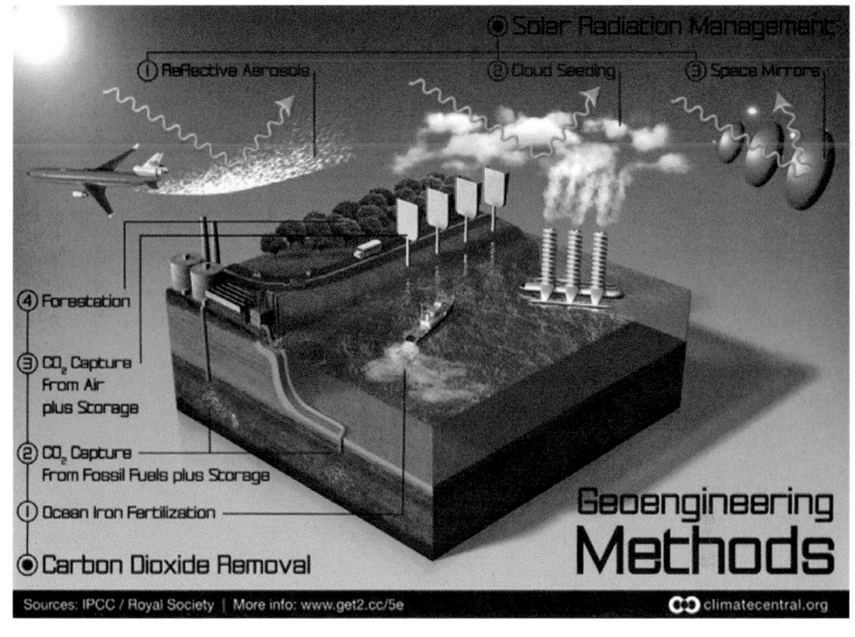


**Figure 25.3 Geoengineering proposals.**

*Source:* © Climate Central. Reproduced with permission.

In Figure 25.3, which illustrates a few of the many proposals that have been discussed for controlling climate change, "foresting" is clearly ecoengineering and space reflectors are geoengineering, but many of the other techniques could fall in either category.

There are so many proposals for reducing atmospheric $CO_2$ (other than by burning less fossil fuel) that it would take a whole book to cover the topic.[27] For ecoengineering, an Oxford website lists, as options,

- afforestation (planting trees);
- biochar (look it up);
- carbon capture and sequestration at the $CO_2$ source;
- ambient air capture;
- ocean fertilization;
- enhanced weathering (look that up too); and
- ocean alkalinity enhancement.

[27] And (surprise!) there are many such books. I'm sorry that I can't recommend one in particular.

For geoengineering in the sense of reducing the amount of sunlight reaching the Earth, the website lists:

- albedo enhancement (albedo is the extent to which a surface reflects light);
- space reflectors; and
- stratospheric aerosols (such as sulfuric acid).

I'll discuss a small sample of these.

## Plants Do It . . .

Probably the most promising proposal is to create *carbon sinks*, environments where $CO_2$ is removed from the atmosphere. Such sinks are highly desirable because it's unlikely we will ever completely eliminate the use of fossil fuels. They already exist: plants do this for a living, so simply increasing the amount of land devoted to plants—afforestation or "foresting"—will have some beneficial effect. We may be able to genetically engineer more carbon-hungry plants. And, since the majority of carbon capture by plants takes place in the ocean, the ocean can be fertilized to encourage growth of these plants.

## Fake Plants Do It Better . . .

Note that Figure 25.3 refers to afforestation. But natural trees are only moderately efficient at separating $CO_2$ into carbon and oxygen. Artificial trees can be many times more efficient, because they are designed to do that and only that.

What are artificial trees? The physicist Klaus Lackner has designed a "species" that is "made from a special resin—a unique plastic that sponges up $CO_2$ from the air in a chemical reaction."[28] Lackner says his robo-trees are 1,000 times as efficient as natural trees at carbon capture. A "treepod" inspired by Lackner's discoveries (and, originally, based on his daughter's eighth-grade science project) was proposed by a Boston group as a streetscape enhancement and is shown in Figure 25.4. It is designed to resemble the dragon blood tree, found in Socotra, Yemen, because of its efficiency in gathering solar energy.

A forest of 100 million of these spacy trees—roughly what it would take to offset all the $CO_2$ emissions currently being produced in the world—would be quite a space hog (and getting the materials to build it would be a challenge). But the economizing effects of large-scale industry haven't begun to take effect yet. I'm jumping way ahead, but they could become as cheap as *faux* Christmas trees, with radical ecological benefits.

---

[28] Cantieri (2016).

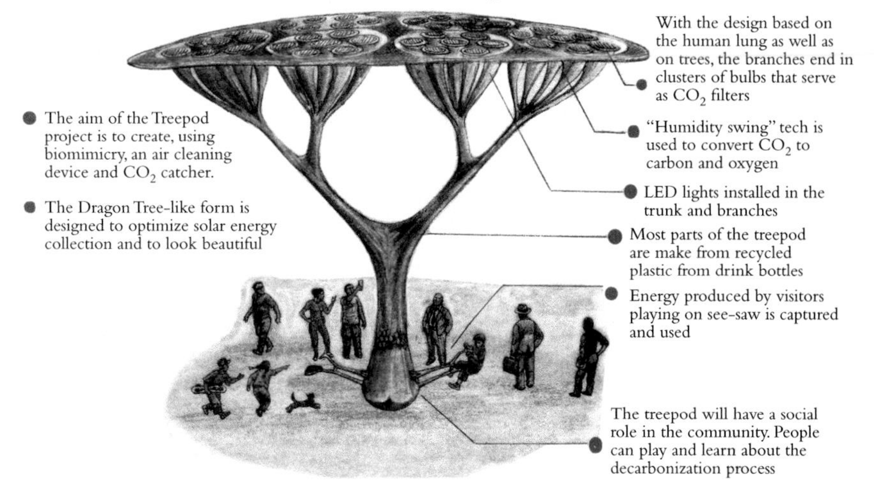


**Figure 25.4 Boston Treepod Party.**

*Source:* © Bogatyr Khan. Reproduced with permission.

## A Giant Sucking Sound

David Keith, a respected Harvard physicist, has established a company in Squamish, British Columbia, Canada, called Carbon Engineering, that has developed a carbon capture plant that is said to do the work of 40 million trees in sucking $CO_2$ out of the air and converting it to compressed $CO_2$, which can be sequestered or converted to fuels. Bill Gates backs the company. Although this negative-emissions technology requires energy to operate, the energy can come from nuclear or other zero-carbon-emissions sources. The company asserts that the technology can be scaled up so as to meet the goal of limiting planetary average warming to 1.5°C above the present level.[29]

## A Pair of Sunglasses for the Earth?

As far as I'm concerned, every eco- and geoengineering method I've discussed so far has at least a veneer of reasonableness. Putting sunglasses on the Earth does not, but since serious people are discussing it, I feel compelled to mention (and lampoon) it. The "sunglasses" consist of, according to the astronomer Roger Angel, 18 trillion

[29] See "Bill Gates-backed"(2019).

reflectors launched into orbit around the Sun in such a way as to block part of the Sun's rays as seen from Earth. There is a point in space, called the first Lagrange point or L1, where the Earth's and Sun's gravity balance each other out and an object at that position will (theoretically) remain in solar orbit there, casting a shadow on the Earth. If the object or group of objects is big enough, it will reduce the amount of solar radiation reaching the Earth enough to cool the planet significantly.

What will the 18 trillion tiny space robots do? "Each would weigh about a gram—the same as a large butterfly," a BBC report says, "and deflect sunlight with a transparent film pierced with tiny holes. . . . The robots would steer themselves into orbit by solar-powered ion propulsion . . . to form a cylindrical cloud 60,000 miles wide. After that they'd need regular nudges from 'shepherd dog satellites' to stop them crashing into each other or being blown off course by the sunlight they're deflecting."[30]

What could possibly go wrong? Even if the plan could somehow be executed, the Law of Unintended Consequences would "get" us eventually . . . probably sooner rather than later. Figure 25.5 is one such

**Figure 25.5 Putting sunglasses on the Earth could have a downside.**
*Source:* AuntSpray/Shutterstock.

[30] Gorvett (2016).

possible consequence: a new Ice Age, coming sooner than we're expected (remember that, despite global warming, we're almost certainly between two ice ages).

## Risks of Ecoengineering and Geoengineering

Geoengineering is often regarded as a last resort, to be used only if climate change becomes unbearable or some other environmental disaster occurs. Some of the more extreme proposals, such as 16 trillion space butterflies ("sunglasses on the Earth"), *should* be a last resort.

But all the proposals have downsides. Do we really want sulfuric acid in the upper atmosphere and iron filings in the ocean? These used to be called *air pollution* and *water pollution*. Planting more forests seems to be the only really safe and easy option, but it will probably not be sufficient, and we will have to try other methods of controlling the Earth's climate. We do not know exactly what to do.

Ecoengineering and geoengineering—as practiced by humans with the intent of affecting the whole planet—are in their infancy, and it's only natural, at this stage, that a wild variety of designs and techniques are being discussed. Only a few will ever become practical, and probably only one or two will become widely used. But it's fun to see how a technology in its formative stages explores the limits of possibility.

## Eco- and Geoengineering: Conclusion

The Earth is likely to be our only home for a very long time—perhaps for as long as the human race continues to exist. Before we try settling Mars or some other place by giving it a breathable atmosphere, a water supply, and a mild climate—a process called *terraforming* that seems impossible to me—let's terraform the Earth.

We've been shaping our environment so that it's more people-friendly, for tens of thousands of years. Sometimes we fail, but the fact that the Earth is home to more than seven billion people, a large majority of whom are eating well and feeling well, is testimony to the unique ability of mankind to shape its own environment in ways that other species can't. Eco- and geoengineering are simply our latest thoughts about how to do that better and on a larger scale. We are born terraformers.

# *Afterword*

## Save the Children (from Apocalyptic Thinking)

An old story has a young man coming home from college to visit his father, who asks him, "What have you learned?"

The young man says, "I've just heard the worst news . . . and it was in astronomy class. The world is going to end in five billion years, when the Sun becomes a supernova."

The father says, "I'm not sure I heard you right. The Sun is going to explode when?"

The son replies, "Five billion years."

The father wipes his forehead and says, "Whew! You had me worried there for a moment. I thought you said five million."

■ ■ ■

The point of the story, of course, is that we can't think sensibly about the very far future. We can barely understand the numbers involved, and, if our future is constrained by astrophysics, we surely can't do much about it. What we can do is make sure we don't destroy the near future.

In this book, we have seen how:

- The population explosion is coming to an end, affording an opportunity to solve problems that once seemed intractable.
- The increase in the wealth and well-being of the world is broadening to include traditionally poor societies, which are closing the gap with traditionally rich ones.
- Richer means greener, as resources and technology become available to solve environmental problems that may be daunting but are almost certainly surmountable.

Thus, the future of humanity offers continued and widespread betterment, punctuated by challenges that we have the riches and knowledge to address. Why do so many people fear it?

Apocalyptic thinking seems to be hard-wired into the human mind. Michael Shermer hypothesizes that this is the case for evolutionary reasons:

> You are a hominid on the plains of Africa 3 million years ago. You hear a rustle in the grass. Is it just the wind or is it a dangerous predator? If you assume it is a predator but it turns out that it is just the wind, you have made what is called a type I error in cognition, also known as a false positive, or believing something is real when it is not. You connected A, the rustle in the grass, to B, a dangerous predator, but no harm.
>
> On the other hand, if you assume that the rustle in the grass is just the wind but it turns out that it is a dangerous predator, you have made a type II error in cognition, also known as a false negative, or believing something is not real when it is. You failed to connect A to B, and in this case you're lunch.[1]

In other words, we have survived because we assume the worst about our environment until we're proven wrong and the danger goes away. This is not a bad guess about why human history is full of stories about the Apocalypse. Here are some examples:

- The Apocalypse itself, in the New Testament book of Revelation.
- The Mesopotamian epic of Gilgamesh, which is two millennia older than the New Testament.
- The end times in Judaism, Islam, Hinduism, and Buddhism.
- Countless prophecies and cults of doom from ancient times to the current day (each time that the world doesn't end, the prophet resets the date of the end of the world).
- A vast body of serious and popular literature, ranging from Mary Shelley's *The Last Man* to the retelling of the Epic of Gilgamesh in *Night of the Living Dead*.

Interestingly, these stories usually end with a new beginning, life organized under some other principle, such as the reign of God or various

[1] Shermer (2011).

gods, the New Man hypothesized in communism and fascism, and the latest wrinkle: transhumanism.[2]

This time *is* different. Modern apocalyptic thinking does not usually involve a better place after the old, bad one is destroyed.

All this does not mean we don't have real problems, including a few that could spell the end of human life as we known it. The boy who cried wolf may have done so one too many times to be believable but sometimes there really is a wolf. The closest we came to destroying civilization was on October 27, 1963, not that long ago. The Cuban missile crisis is sometimes thought of as proof that we will always pull back from the brink of destruction.

But it's not proof of anything. Although Kennedy and Khrushchev should be commended for their restraint, it's possible that the person who actually averted a nuclear world war was the brave Soviet submarine commander, Vasili Arkhipov.[3] Unanimity among the three officers in charge of his submarine would have been required to launch a nuclear missile aimed at the United States as a response to what appeared to the officers to be an American attack on the sub; Arkhipov cast the lone vote not to.[4] We do know that Fidel Castro wanted to use the weapons against the United States, even if the Soviet Union was using them as a bluffing tactic.

The challenges we face now are far less immediate: climate change, nuclear proliferation, pandemics. The *Bulletin of the Atomic Scientists* notwithstanding, we are not a few minutes away from almost complete destruction. We are getting richer and greener, but we have engineering challenges to overcome. Some of them could be daunting.

In the very, very long run we are, of course, subject to the laws of astrophysics. The Sun *will* explode. Long before then, the human race

---

[2] Transhumanism is a philosophy that suggests the human race can direct its own evolution into a superior species that incorporates massive computational power and artificial intelligence into its biological existence. Aspects of transhumanism, with its emphasis on life extension and, according to some adherents, immortality, are reminiscent of traditional religions.

[3] It could be argued, based on this story, that Arkhipov, not Churchill, Hitler, Mao, Roosevelt, Gavrilo Princip, or Einstein, was the most influential person of the twentieth century. (Princip was the man who started World War I, and indirectly World War II as well, by assassinating Archduke Franz Ferdinand of Austria and his wife.)

[4] Something similar happened again in 1983, when Stanislav Petrov, an officer of the Soviet Air Defense Forces, disobeyed an order to launch a retaliatory strike against the United States after it was falsely believed, due to warnings from a nuclear early-warning system, that the United States had attacked. This incident occurred in the wake of the shooting down of Korean Air Lines flight 007, a passenger airliner traveling from the United States, over Soviet airspace.

will have evolved into something unrecognizably different or will have disappeared for other reasons. I cannot fathom the next billion or even the next million years. The next thousand are hard enough to grasp. The next few hundred are probably the most that we can think about sensibly. Let's put our aspirations and fears within that frame.

■ ■ ■

Let's not teach our children that apocalyptic thinking is right thinking. It has always been wrong as a forecast, and it will continue to be wrong. Some young people are angry with their parents for bringing them into the world. We should stop terrifying them into not wanting to be alive. We should stop frightening them into not wanting to procreate. That is child mistreatment. We should be ashamed that we—some of us, anyway—have led our children to think this way.

We should, instead, teach them how to identify and solve problems. Specifically, they need to learn how to tell real problems from illusions and chimeras, and how to distinguish problems that can be solved from difficulties we're simply going to have to live with.

We should explain that apocalyptic thinking is a neural mistake based on our need to survive in a cruelly hostile environment that doesn't exist anymore. We are descended, over billions of years, from animal ancestors that could not craft an environment friendly to their own needs. We are not that species. To an extent unprecedented in the history of life on Earth, we can control our fate.

Most of all, let's teach them to enjoy and celebrate life, their own and that of future generations. The mind, the body, and the spirit (I do not know what the spirit is, but the word does seem to have some meaning) are our primary gifts from Nature. The others are the natural world, each other, and the works we humans have created and preserved over time. Let us ask our children to respect the past and embrace the future, not with fear, but with courage and the desire to right what might be wrong with it.

Remind our children that they are members of the only species that can do any of these things.

What a piece of work is man.

# *Reader's Guide: Annotated Suggestions for Further Learning*

## Basics: Betterment Over Time

Norberg, Johan. 2016. *Progress: Ten Reasons to Look Forward to the Future.* London: Oneworld Publications.

> Readers seeking a quick, informative read on the issues in my book should start with Norberg's cheerful take on the state of the world and its future prospects. Norberg, a think-tank scholar and free-market advocate, features chapters on "food, sanitation, life expectancy, poverty, violence, the environment, literacy, freedom, equality, [and] the conditions of childhood" (as reported by *Kirkus Reviews*). His outline inspired mine, although it's hard to see how any book on this topic could be organized very differently, and Steven Pinker's outline (see below) is similar.

Ridley, Matt. 2010. *The Rational Optimist: How Prosperity Evolves.* New York: HarperCollins.

> The book that started me on this journey (Wattenberg's earlier work on population aside). Drawing on economics, history, political philosophy, and many branches of science, the polymathic science writer and member of the House of Lords presents a powerful case that the future will be better than the present, just as the present time is dramatically

wealthier, healthier, safer, more fun, and more peaceful than at any time in history. He credits freedom, and in particular capitalism, for this Great Enrichment.

Rosling, Hans, Anna Rosling Rönnlund, and Ola Rosling. 2018. *Factfulness: Ten Reasons We're Wrong About the World—and Why Things Are Better Than You Think.* New York: Flatiron Books.

Rosling, a medical doctor, was famous for his spectacular TED talks, which describe the improvement in living conditions since the Industrial Revolution using inventive graphics and boundless enthusiasm. Rosling passed away at a young age (he was 68—how our expectations have changed!), and his last book, co-authored with his son and daughter-in-law, is a poignant memorial to his work. It captures the Rosling family's love of data and hope for the future, although there's really nothing quite like seeing Hans jump around onstage, pointing to bubbles representing countries rising from being "poor and sick" to "rich and healthy."

## Fewer

Simon, Julian. (1981) 1996. *The Ultimate Resource* and *The Ultimate Resource 2*. Princeton, NJ: Princeton University Press.

The first serious contribution to the New Optimism genre. *The Ultimate Resource*, in two very different editions, is a now dated but groundbreaking work that argues for continued population growth. Think of all the Einsteins and Beethovens who would otherwise never be born. Simon lets others worry about the downside.

Wattenberg, Ben J. 2005. *Fewer: How the New Demography of Depopulation Will Shape Our Future*. Chicago: Ivan R. Dee.

Wattenberg describes the problems that come with the end of the population explosion. We're old, we need a lot of care, we lack vitality and a sense of possibility. Our governments are financially strained. Mentally, we live in the past. An aging society really does have a downside. Not all of Wattenberg's book is negative—he proposes solutions and recognizes the upside of population stabilization. At any rate, *Fewer* convinced me that the population explosion really is coming to an end, and supplied part of the title for my book.

## Richer

Acemoglu, Daron, and James A. Robinson. 2013. *Why Nations Fail: The Origins of Power, Prosperity, and Poverty.* New York: Crown Publishing.

A cautionary tale, well written, a page-turner, reminding us that progress is not automatic. Betterment is hard won and easily destroyed, and we need good institutions—which many countries do not have—to avoid the latter.

Brynjolfsson, Erik, and Andrew McAfee. 2016. *The Second Machine Age: Work, Progress, and Prosperity in a Time of Brilliant Technologies.* New York: W. W. Norton.

The authors both celebrate technology and express concern about its effects on jobs, economic equality, and human welfare. You should explore the downside of progress as well as the upsides that I've dwelled on. Brynjolfsson and McAfee, more pessimistic than most tech enthusiasts, do a literate and entertaining job of presenting both sides.

Deaton, Angus. 2013. *The Great Escape: Health, Wealth, and the Origins of Inequality.* Princeton, NJ: Princeton University Press.

A history of the Industrial Revolution and its aftermath, the great escape from poverty that was not possible until the last 250 years. Deaton won the 2015 Nobel Prize in Economics for helping us understand the economics of happiness, a novel concept that has become wildly popular in recent years.

Diamandis, Peter, and Steven Kotler. 2012. *Abundance: The Future Is Better Than You Think.* New York: Free Press/Simon & Schuster.

The authors, an inventor-entrepreneur and a reporter, describe technologies large and small (especially small) that enrich the poor and change the world for the better a little at a time. While the authors sometimes come across as dreamers, their coverage of technological issues is balanced and includes the costs as well as the benefits.

Gould, Stephen Jay. 1993. *Eight Little Piggies: Reflections in Natural History.* New York: W. W. Norton.

In terms of actually writing my book, I include Gould here only because of a single chapter (his chapter 14, "Shoemaker and Morning Star") in a single volume (*Eight Little Piggies*) of his Reflections in Natural History series. But Gould is the science writer that all science writers want to be when they grow up. Everything he wrote is worth reading.

Maddison, Angus. 2007. *Contours of the World Economy, 1–2030 AD; Essays in Macroeconomic History.* Oxford, UK: Oxford University Press.

This is a technical book for readers who want to look really, really deeply into economic development issues and data. It is the basis of much of my research, but hard to recommend as further reading.

McCloskey, Deirdre. 2006. *The Bourgeois Virtues: Ethics for an Age of Commerce.* Chicago: University of Chicago Press.

The most compelling of McCloskey's Bourgeois trilogy, *The Bourgeois Virtues* sets forth the theory that the Great Enrichment (her phrase) was set in motion by *words*, specifically those that supported the "dignity and liberty of the bourgeoisie" so that business could come respectably to the forefront of human action. It is a serious book of economic and political philosophy for serious people. The rest of McCloskey's Bourgeois trilogy, *Bourgeois Dignity* and *Bourgeois Equality*, are also recommended (budget a whole summer).

Pinker, Steven. 2018. *Enlightenment Now!: The Case for Reason, Science, Humanism, and Progress*. New York: Viking.

He should have worked harder on the title, but this magisterial volume covers much of the same territory as *Fewer, Richer, Greener* but with a much more scholarly tone. Pinker argues that Enlightenment values are responsible for mankind's past successes and asks us to renew our commitment to them. He peppers his stew with a good bit of philosophy, a welcome change from the facts and figures that often characterize books on economic progress.

Pinker, Steven. 2011. *The Better Angels of Our Nature: Why Violence Has Declined*. New York: Viking.

In a precursor to *Enlightenment Now*, Pinker makes the argument, novel and astonishing at the time (it has since become received wisdom) that violence has steadily declined over time and that we live in the most peaceful time in history. This thesis applies to both state-sponsored violence (wars) and crime. Although Pinker presents a wealth of data supporting his case, his work has been criticized by serious thinkers such as Nassim Taleb and John Gray, who question either his philosophical approach or his statistical methods, or both.

Winchester, Simon. 2018. *The Perfectionists: How Precision Engineers Created the Modern World.* New York: HarperCollins.

Where inventions come from, told by a master. The book focuses on the Victorian era, when improving technology became a popular obsession. Read anything by Winchester—for that matter, read everything. *The Professor and the Madman,* about the making of the Oxford English Dictionary and the very peculiar doctor who wrote some of it, is his best-known work, but they are all that good.

## Greener

Brand, Stewart. 2009. *Whole Earth Discipline: An Ecopragmatist Manifesto.* New York: Viking.

Author of the *Whole Earth Catalogue,* business and social entrepreneur, *über*-environmentalist, Renaissance man, and counterculture hero, Brand has written a book that is a manifesto for a new and more practical kind of environmentalism, one that embraces urbanism, nuclear power, genetic engineering, and geoengineering in pursuit of a better and greener world.

Lomborg, Bjørn. 2001. *The Skeptical Environmentalist: Measuring the Real State of the World.* Cambridge, UK: Cambridge University Press.

Lomborg challenges radical environmentalism with a moderate agenda that embraces economic considerations and champions cost-benefit analysis. His Copenhagen Consensus Center seeks to apply welfare economics and cost-benefit analysis to a variety of world problems. A climate "lukewarmist" when he wrote this book, he's become more focused on climate change in recent years. Originally part of the Copenhagen Business School, the center is now an independent nonprofit.

Mann, Charles C. 2018. *The Wizard and the Prophet.* New York: Alfred A. Knopf.

Two views of the proper relation between humans and the environment, seen through biography. The wizard, Norman Borlaug, whose innovations fed a billion people or more, was half-seriously described by magician-comedians Penn and Teller as the greatest person in history. The prophet, William Vogt, was the founder of "apocalyptic environmentalism" long before it became an influential movement. Let the best man win.

## Lots of Data

### Essentials

Gapminder (Hans Rosling), www.gapminder.org. Historical and current data on human development, presented with inspiring graphic design values reminiscent of Edward Tufte; data are downloadable and presentation graphics can be created online.

Maddison Project, www.ggdc.net/maddison/maddison-project/data.html. The history of the world economy, country by country, spanning two millennia. Downloadable.

Roser, Max, et al. Our World in Data, www.ourworldindata.org. Everything you always wanted to know about human development. Data are downloadable and presentation graphics can be created online.

### Treats

Geocurrents, www.geocurrents.info. "The peoples, places, and languages shaping current events." Physical and human geography are dying fields, with geographic information systems being the only growth industry in the general vicinity; let's revive the old disciplines.

Historical maps, www.davidrumsey.com/view. Addictive.

https://tradingeconomics.com. Capital markets and national accounts data by country.

www.edwardtufte.com/tufte. How to make art out of data and get your point across a zillion times more clearly. Leonardo da Vinci and Charles Joseph Minard (Google him) would be proud.

www.marginalrevolution.com. The mother of all economics blogs, curated by the hyper-prolific Tyler Cowen with co-author Alex Tabarrok. Links to every other economics blog worth reading. I check it every day. Not a data source specifically, but it really is a treat—intellectual candy.

www.measuringworth.com. Some economic data back to the year 1270! With convenient online calculators and downloads. A non-profit with a board of advisors that includes several of the people discussed in this book.

OECD, https://data.oecd.org. The OECD is the rich countries' club, but the website has data for all countries.

The amazing reading list compiled by Pseudoerasmus, an anonymous, and very good, economics blogger claiming to live in Siberia. Rehosted by the economics professor and blogger Brad DeLong at www.bradford-delong.com/2016/08/pseudoerasmuss-economic-history-reading-list.html.

UN Population Division, www.un.org/en/development/desa/population.

World Bank global development data, https://data.worldbank.org.

# *References*

## Preface

McCloskey, Deirdre N. 2019. *Why Liberalism Works: How True Liberal Values Produce a Freer, More Equal, Prosperous World for All*. New Haven, CT: Yale University Press.

## Chapter 1: Right Here, Right Now

Bolt, Jutta, Robert Inklaar, Herman de Jong, and Jan Luiten van Zanden. 2018. "Rebasing 'Maddison': New Income Comparisons and the Shape of Long-run Economic Development," Maddison Project Working paper 10. (This paper incorporates, through a web link, the Maddison Project Database, version 2018, which was used by Our World in Data to construct Figure 1.1. The link is www.rug.nl/ggdc/historicaldevelopment/maddison/releases/maddison-project-database-2018.)

McCloskey, Deirdre N. 2010: *Bourgeois Dignity: Why Economics Can't Explain the Modern World*. Chicago: University of Chicago Press.

Obama, Barack. 2016. "Barack Obama: Now Is the Greatest Time to Be Alive." *Wired*, October 12, www.wired.com/2016/10/president-obama-guest-edits-wired-essay/.

Ridley, Matt. 2010. *The Rational Optimist: How Prosperity Evolves*. New York: HarperCollins.

Rosling, Hans, with Ola Rosling and Anna Rosling Rönnlund. 2018. *Factfulness: Ten Reasons We're Wrong About the World—and Why Things Are Better Than You Think*. New York: Flatiron Books (Macmillan).

Stevenson, Katherine Cole. 1986. *Houses by Mail: A Guide to Houses from Sears, Roebuck and Company*. National Trust for Historic Preservation. New York: John Wiley & Sons.

Ward, J. H., and Stevenson, K. C. 1996. *Houses by Mail: A Guide to Houses from Sears, Roebuck and Company.* New York: John Wiley & Sons.

## Chapter 2: The Population Explosion, Malthus, and the Ghost of Christmas Present

Abbey, Edward. 1977. *The Journey Home: Some Words in Defense of the American West.* New York: Plume (Penguin), 183.

Bowyer, Jerry. 2012. "Malthus and Scrooge: How Charles Dickens Put Holly Branch Through the Heart of the Worst Economics Ever." *Forbes.com* (December 24), www.forbes.com/sites/jerrybowyer/2012/12/24/malthus-and-scrooge-how-charles-dickens-put-holly-branch-through-the-heart-of-the-worst-economics-ever/#25635d16672d.

Charles, Enid. 1934. *The Twilight of Parenthood: A Biological Study of the Decline of Population Growth.* London: Watts and Company.

Dickens, Charles. 1843. *A Christmas Carol.* London: Chapman & Hall. Full text online at www.gutenberg.org/files/46/46-hit/46-h.htm.

Hobbes, Thomas. 1651. *Leviathan or The Matter, Forme and Power of a Common-Wealth Ecclesiasticall and Civil.*

Malthus, Thomas R. (1798) 1993. *An Essay on the Principle of Population.* Oxford, UK: Oxford World's Classics reprint. Chapter VII, 44.

McDonald, Randy. 2004. "The Fewer the Better? Wattenberg's Fewer." *A Bit More Detail* (blog, November 18). https://abitmoredetail.wordpress.com/2004/11/18/review-the-fewer-the-better-wattenbergs-fewer.

"Pseudoerasmus" (pseudonym). 2014. "The Little Divergence," blog post (June 12), https://pseudoerasmus.com/2014/06/12/the-little-divergence.

Say, Jean Baptiste. 1821. *Letters to Mr. Malthus, and A Catechism of Political Economy* (in translation). London: Sherwood, Neely & Jones. Full text online at http://oll.libertyfund.org/titles/say-letters-to-mr-malthus-and-a-catechism-of-political-economy.

Stephen, Sir Leslie. 1900. *The English Utilitarians, Vol. 1.* New York: G. P. Putnam's Sons.

## Chapter 3: The Demographic Transition: Running Out of and Into People

Gould, Stephen Jay. 1993. "Shoemaker and Morning Star," in *Eight Little Piggies.* New York: W. W. Norton.

McCloskey, Deirdre N. 2016. "The Economic Sky Will Not Fall." *Prospect Magazine*, London (March). http://deirdremccloskey.org/docs/pdf/McCloskey_ProspectMagazine2016.pdf.

Nordhaus, Ted. 2018. "The Earth's Carrying Capacity for Human Life Is Not Fixed." *Aeon* (online, July 5). https://aeon.co/ideas/the-earths-carrying-capacity-for-human-life-is-not-fixed.

Roser, Max. 2017. 'Fertility Rate," Our World in Data (online), https://ourworldindata.org/fertility-rate#fertility-is-first-falling-with-development-and-then-rising-with-development, first published in 2014, revised in 2017. The source to which Roser refers is Myrskylä, Mikko, Hans-Peter Kohler, and Francesco C. Billari. 2009. "Advances in Development Reverse Fertility Declines." *Nature* 460 (August 6): 741–743.

Roser, Max, and Esteban Ortiz-Ospina. 2017. "World Population Growth," in Max Roser, editor, Our World in Data (online), https://ourworldindata.org/world-population-growth#demographic-transition (first published in 2013; updated April 2017).

## Chapter 4: Having Fewer Children: "People Respond to Incentives"

Becker, Gary S. 1960. "An Economic Analysis of Fertility." In National Bureau of Economic Research (special conference series), *Demographic and Economic Change in Developed Countries*. New York: Columbia University Press, www.nber.org/chapters/c2387.pdf.

Doepke, Matthias. 2015. "Gary Becker on the Quantity and Quality of Children." *Journal of Demographic Economics* 81: 59–66.

*The Economist*. 2017. "Gary Becker's Concept of Human Capital" (August 3). www.economist.com/economics-brief/2017/08/03/gary-beckers-concept-of-human-capital.

Kirk, Dudley. 1996. "Demographic Transition Theory." *Population Studies* 50: 361–387.

Landry, Adolphe. (1934) 1982. *La révolution démographique: études et essais sur les problèmes de la population*. Paris: Institut National d'Études Démographiques.

Landsburg, Steven E. (1993) 2012. *The Armchair Economist: Economics and Everyday Life*. 2nd ed. New York: Free Press.

Lee, Ronald, and Andrew Mason. 2014. "Are Our Low Birthrates and Aging Population Threatening the Economy?" *Los Angeles Times* (November 12). www.latimes.com/opinion/op-ed/la-oe-lee-birthrate-standard-of-living-20141113-story.html.

MacArthur, Robert H., and Edward O. Wilson. (1967) 2001. *The Theory of Island Biogeography*. Reprint ed. Princeton, NJ: Princeton University Press.

Notestein, Frank W. 1945. "Population: The Long View," in Theodore Schultz, editor, *Food for the World*. Chicago: University of Chicago Press.

Notestein, Frank W. 1953. "Economic Problems of Population Change." *Proceedings of the Eighth International Conference of Agricultural Economists* (New York), pp. 13–31.

Thomson, Adam. 2013. "Demographics: Birth Rate Fall and Prospect of Longer Life Cloud Mexico's Future." *Financial Times* (June 3). www.ft.com/content/4b531a9e-c766-11e2-9c52-00144feab7de.

Weller, Chris. 2017. "'This Is Death to the Family': Japan's Fertility Crisis Is Creating Economic and Social Woes Never Seen Before." *Business Insider* (online, May 21), www.businessinsider.com/japan-fertility-crisis-2017-4.

"Why Ethiopian Women Are Having Fewer Children Than Their Mothers." 2015. BBC News (November 6). www.bbc.com/news/world-africa-34732609.

## Chapter 5: Age Before Beauty: Life in an Aging Society

Collins, Patrick J., Huy D. Lam, and Josh Stampfli. 2015. *Longevity Risk and Retirement Income Planning*. Charlottesville, VA: CFA Institute Research Foundation. www.cfainstitute.org/media/documents/book/rf-lit-review/2015/rflr-v10-n2-1-pdf.ashx.

Dawkins, Richard. 1976. *The Selfish Gene*. Oxford, UK: Oxford University Press.

McLellan, Faith. 2007. "Medicalisation: A Medical Nemesis." *The Lancet* 369 (9562, February): 627–628. www.thelancet.com/journals/lancet/article/PIIS0140-6736(07)60293-1/fulltext.

Papanicolas, Irene, Liana R. Woskie, and Ashish K. Jha. 2018. "Health Care Spending in the United States and Other High-Income Countries." *Journal of the American Medical Association* 319 (10): 1024–1039; abstract posted March 13, 2018 at https://jamanetwork.com/journals/jama/article-abstract/2674671.

Sexauer, Stephen C., and Laurence B. Siegel. 2013. "A Pension Promise to Oneself." *Financial Analysts Journal* 69 (6, November/December). www.cfapubs.org/doi/pdf/10.2469/faj.v69.n6.4.

Siegel, Laurence B., and M. Barton Waring. 2017. "The Only Saving Rate Article You Will Ever Need—*Using Just a Handheld Financial Calculator!*" *Journal of Investing* 26 (1, Spring): 57–69.

Totten, Thomas L., and Laurence B. Siegel. 2019. "Combining Conventional Investing with a Lifetime Income Guarantee: A Blueprint for Retirement Security." *The Journal of Retirement* 6 (4, Spring): 45–59.

Vonnegut, Jr., Kurt. 1974. *Wampeters, Foma & Granfalloons.* New York: Delacorte Press.

Waring, M. Barton, and Laurence B. Siegel. 2015. "The Only Spending Rule Article You Will Ever Need." *Financial Analysts Journal* 71(1, January/February). www.cfapubs.org/doi/pdf/10.2469/faj.v71.n1.2.

## Chapter 6: Before the Great Enrichment: The Year 1 to 1750

Buringh, Eltjo, and Jan Luiten van Zanden. 2009. "Charting the 'Rise of the West': Manuscripts and Printed Books in Europe, A Long-Term Perspective from the Sixth through Eighteenth Centuries." *Journal of Economic History* 69 (2): 409–445 (416–417, tables 1 and 2).

DeLong, J. Bradford. 2014. Blog post, www.bradford-delong.com/2014/05/estimates-of-world-gdp-one-million-bc-present-1998-my-view-as-of-1998-the-honest-broker-for-the-week-of-may-24-2014.html.

Karayalcin, Cem. 2016. "Property Rights and the First Great Divergence: Europe 1500–1800." *International Review of Economics & Finance* 42 (March 2016): 484–498.

Koot, Gerard M. "The Little Divergence and the Birth of the first Modern Economy, or When and Why Did Northwestern Europe Become Much Richer than Southern and Eastern Europe." www1.umassd.edu/euro/resources/worldeconomy/4.pdf.

Locke, John. 1689. [Thomas Hollis edition of 1764]. Two treatises [of Civil Government]. Republished online by the Liberty Fund at http://oll.libertyfund.org/pages/john-locke-two-treatises-1689.

Maddison, Angus. 2007. *Contours of the World Economy 1–2030 AD: Essays in Macro-Economic History.* Oxford, UK: Oxford University Press.

McCloskey, Deirdre N. 2016. "The Economic Sky Will Not Fall." *Prospect Magazine*, London (March). http://deirdremccloskey.org/docs/pdf/McCloskey_ProspectMagazine2016.pdf.

Moore, Robert I. 2015. "The First Great Divergence?" *Medieval Worlds* 1 (1). www.medievalists.net/2015/07/the-first-great-divergence/.

Our World in Data. 2019. https://ourworldindata.org/economic-growth.

Owen, Virginia Lee. 1989. "The Economic Legacy of Gothic Cathedral Building: France and England Compared." *Journal of Cultural Economics* 13 (1, June): 89–100.

Van Zanden, Jan Luiten. 2009. *The Long Road to the Industrial Revolution.* Boston: Brill.

## Chapter 7: The Great Enrichment: 1750 to Today

Fuller, R. Buckminster. 1940. "World Energy" (cover illustration), *Fortune* (February). https://placesjournal.org/article/hubberts-peak-eneropa-and-the-visualization-of-renewable-energy/ on September 5, 2018.

Long, Heather, and Leslie Shapiro. 2018. "Does $60,000 Make You Middle-Class or Wealthy on Planet Earth?" *Washington Post*. www.washingtonpost.com/business/2018/08/20/does-make-you-middle-class-or-wealthy-planet-earth/?noredirect=on&utm_term=.05eb5b070671.

McCloskey, Deirdre N. 2004. "Review of *The Cambridge Economic History of Modern Britain*, edited by Roderick Floud and Paul Johnson." *Times Higher Education Supplement* (January 15), www.deirdremccloskey.com/articles/floud.php.

Ridley, Matt. 2010. *The Rational Optimist: How Prosperity Evolves*. New York: HarperCollins.

Tabarrok, Alex. 2010. "Soviet Growth & American Textbooks," blog post (January 4), https://marginalrevolution.com/marginalrevolution/2010/01/soviet-growth-american-textbooks.html.

## Chapter 8: Food

Alter, Lloyd. 2016. "Vertical Farms: Wrong on So Many Levels." *TreeHugger* (online, February 22). www.treehugger.com/green-architecture/vertical-farms-wrong-so-many-levels.html.

Blackburn-Dwyer, Brandon. 2016. "Farms Grow Up: Why Vertical Farming May Be Our Future." *Global Citizen* (online, August 11). www.globalcitizen.org/en/content/farms-grow-up-why-vertical-farming-may-be-our-futu/.

Brand, Stewart C. 2018. "Charles C. Mann: The Wizard and the Prophet," blog post (January 22). http://longnow.org/seminars/02018/jan/22/wizard-and-prophet/.

Despommier, Dickson. 2011. *The Vertical Farm: Feeding the World in the 21st Century*. New York: St. Martin's Press.

Despommier, Dickson. Undated. "The Vertical Essay." *The Vertical Farm: Feeding the World in the 21st Century* (website). www.verticalfarm.com/?page_id=36.

Diamandis, Peter H., and Steven Kotler. 2012. *Abundance: The Future Is Better Than You Think*. New York: Free Press.

Floud, Roderick, Robert W. Fogel, Bernard Harris, and Sok Chul Hong. 2011. *The Changing Body: Health, Nutrition, and Human Development in the Western World Since 1700*. Cambridge, UK: Cambridge University Press.

Gates, Bill. 2014. "Why Does Hunger Still Exist in Africa?" *Gates Notes* (blog, August 1). www.gatesnotes.com/Development/Why-Does-Hunger-Still-Exist-Africas-Table-Day-One.

Gombrich, Ernst. 2005. *A Little History of the World*. New Haven, CT: Yale University Press. Originally published in 1935 as *Eine kurz Weltgeschichte für junger Leser*.

Lacey, Richard, and Danny Danziger. 1999. *The Year 1000: What Life Was Like at the Turn of the First Millennium, An Englishman's World*. Boston: Little, Brown.

Massing, Michael. 2003. "Does Democracy Avert Famine? *New York Times* (March 1), www.nytimes.com/2003/03/01/arts/does-democracy-avert-famine.html.

Norberg, Johan. 2017. *Progress: Ten Reasons to Look Forward to the Future*. London: Oneworld.

Pinker, Steven. 2017. *Enlightenment Now: The Case for Reason, Science, Humanism, and Progress*. New York: Penguin Random House.

Prinz, Daniel. 2015. "Robert Fogel: The Escape from Hunger and Premature Death, 1700–2100. Chapters 1 and 2" (class notes, September 4). http://scholar.harvard.edu/files/dprinz/files/fogel_the_escape_from_hunger_an_premature_death_1700_2100.pdf.

*Science Daily*. 2004. "Men From Early Middle Ages Were Nearly as Tall as Modern People" (September 2). www.sciencedaily.com/releases/2004/09/040902090552.htm.

Sen, Amartya. 1999. "Democracy as a Universal Value." *Journal of Democracy* 10 (3): 3–17.

Sen, Amartya. 2000. *Development as Freedom*. New York: Anchor Books.

Senthilingam, Meera, 2017. "The Tech Solutions to End Global Hunger." CNN (online, February 24). www.cnn.com/2017/02/23/health/tech-apps-solving-global-hunger-famine/index.html.

Stone, Daniel. 2014. "The Other Other White Meat." *National Geographic* (online, April 30). https://news.nationalgeographic.com/news/special-features/2014/04/140430-other-white-meat-fish-aquaculture-cobia/.

Williams, William Carlos. 1962. "The Kermess," in *Pictures from Brueghel and Other Poems*. New York: New Directions. http://english.emory.edu/classes/paintings&poems/kermess.html.

World Health Organization. 2018. "Malnutrition" (February 16). www.who.int/news-room/fact-sheets/detail/malnutrition.

## Chapter 9: Health and Longevity

Coleman, Thomas S. 2019. "Causality in the Time of Cholera: John Snow as a Prototype for Causal Inference." Unpublished manuscript. https://papers.ssrn.com/abstract=3262234.

Coleman, Marilyn, Lawrence H. Ganong, and Kelly Warzinik. 2007. *Family Life in 20th-century America.* Westport, CT: Greenwood Press.

Goetzmann, William N. 2016. *Money Changes Everything: How Finance Made Civilization Possible.* Princeton, NJ: Princeton University Press.

Johnson, Steven. 2006. *The Ghost Map: The Story of London's Most Terrifying Epidemic—and How It Changed Science, Cities, and the Modern World.* New York: Riverhead (Penguin).

Kimura, Shigeyoshi. 1987. "100-Year-Old Conquers Mount Fuji." Associated Press News (August 2). www.apnews.com/7bca40602fb4c2b2b9468f2815d530c7.

Peterson, Tim. 2017. "Healthspan Is More Important Than Lifespan, So Why Don't More People Know About It?" (May 30). Institute for Public Health, Washington University in St. Louis. https://publichealth.wustl.edu/heatlhspan-is-more-important-than-lifespan-so-why-dont-more-people-know-about-it/.

Robine, J.- M., and M. Allard. 2019. "Jeanne Calment: Validation of the Duration of Her Life." Max Planck Institute for Demographic Research (Rostock, Germany). https://www.demogr.mpg.de/books/odense/6/09.htm.

Rothman, Lily. 2016. "This Is What Happened to the First American Treated with Penicillin." *Time* (March 14). http://time.com/4250235/penicillin-1942-history/.

Saxon, Wolfgang. 1999. "Anne Miller, 90, First Patient Who Was Saved by Penicillin." *New York Times* (June 9). www.nytimes.com/1999/06/09/us/anne-miller-90-first-patient-who-was-saved-by-penicillin.html.

Snow, John. 1855. *On the Mode of Communication of Cholera.* London: Wilson & Ogilvy. http://archive.org/details/b28985266.

Twain, Mark (Samuel L. Clemens). 1889. "Letter to Walt Whitman." Beinecke Rare Book & Manuscript Library, Yale University (New Haven, CT). https://brbl-dl.library.yale.edu/vufind/Record/3530736.

Uhlenberg, Peter. 1996. "Mortality Decline in the Twentieth Century and Supply of Kin Over the Life Course." *The Gerontologist* 36 (5, October 1): 681–685.

## Chapter 10: Energy: A BTU Is a Unit of Work You Don't Have to Do

Fuller, R. Buckminster. 1940. "World Energy" (magazine cover drawing), *Fortune* (February).
"Great Pyramid Tombs Unearth 'Proof' Workers Were Not Slaves" (unattributed). 2010. *Guardian* (January 11). www.theguardian.com/world/2010/jan/11/great-pyramid-tombs-slaves-egypt.
Hubbert, M. King. 1956. "Nuclear Energy and the Fossil Fuels," presented at the American Petroleum Institute, San Antonio, TX (March 7–9). https://web.archive.org/web/20080527233843/http://www.hubbertpeak.com/hubbert/1956/1956.pdf.
Pinker, Steven. 2017. *Enlightenment Now: The Case for Reason, Science, Humanism, and Progress*. New York: Penguin Random House.
Ridley, Matt. 2010. *The Rational Optimist: How Prosperity Evolves*. New York: Harper.
Smil, Vaclav. 2010. "Science, Energy, Ethics, and Civilization," in R. Y. Chiao et al., eds., *Visions of Discovery: New Light on Physics, Cosmology, and Consciousness*. Cambridge, UK: Cambridge University Press. http://vaclavsmil.com/wp-content/uploads/docs/smil-articles-science-energy-ethics-civilization.pdf.
Smil, Vaclav. 2014. "The Long, Slow Rise of Solar and Wind Energy: The Great Hope for a Quick and Sweeping Transition to Renewable Energy Is Wishful Thinking." *Scientific American* (January). http://vaclavsmil.com/wp-content/uploads/scientificamerican0114-521.pdf.

## Chapter 11: Cities

Adler, Jonathan H. 1993. "Poplar Front: The Rebirth of America's Forests." *Policy Review* (Spring): 84–87.
Follett, Chelsea. 2018. "The Remarkable Fall in China's Suicide Rate." *Human Progress* (online), May 18. https://humanprogress.org/article.php?p=1291.
Glaeser, Edward. 2011. *Triumph of the City: How Our Greatest Invention Makes Us Richer, Smarter, Greener, Healthier, and Happier*. New York: PenguinRandomHouse.
Hodges, Glenn. 2011. "America's Forgotten City." *National Geographic* (January). www.nationalgeographic.com/magazine/2011/01/americas-forgotten-city/.

Ingraham, Christopher. 2018. "How 2,000-year-old Roads Predict Modern-day Prosperity." *Washington Post* (August 7). https://washingtonpost.com/amphtml/business/2018/08/06/how-year-old-roads-predict-modern-day-prosperity/.

Jacobs, Jane. 1961. *The Death and Life of Great American Cities*. 50th Anniv. ed., 2011. New York: Modern Library. Random House.

Leite, Aureliano, C.W. Minkel, and R. M. Schneider. Undated. "São Paulo." *Encyclopedia Britannica*, accessed at www.britannica.com/place/Sao-Paulo-Brazil on August 24, 2018.

Mumford, Lewis. 1961. *The City in History: Its Origins, Its Transformations, and Its Prospects*. New York: Harcourt, Brace & World.

Ridley, Matt. 2010. *The Rational Optimist: How Prosperity Evolves*. New York: Harper.

Siegel, Laurence B. 2017. "The Rules of Growth: Organisms, Cities and Companies." *Advisor Perspectives* (October 30). www.advisorperspectives.com/articles/2017/10/30/the-rules-of-growth-organisms-cities-and-companies.

Stroud, Ellen. 2012. *Nature Next Door: Cities and the Rebirth of Northeastern Forests*. Seattle: University of Washington Press.

UN News. 2014. "More Than Half of World's Population Now Living in Urban Areas, UN Survey Finds." (July 10).

West, Geoffrey. 2017. *Scale: The Universal Laws of Life and Death in Organisms, Cities and Companies*. New York: Penguin.

*Wired*. 2009. "Stewart Brand: Save the Slums" (September 21). www.wired.com/2009/09/ff-smartlist-brand/.

## Chapter 12: Education: The Third Democratization

Adler, Mortimer. 1940. *How to Read a Book*. Updated in 1972 by Adler and Charles Van Doren. New York: Simon & Schuster.

Banning, Margaret Culkin. 1933. "What a Young Girl Should Know." *Harper's* (December). Reprinted in Gehlmann, John, editor. (1950) 1961. *The Challenge of Ideas: An Essay Reader*. New York: Odyssey Press.

Case, Anne, and Christina Paxson. 2008. "Stature and Status: Height, Ability, and Labor Market Outcomes." *Journal of Political Economy* 116 (3, June): 499–532.

CredeCalhoun (pseudonym). 2018. "What Is Chautauqua?" *Voices from the Earth* (online). https://voicesfromtheearth.org/2018/02/01/what-is-chautauqua/.

Crispe, Imogen. 2019. "Coding Bootcamp Cost Comparison." Course Report (online), April 1. www.coursereport.com/blog/coding-bootcamp-cost-comparison-full-stack-immersives.

Csikszentmihalyi, Mihaly. 1990. *Flow: The Psychology of Optimal Experience*. New York: Harper & Row.

Friedman, Milton. 1968. "The Higher Schooling in America." *The Public Interest* 11 (Spring): 108–112.

Heckman, James J., and Paul A. LaFontaine. 2010. "The American High School Graduation Rate: Trends and Levels." *Review of Economics and Statistics* 92 (2, May): 244–262. www.ncbi.nlm.nih.gov/pmc/articles/PMC2900934/.

Holmquist, Annie. 2016. "Middle School Reading Lists 100 Years Ago vs. Today." *Intellectual Takeout* (July 19). www.intellectualtakeout.org/blog/middle-school-reading-lists-100-years-ago-vs-today.

Hoover, Gary E. 2016. "Why Higher Education Is Too Important to Be Free." https://hooversworld.com/why-higher-education-is-too-important-to-be-free/.

Hoover, Gary E. 2017. *The Lifetime Learner's Guide to Reading and Learning*. Philadelphia: Assiduity Publishing House.

Housman, A. E. 1896. "Terence, this is stupid stuff," in *A Shropshire Lad*. London: Kegan Paul, Trench, Trübner, & Co.

Littell, Robert. 1933. "What the Young Man Should Know." *Harper's* (March). https://archive.harpers.org/1933/03/pdf/HarpersMagazine-1933-03-0018458.pdf

Lerman, Robert. 2016. "Let's Help Intermediaries Expand U.S. Apprenticeships." *American Institute for Innovative Apprenticeship* (July 7). https://innovativeapprenticeship.org/author/blerman/

Longfellow, Henry Wadsworth. 1856. *Evangeline: A Tale of Acadie*. New York: E.P. Dutton & Co.

Lynch, Matthew. 2016. "What's Wrong with MOOCs and Why Aren't They Working? The Tech Edvocate (June 12). www.thetechedvocate.org/whats-wrong-with-moocs-and-why-arent-they-working/.

Nguyen, Son. 2014. "A Booth Education, for Free? With Professor Cochrane's Online Class, It's Already Started." Chicago Business, University of Chicago Booth School of Business (May 5). www.chibus.com/news/2014/5/5/a-booth-education-for-free-with-professor-cochranes-online-class-its-already-started.

Wortis, Edward Irving (pseudonym: "Avi"). 1991. *Nothing But the Truth: A Documentary Novel*. London: Orchard Books.

## Chapter 13: Conflict, Safety, and Freedom

Bonnefoy, Pascale, and Patrick J. Lyons. 2015. "Why Chile's Latest Big Earthquake Has a Smaller Death Toll." *New York Times* (September 17).

Dawkins, Richard. 2006. *The God Delusion*. New York: Bantam Books.

Economist Intelligence Unit. 2018. "Democracy Index 2017: Free Speech Under Attack." https://pages.eiu.com/rs/753-RIQ-438/images/Democracy_Index_2017.pdf.

Edesess, Michael, 2019. "Klarman, Fink and Grantham and the Threat to American Capitalism." *Advisor Perspectives* (online, February 18). https://www.advisorperspectives.com/articles/2019/02/18/klarman-fink-and-grantham-and-the-threat-to-american-capitalism.

Gibbons, Ann. 2018. "Why 536 Was 'the Worst Year to Be Alive.'" *Science*, American Association for the Advancement of Science (November 15). https://www.sciencemag.org/news/2018/11/why-536-was-worst-year-be-alive.

Gray, John. 2015. "John Gray: Steven Pinker Is Wrong About Violence and War." *Guardian* (March 13). www.theguardian.com/books/2015/mar/13/john-gray-steven-pinker-wrong-violence-war-declining.

Mamet, David. 2011. *The Secret Knowledge: On the Dismantling of American Culture*. New York: Penguin, 84.

Monkkonen, Eric H. 2001. *Murder in New York City*. Oakland: University of California Press.

Norberg, Johan. 2017. *Progress: Ten Reasons to Look Forward to the Future*. London: Oneworld, 158.

Pinker, Steven. 2011. *The Better Angels of Our Nature: A History of Violence and Humanity*. New York: Penguin.

Real-McKeighan, Tammy. 2010. "Pony Express Riders Faced Many Dangers." Fremont Tribune (July 19). https://fremonttribune.com/news/local/pony-express-riders-faced-many-dangers/article_ef8e9c90-934e-11df-b50b-001cc4c03286.html.

Roser, Max. 2018. "Democracy." In Our World in Data (online). https://ourworldindata.org/democracy.

Roth, Randolph. 2009. *American Homicide*. Cambridge, MA: Harvard University Press (Belknap Press).

Wells, Herbert George. 1895. *The Time Machine*. London: William Heinemann.

Wells, Herbert George. (1901) 1914. *Anticipations [of the Reaction of Mechanical and Scientific Progress upon Human Life and Thought]*. London: Chapman and Hall.

## Chapter 14: The Alleviation of Poverty

Adegoke, Yinka. 2018. "The Household Electrification Rate in Sub-Saharan Africa is the Lowest in the World." *Quartz* (May 7). https://qz.com/africa/1271252/world-bank-recommendations-on-electricity-in-sub-saharan-africa/.

Adler, David E., and Laurence B. Siegel, eds. 2019. *The Productivity Puzzle: Restoring Economic Dynamism*. Charlottesville, VA: CFA Institute Research Foundation.

Breznitz, Dan, and Peter L. Cowhey. 2012. *Reviving America's Forgotten Innovation System: Fostering U.S. Growth through Incremental Product and Process Innovation*. Cambridge, MA: MIT Press.

Cowen, Tyler. 2018. *Stubborn Attachments: A Vision for a Society of Free, Prosperous, and Responsible Individuals*. San Francisco: Stripe Press.

De Soto, Hernando. 2000. *The Mystery of Capital: Why Capitalism Triumphs in the West and Fails Everywhere Else*. New York: Basic Books.

Diamandis, Peter H., and Steven Kotler. 2012. *Abundance: The Future is Better Than You Think*. New York: Free Press (Simon & Schuster).

Falk, Michael S. 2016. *Let's All Learn How to Fish . . . to Sustain Long-Term Economic Growth*. Charlottesville, VA: CFA Institute Research Foundation. www.cfainstitute.org/en/research/foundation/2016/lets-all-learn-how-to-fish—-to-sustain-long-term-economic-growth.

Friedman, Milton. 1962. *Capitalism and Freedom*. Chicago: University of Chicago Press.

Giannarelli, Linda, and C. Eugene Steuerle. 1995. *The Twice-Poverty Trap: Tax Rates Faced by AFDC Recipients*. Washington, DC: Urban Institute.

Gittings, John. 2005. *The Changing Face of China*. Oxford University Press, Oxford.

"Granny Cloud." Undated. School in the Cloud (Newcastle, UK). www.theschoolinthecloud.org/people/the-granny-cloud/.

Jacobs, Jane. 1969. *The Economy of Cities*. New York: Random House.

Kharas, Homi, and Kristofer Hamel. 2018. "A Global Tipping Point: Half the World Is Now Middle Class or Wealthier." Brookings (blog, September 27). www.brookings.edu/blog/future-development/2018/09/27/a-global-tipping-point-half-the-world-is-now-middle-class-or-wealthier/.

"King Gustaf of Sweden." 1938. *Life* (July 11), p. 31. https://books.google.com/books?id=f08EAAAAMBAJ&lpg=PP1&ots=ADyf8WOb5i&pg=PA31#v=onepage&q&f=true on January 5, 2019.

Landes, David. 1998. *The Wealth and Poverty of Nations: Why Some Are So Rich and Some Are So Poor.* New York: W. W. Norton.

North, Douglass. 1991. "Institutions." *Journal of Economic Perspectives* 5(1, Winter): 97–112.

Pinker, Steven. 2017. *Enlightenment Now: The Case for Reason, Science, Humanism, and Progress.* New York: Penguin Random House.

Weber, Max. (1904–1905) 1958. *The Protestant Ethic and the Spirit of Capitalism.* Translated from German. New York: Scribner.

## Chapter 15: Robots Don't Work for Free: A Meditation on Technology and Jobs

Adler, David E., and Laurence B. Siegel, eds. 2019. *The Productivity Puzzle: Restoring Economic Dynamism.* Charlottesville, VA: CFA Institute Research Foundation.

American Federation of State, County & Municipal Employees. Undated. "Dr. Martin Luther King, Jr. on Labor" (collection of quotes). www.afscme.org/union/history/mlk/dr-martin-luther-king-jr-on-labor.

Armstrong, Thomas. Undated. "Multiple Intelligences." American Institute for Learning and Development (online). www.institute4learning.com/resources/articles/multiple-intelligences/.

Ausubel, Jesse H. 2004. "Will the Rest of the World Live Like America?" *Technology in Society* 26: 343–360, https://phe.rockefeller.edu/PDF_FILES/LiveLikeAmerica.pdf, p. 346.

Autor, David H. 2016. "Why Are There Still So Many Jobs? The History and Future of Workplace Automation and Anxiety." *MIT Initiative on the Digital Economy Research Brief* 2016.07. http://ide.mit.edu/sites/default/files/publications/IDE_Research_Brief_v07.pdf.

Breznitz, Dan, and Peter L. Cowhey. 2019. "Reviving America's Forgotten Innovation System: Fostering U.S. Growth through Incremental Product and Process Innovation," in David E. Adler and Laurence B. Siegel, eds., *The Productivity Puzzle: Restoring Economic Dynamism.* Charlottesville, VA: CFA Institute Research Foundation.

Brynjolfsson, Erik, and Andrew McAfee. 2016. *The Second Machine Age: Work, Progress, and Prosperity in a Time of Brilliant Technologies.* New York: W. W. Norton.

Coen, Robert M. 1973. "Labor Force and Unemployment in the 1920's and 1930's: A Re-Examination Based on Postwar Experience." *Review of Economics and Statistics* 55 (1): 46–55.

Duffin, Erin. 2019. "Percentage of White, Non-Hispanic Married-couple Families in the U.S. Who Live Below the Poverty Level from 1990 to 2017." *Statista* (online, April 29). www.statista.com/statistics/205033/percentage-of-poor-white-married-couple-families-in-the-us/.

Gardner, Howard. (1983) 2011. *Frames of Mind: The Theory of Multiple Intelligences*. New York: Basic Books.

Henrie, Brad. 2018. "Timeline of Industrial Automation" (July 25). https://automationexperts.ca/2018/07/25/timeline-of-industrial-automation/.

Lebergott, Stanley. 1964. *Manpower in Economic Growth: The American Record since 1800*. New York: McGraw-Hill.

McCloskey, Deirdre N. 2017. "The Myth of Technological Unemployment." *Reason* (August/September). http://reason.com/archives/2017/07/11/the-myth-of-technological-unem.

Perry, Mark J. 2012. "Phenomenal Gains in Manufacturing Productivity." AEIdeas, American Enterprise Institute for Public Policy Research. www.aei.org/publication/phenomenal-gains-in-manufacturing-productivity/.

Romer, Christina. 1986. "Spurious Volatility in Historical Unemployment Data." *Journal of Political Economy* 94(1): 1–37.

Sexauer, Stephen C., and Laurence B. Siegel. 2017. "The Age of Experts: A Review of Marc Levinson's An Extraordinary Time." *Business Economics* (October 26).

Weaver, Caroline. 2017. "A Brief History of the Ferrule." CW Pencil Enterprise (September 19). https://cwpencils.com/blogs/news/a-moment-in-pencil-history-the-ferrule.

## Chapter 16: The Mismeasurement of Growth: Why You Aren't Driving a Model T

Adler, David E., and Laurence B. Siegel, eds. 2019. *The Productivity Puzzle: Restoring Economic Dynamism*. Charlottesville, VA: CFA Institute Research Foundation.

Brynjolfsson, Erik, and Andrew McAfee. 2016. *The Second Machine Age: Work, Progress, and Prosperity in a Time of Brilliant Technologies*. New York: W. W. Norton.

Christensen, Clayton, Michael E. Raynor, and Rory McDonald. 2015. "What Is Disruptive Innovation?" *Harvard Business Review* (December). https://hbr.org/2015/12/what-is-disruptive-innovation.

DeLong, J. Bradford. 2017. "Rethinking Productivity Growth." *Project Syndicate* (online, March 3). www.project-syndicate.org/commentary/rethinking-productivity-growth-by-j–bradford-delong-2017-03.

Feldstein, Martin. 2017. "Underestimating the Real Growth of GDP, Personal Income, and Productivity." *Journal of Economic Perspectives* (Spring).

Le Page, Michael. 2017. "We're Nearly Ready to Use CRISPR to Target Far More Diseases." *New Scientist* (October 2). www.newscientist.com/article/2149129-were-nearly-ready-to-use-crispr-to-target-far-more-diseases/.

Meinig, Donald W. 2004. *The Shaping of America: A Geographical Perspective on 500 Years of History*, Volume 4: Global America, 1915–2000. New Haven, CT: Yale University Press.

NASA Explores. 2014. "Space Station 3-D Printer Builds Ratchet Wrench to Complete First Phase of Operations." (December 22). www.nasa.gov/mission_pages/station/research/news/3Dratchet_wrench.

Organisation for Economic Co-operation and Development. 2006. "Beyond GDP: Measuring Progress, True Wealth, and the Well-Being of Nations" (conference description). www.oecd.org/site/worldforum06/38433373.pdf.

Perry, Mark J. 2015. "Today Is Manufacturing Day, So Let's Recognize America's World-Class Manufacturing Sector and Factory Workers." AEIdeas, American Enterprise Institute for Public Policy Research (October 1). www.aei.org/publication/october-2-is-manufacturing-day-so-lets-recognize-americas-world-class-manufacturing-sector-and-factory-workers/.

Surowiecki, James. 2013. "Gross Domestic Freebie." *The New Yorker* (November 25). www.newyorker.com/magazine/2013/11/25/gross-domestic-freebie.

Thompson, William H. 1989. "Transportation in Iowa: A Historical Summary." Iowa Department of Transportation. https://iowadot.gov/history/transportationiniowa, Chapter 5, The Lincoln Highway.

Triplett, Jack E., and Barry Bosworth, 2004. "Productivity in the US Services Sector: New Sources of Economic Growth." Washington, DC: Brookings Institution Press.

Walsh, Margaret. Undated. "Gender and the Automobile in the United States." *Automobile in American Life and Society* (website), University of Michigan. www.autolife.umd.umich.edu/Gender/Walsh/G_Overview.htm.

## Chapter 17: *The Discreet Charm of the Bourgeoisie:* Deirdre McCloskey, Capitalism, and Christian Ethics


Adams, John. 1780. "Letter from John Adams to Abigail Adams, post 12 May 1780." Adams Family Papers, Massachusetts Historical Society. www.masshist.org/digitaladams/archive/doc?id=L17800512jasecond.

Feireiss, Kristin, ed. 2003. *Curves and Spikes: Kunsthaus und Stadthalle für Graz*. Berlin: Aedes.

Goldberg, Jonah. 2018. "Karl Marx's Jew-Hating Conspiracy Theory." *Commentary* (March). www.commentarymagazine.com/articles/karl-marxs-jew-hating-conspiracy-theory/.

Kristol, Irving. 1979. "The Adversary Culture of the Intellectuals," in Michael Novak, ed., *The Moral Basis of Democratic Capitalism*. Washington, DC: American Enterprise Institute.

Lovecraft, H. P. (1928) 2016. *The Call of Cthulhu*. Auckland, New Zealand: Floating Press (ebook).

McCloskey, Deirdre N. 1995. "Some News That At Least Will Not Bore You." *Eastern Economic Journal* (Fall).

McCloskey, Deirdre N. 2016. *Bourgeois Equality: How Ideas, Not Capital or Institutions, Enriched the World*. Chicago: University of Chicago Press.

Mencken, Henry Louis. 1917. "Dreiser," in *A Book of Prefaces*, New York: Alfred A. Knopf. Full text at http://www.gutenberg.org/files/19355/19355-h/19355-h.htm.

Mokyr, Joel. 1993. "Editor's Introduction," in Joel Mokyr, ed., *The New Economic History and the Industrial Revolution*, Boulder, CO: Westview Press. www.unsa.edu.ar/histocat/haeconomica07/mokyr.pdf.

Mokyr, Joel. 2017. "The Persistence of Technological Creativity and the Great Enrichment: Reflections on the 'Rise of Europe.'" *Voxeu.org* (February 2). https://voxeu.org/article/technological-creativity-and-great-enrichment-reflections-rise-europe.

Rennix, Brianna, and Nathan J. Robinson. 2017. "Why You Hate Contemporary Architecture." *Current Affairs* (October 31). www.currentaffairs.org/2017/10/why-you-hate-contemporary-architecture.

Rostow, Walt W. 1960. *The Stages of Economic Growth: A Non-Communist Manifesto*. Cambridge, UK: Cambridge University Press.

Toynbee, Arnold. (1884) 1969. *Lectures on the Industrial Revolution,* reprinted as *Toynbee's Industrial Revolution*, Newton Abbot, UK: David & Charles.

VoxEU.org. 2017. *The Long Economic and Political Shadow of History*. Volume 1, ed. Stelios, Michalopoulos, and Elias Papaioannou.

## Chapter 18: Simon and Ehrlich: Cornucopianism versus the Limits to Growth

Bailey, Ronald. 2010. "Cracked Crystal Ball: Environmental Catastrophe Edition." *Reason.com* (blog, December 30). http://reason.com/blog/2010/12/30/cracked-crystal-ball-environme.

Berlin, Isaiah. 1953. *The Hedgehog and the Fox: An Essay on Tolstoy's View of History*. London: Weidenfield and Nicolson.

Simon, Julian L. 1998. "Epilogue: My Critics and I." In *The Ultimate Resource II: People, Materials, and Environment* (online book draft), www.juliansimon.com/writings/Ultimate_Resource/TEPILOG.txt.

Tierney, John. 1990. "Betting on the Planet." *New York Times*, archived at www.nytimes.com/1990/12/02/magazine/betting-on-the-planet.html.

## Chapter 19: Obstacles

Alexander, Scott (pseudonym). 2015. "How Bad are Things?" *Slate Star Codex: The Joyful Reduction of Uncertainty* (blog). http://slatestarcodex.com (post of December 24).

Bastiat, [Claude] Frédéric. 1850. "That Which Is Seen, and That Which Is Not Seen." Translator unknown. Italics in the translated original. https://mises.org/library/which-seen-and-which-not-seen.

Clifton, Jon. 2017. "Coming to America." Gallup Blog (June 28). https://news.gallup.com/opinion/gallup/212687/coming-america.aspx.

Forbes, Robert James. 1966. *Studies in Ancient Technology: Volume 6—Heat and Heating, Refrigeration, Light*. Leiden: E. J. Brill.

Gould, David M., and William C. Gruben. 1997. "The Role of Intellectual Property Rights in Economic Growth," in Satya Dev Gupta and Nanda K. Choudhry, eds., *Dynamics of Globalization and Development*, New York: Springer, 209–241.

Hazlitt, Henry. (1946) 1988. *Economics in One Lesson: The Shortest and Surest Way to Understand Basic Economics*. New York: Three Rivers Press (Random House).

Postrel, Virginia. 2018. "Before Drug Prohibition, There Was the War on Calico." *Reason* (July). https://reason.com/archives/2018/06/25/before-drug-prohibition-there.

Ricardo, David. 1817. *On the Principles of Political Economy and Taxation*. London: John Murray.

Wolfe, Tom. 2000. "The Great Relearning," in *Hooking Up*. New York: G. P. Putnam's Sons. "The Great Relearning" originally appeared in *The American Spectator*, December 1987.

## Chapter 20: "He Shall Laugh": Why Weren't Our Ancestors Miserable All the Time?

Feynman, Richard P., with Ralph Leighton. 1985. *Surely You're Joking, Mr. Feynman! (Adventures of a Curious Character)*. New York: W. W. Norton.

Frederick, Shane. 2007. "Hedonic Treadmill," in Roy Baumeister and Kathleen D. Vohs, eds., *Encyclopedia of Social Psychology*. Thousand Oaks, CA: SAGE Publications.

Kahneman, Daniel, Edward Diener, and Norbert Schwarz, eds. 1999. *Well-Being: Foundations of Hedonic Psychology*. New York: Russell Sage Foundation.

Kahneman, Daniel, and Angus Deaton. 2010. "High Income Improves Evaluation of Life But Not Emotional Well-being." *Proceedings of the National Academy of Sciences* 107:38 (September 21): 16489–16493. https://doi.org/10.1073/pnas.1011492107.

Ortiz-Ospina, Esteban, and Max Roser. 2013. "Happiness and Life Satisfaction." Our World in Data (online). www.ourworldindata.org/happiness-and-life-satisfaction. Posted 2013; revised 2017.

Parker, Robert B. 1999. *Family Honor.* New York: G. P. Putnam's Sons.

## Chapter 21: Prologue: Why Poor Is Brown and Rich Is Green

Carlin, Dan. 2018. *Hardcore History 62: Supernova of the East I* (podcast, July 14). www.dancarlin.com/product/hardcore-history-62-supernova-in-the-east-i/ (at 22:40).

Grossman, Gene M., and Alan B. Krueger. 1991. "Environmental Impact of a North American Free Trade Agreement." NBER Working Paper 3914 (November).

Grossman, Gene M., and Alan B. Krueger. 1995. "Economic Growth and the Environment." *Quarterly Journal of Economics* 110 (2, February): 353–377.

Warsh, David. 2014. "Nobel Prizes and Macro vs. Growth." *New England Diary* (online, July 9). https://newenglanddiary.squarespace.com/blog/david-warsh-nobel-prizes-macro-vs-growth.

## Chapter 22: A Skeptical Environmentalist: The Greening World of Bjørn Lomborg


Aldy, Joseph E. 2004. "An Environmental Kuznets Curve Analysis of U.S. State-Level Carbon Dioxide Emissions." Department of Economics, Harvard University (August 9). https://sites.hks.harvard.edu/m-rcbg/repsol_ypf-ksg_fellows/Papers/Aldy/Aldy%20States%20EKC%20Paper.pdf.

Federal Reserve Bank of Saint Louis. 2014. "How Much Do Americans Drive?" *The FRED Blog* (June 9). https://fredblog.stlouisfed.org/2014/06/how-much-do-americans-drive/.

Hammond, Alexander C. R. 2018. "No, We Are Not Running Out of Forests." *Humanprogress.org* (online, May 24). https://humanprogress.org/article.php?p=1295.

Hazlitt, Henry. 1946. *Economics in One Lesson*. New York: Harper & Brothers.

Hickel, Jason. 2015. "The Hunger Numbers: Are We Counting Right?" *Guardian* (July 17). www.theguardian.com/global-development-professionals-network/2015/jul/17/the-hunger-numbers-are-we-counting-right.

Huber, Peter W. 1999. *Hard Green: Saving the Environment from the Environmentalists, A Conservative Manifesto*. New York: Basic Books.

Lomborg, Bjørn. 2001. *The Skeptical Environmentalist: Measuring the Real State of the World*. Cambridge, UK: Cambridge University Press.

Lomborg, Bjørn. 2004. "Need for Economists to Set Global Priorities." Correspondence, *Nature* 431 (September 2): 17.

Moore, Matthew. 2010. "Climate 'Sceptic' Bjørn Lomborg Now Believes Global Warming Is One of World's Greatest Threats." *Telegraph* (August 21). www.telegraph.co.uk/news/earth/environment/globalwarming/7972383/Climate-sceptic-Bjorn-Lomborg-now-believes-global-warming-is-one-of-worlds-greatest-threats.html.

Tierney, John. 2009. "Use Energy, Get Rich, and Save the Planet." *New York Times* (April 20). www.nytimes.com/2009/04/21/science/earth/21tier.html.


## Chapter 23: Dematerialization: Where Did My Record Collection Go?


Assadourian, Erik. 2012. "The Path to Degrowth in Overdeveloped Countries." In *State of the World 2012: Moving Toward Sustainable Prosperity*. Washington, DC: Worldwatch Institute.

Ausubel, Jesse H. 2015. "The Return of Nature: How Technology Liberates the Environment." *The Breakthrough* (Spring), https://thebreakthrough.org/index.php/journal/past-issues/issue-5/the-return-of-nature.

Bailey, Ronald. 2014. "Modernity Means More Stuff." *Reason* (February 21). http://vaclavsmil.com/wp-content/uploads/Modernity-Means-More-Stuff-Reason1.pdf.

Cichon, Steve. 2014. "Everything from 1991 Radio Shack Ad I Now Do with My Phone." *Trending Buffalo* (online, January 14). http://www.trendingbuffalo.com/life/uncle-steves-buffalo/everything-from-1991-radio-shack-ad-now/#prettyPhoto[mixed]/1/.

Commonwealth Scientific and Industrial Research Organisation (CSIRO). Undated. "Improving Safety and Productivity in Coal Mines." www.csiro.au/en/About/Our-impact/Our-impact-in-action/Industry-and-defence/Coal.

Diamandis, Peter H., and Steven Kotler. 2012. *Abundance: The Future Is Better Than You Think*. New York: Free Press/Simon & Schuster.

Dvorsky, George. 2012. "1846: The Year We Hit Peak Sperm Whale Oil." *Gizmodo* (online, July 31). https://io9.gizmodo.com/5930414/1846-the-year-we-hit-peak-sperm-whale-oil.

Eastwood, Brian. 2018. "Why Economic Growth Hasn't Ruined the Planet." Ideas Made to Matter (online, May 25), MIT Sloan School of Management. http://mitsloan.mit.edu/newsroom/articles/why-economic-growth-hasnt-ruined-the-planet/.

Fuller, R. Buckminster. (1938) 2000. *Nine Chains to the Moon*. New York: Anchor/Doubleday.

Goodall, Chris. 2011. "'Peak Stuff': Did the UK reach a maximum use of material resources in the early part of the last decade?" www.carboncommentary.com (blog) (October 13). http://static.squarespace.com/static/545e40d0e4b054a6f8622bc9/t/54720c6ae4b06f326a8502f9/1416760426697/Peak_Stuff_17.10.11.pdf.

Kadlec, Dan. 2015. "What Millennials Can Teach Boomers About Happiness." *Money* (September 15). http://time.com/money/4030036/millennials-boomers-buying-experiences/.

McPhee, John. 1971. *Encounters with the Archdruid: Narratives About a Conservationist and Three of His Natural Enemies.* New York: Farrar, Straus and Giroux.

Peeters, Paul, et al. 2016. "Are Technology Myths Stalling Aviation Climate Policy?" *Transportation Research Part D*, http://daneshyari.com/article/preview/1065643.pdf, pp. 4–5.

Pine, B. Joseph II, and James H. Gilmore. 1998. "Welcome to the Experience Economy." *Harvard Business Review* (July–August). https://hbr.org/1998/07/welcome-to-the-experience-economy.

Pine, B. Joseph II, and James H. Gilmore. 2011. *The Experience Economy*, updated ed. Boston: Harvard Business School.

Morris, David Z. 2016. "Today's Cars Are Parked 95% of the Time." *Fortune* (March 13). http://fortune.com/2016/03/13/cars-parked-95-percent-of-time/.

Perry, Mark J. 2016. "1991 Radio Shack Ad: 13 electronic products for $5k (and 290 hrs. work) can now be replaced with a $200 iPhone (10 hrs.)" *AEI* (January 25). www.aei.org/publication/1991-radio-shack-ad-13-electronic-products-for-5k-and-290-hrs-work-can-now-be-replaced-with-a-200-iphone-10-hrs/.

Shellenberger, Michael. 2018. "If Solar Panels Are So Clean, Why Do They Produce So Much Toxic Waste?" *Forbes.com* (May 23). www.forbes.com/sites/michaelshellenberger/2018/05/23/if-solar-panels-are-so-clean-why-do-they-produce-so-much-toxic-waste/#19cc80f4121c.

Smil, Vaclav. 2014. *Making the Modern World: Materials and Dematerialization*. Chichester, West Sussex (UK): Wiley.

Smil, Vaclav. 2016. ""What I See When I See a Wind Turbine." *IEEE Spectrum* (March). https://spectrum.ieee.org/energy/renewables/to-get-wind-power-you-need-oil.

Saiidi, Uptin. 2016. "Millennials Are Prioritizing 'Experiences' over Stuff." CNBC (May 5). www.cnbc.com/2016/05/05/millennials-are-prioritizing-experiences-over-stuff.html.

## Chapter 24: "We Are as Gods": The Fertile Mind of Stewart Brand

Brand, Stewart. 1968. *Whole Earth Catalog: Access to Tools*. Menlo Park, CA: Portola Institute.

Brand, Stewart. 1994. *How Buildings Learn: What Happens After They're Built*. New York: Penguin.

Brand, Stewart. 2009. *Whole Earth Discipline: An Ecopragmatist Manifesto*. New York: Viking Penguin.

"Environmental Heresies: An Interview with Stewart Brand." 2008. (Interviewer unattributed.) *Conservation Magazine*, University of Washington (July 29). www.conservationmagazine.org/2008/07/environmental-heresies/.

Lewis, Michael. 2011. *Boomerang*. New York: W. W. Norton.
Siegel, Laurence B. 2012. "Fewer, Richer, Greener: The End of the Population Explosion and the Future for Investors." *Financial Analysts Journal* 68 (6, November/December).

## Chapter 25: Ecomodernism: A Way Forward

Agaba, John. 2019. "Why South Africa and Sudan Lead the Continent in GMO Crops." Cornell Alliance for Science (January 15). https://allianceforscience.cornell.edu/blog/2019/01/south-africa-sudan-lead-continent-gmo-crops/.
Beckmann, Petr. 1976. *The Health Hazards of Not Going Nuclear*. Boulder, CO: Golem Press. Out of print.
"Bill Gates-Backed Carbon Capture Plant Does the Work of 40 Million Trees" (video). 2019. CNBC. www.youtube.com/watch?v=XHX9pmQ6m_s.
Brand, Stewart. 2009. "Rethinking Green" (video, October 9). http://longnow.org/seminars/02009/oct/09/rethinking-green/.
Calthorpe, Peter. 1985. "Redefining Cities." *Whole Earth Review* (March): 1.
Cantieri, Janice. 2016. "Artificial Trees Could Offset Carbon Dioxide Emissions." *Climate Change*. Medill School of Journalism, Northwestern University. http://climatechange.medill.northwestern.edu/2016/11/29/artificial-trees-might-be-needed-to-offset-carbon-dioxide-emissions/.
Centers for Disease Control. 2009. "Underground Coal Mining Disasters and Fatalities—United States, 1900–2006" (January 2). www.cdc.gov/mmwr/preview/mmwrhtml/mm5751a3.htm.
Gantz, Valentino N., et al. 2015. "Highly efficient Cas9-mediated gene drive for population modification of the malaria vector mosquito *Anopheles stephensi*." *Proceedings of the National Academy of Sciences* (December 8), 112(49).
Gorvett, Zaria. 2016. "How a Giant Space Umbrella Could Stop Global Warming." BBC (26 April). www.bbc.com/future/story/20160425-how-a-giant-space-umbrella-could-stop-global-warming.
London, Mark, and Brian Kelly. 2007. *The Last Forest*. New York: Random House, p. 244.
Millis, John P. 2018. "Could Matter-Antimatter Reactors Work?" *ThoughtCo.* (August 14). www.thoughtco.com/matter-antimatter-power-on-star-trek-3072119

Murray, R. B. 1977. "[Review of] The Health Hazards of Not Going Nuclear, Petr Beckmann, Golem Press, 1976." *International Journal of Energy Research* 1 (2): 18.

Owen, David. 2004. "Green Manhattan." *The New Yorker* (October 18).

Oxford Geoengineering Program. 2018. "What Is Geoengineering?" Author unattributed. Oxford Martin School, University of Oxford, Oxford, UK. www.geoengineering.ox.ac.uk/what-is-geoengineering/what-is-geoengineering/indexd41d.html

Raby, John. 2018. "Coal Mining Deaths Surge After Hitting Record Low." *USA Today* (January 2). www.usatoday.com/story/money/business/2018/01/02/coal-mining-deaths-surge-2017-after-hitting-record-low/998410001/.

Ridl, Sophia, Kyla Retzer, and Ryan Hill. 2017. "Oil and Gas Extraction Worker Fatalities 2014." Centers for Disease Control (August). www.cdc.gov/niosh/docs/2017-193/2017-193.pdf?.

Shellenberger, Michael. 2019. "It Sounds Crazy, but Fukushima, Chernobyl, and Three Mile Island Show Why Nuclear Is Inherently Safe." *Forbes* (March11),www.forbes.com/sites/michaelshellenberger/2019/03/11/it-sounds-crazy-but-fukushima-chernobyl-and-three-mile-island-show-why-nuclear-is-inherently-safe/.

Vidyasagar, Aparna. 2018. "What Is CRISPR?" *LiveScience* (online, April 20). www.livescience.com/58790-crispr-explained.html.

Waldrop, M. Mitchell. 2019. "Nuclear Goes Retro—with a Much Greener Outlook." *Knowable* (February 22). www.knowablemagazine.org/article/technology/2019/nuclear-goes-retro-much-greener-outlook.

## Afterword

Shermer, Michael. 2011. "The End Is Always Nigh in the Human Mind." *New Scientist* (June 1). www.newscientist.com/article/mg21028156-300-the-end-is-always-nigh-in-the-human-mind/.

# Index

## B

## C

# D

# E

## F

# I

# J



# K



# L

# M

# O



# P

# Q



# R

# S

## T

# U

# V



# W

# X



# Y



# Z